Model-Based Enterprise

Model-Based Enterprise describes Model-Based Enterprise (MBE) and Model-Based Definition (MBD) in detail, focusing on how to obtain significant business value from MBE.

This book presents MBE from technical and business perspectives, focusing on process improvement, productivity, quality, and obtaining greater value from our information and how we work. The evolution of MBD and MBE, from computer-aided design (CAD) topics to current approaches and to their future roles, is discussed. Following the progression from manual drawings to 2D CAD, 3D CAD, and to digital data and digital information models, MBE is presented as *the* method to achieve productivity and profitability by understanding the cost of how we work and refining our approaches to creating and using information. Many MBD and MBE implementations have changed how we work but yield little real business value – processes changed, engineering drawings were replaced with 3D models, but the organization achieved minor benefits from their efforts. This book provides methods to become an MBE and achieve the full value possible from digital transformation.

Model-Based Enterprise is essential reading for anyone who creates or uses product-related information in original equipment manufacturers (OEMs) and suppliers, in the private sector, and in government procurement and development activities. This book is also essential for students in all engineering disciplines, manufacturing, quality, information management, product lifecycle management (PLM), and related business disciplines.

Model-Based Enterprise
Achieving Lasting Value with MBD and MBE

Bryan R. Fischer

CRC Press
Taylor & Francis Group
Boca Raton London New York

CRC Press is an imprint of the
Taylor & Francis Group, an **informa** business

Designed cover image: Bryan R. Fischer

First edition published 2025
by CRC Press
2385 NW Executive Center Drive, Suite 320, Boca Raton FL 33431

and by CRC Press
4 Park Square, Milton Park, Abingdon, Oxon, OX14 4RN

CRC Press is an imprint of Taylor & Francis Group, LLC

Library of Congress Cataloging-in-Publication Data
Names: Fischer, Bryan R., 1960- author.
Title: Model-based enterprise : achieving lasting value with MBD and MBE /
Bryan R. Fischer.
Description: First edition. | Boca Raton : CRC Press, 2025. | Includes
bibliographical references and index.
Identifiers: LCCN 2024032698 (print) | LCCN 2024032699 (ebook) | ISBN
9781032067728 (hardback) | ISBN 9781032067759 (paperback) | ISBN
9781003203797 (ebook)
Subjects: LCSH: Application software. | Business--Data processing. |
Computer simulation. | Database management.
Classification: LCC QA76.9.C65 F56 2025 (print) | LCC QA76.9.C65 (ebook)
| DDC 003.3--dc23/eng/20240823
LC record available at https://lccn.loc.gov/2024032698
LC ebook record available at https://lccn.loc.gov/2024032699

ISBN: 978-1-032-06772-8 (hbk)
ISBN: 978-1-032-06775-9 (pbk)
ISBN: 978-1-003-20379-7 (ebk)

DOI: 10.1201/9781003203797

Typeset in Times
by KnowledgeWorks Global Ltd.

Contents

Preface

Thank you for your interest in my book and my take on MBD and MBE. I hope you find the material interesting, useful, and engaging. Note that I've taken a bit of a novel approach to MBD and MBE in this text.

The ideas in this book are different from what a reader familiar with traditional ideas about MBD and MBE would expect. This difference is intentional. My understanding of MBD and MBE has increased over the years, not only in terms of what MBD and MBE are and why they are necessary business practices but also in terms of what constitutes success in MBE and why so many companies miss the mark. Thus, this book represents a much more business-oriented approach to these topics. If companies want to achieve value from MBD and MBE, then the topics must be understood and communicated in ways that focus on the cost of information and business value obtained from it. The core idea in this book is that we need to relentlessly focus on and optimize how information is used, as this ends up being the most important factor in achieving value from MBD and MBE. Ignoring the many technical reasons for MBE, the main reason to become an MBE is to achieve the greatest value from our information at the lowest cost (there are many types of cost, such as financial, labor, CAPEX, OPEX, time and schedule, skills required, HR costs, burden, overhead, opportunity costs, etc.).

We need to adopt a page out of tech's playbook. We need to stop creating information to be used manually by people — we need to focus on creating information to be used directly in software and computation, and we need to use this information directly in software. This is a form of automation. While automation makes some people in the discrete manufacturing sector uncomfortable, it is commonplace in tech (e.g. leading Silicon Valley software and related technology companies). People in tech think about the information they create, gather, and work with as an asset that will be used in automated processes to provide value to users and to obtain value for the organization. We need to adapt this sort of thinking in the discrete manufacturing sector, in which organizations such as OEMs and their supply chain design and manufacture electromechanical products and systems. We need to focus on our information, the value we obtain from our information, and the cost of our information. Information has its own lifecycle, and our information incurs costs and adds value throughout its lifecycle. The value obtained and the costs incurred depend very much on how we create information, how we use information, and how well we coordinate these things.

In this book I do not focus on CAD or how to do this or that in software. The CAD and software parts of MBD and MBE are the easy part. To be clear, MBD and MBE are not about CAD. MBD and MBE are about productivity. MBD and MBE are about information. MBD and MBE are about optimizing how information is created and how it is used. MBD and MBE are about optimizing operations and automation. MBD and MBE are about optimizing information to be used directly in software and optimizing processes to use that information directly, without a human intermediary. CAD and software play important roles in this, but software is merely one of the many tools in the MBE toolbox.

This text is not a step-by-step how-to guide for people wanting to copy what some other company has done. In fact, I strongly discourage companies from copying what other companies claim to have done, for most companies haven't focused on the right things to achieve maximum business value from MBE, and thus they are not achieving adequate value for their efforts.

This text is about change, progress, and transformation. While I provide many examples, my goal is to provide content that leads readers to a clear understanding of what MBD and MBE are, what their value propositions are, why they are important, what our goals should be, and how to get there. This book is about business strategy as much as it is about the discipline of MBE. I've talked to many people who are implementing MBE, and when I ask them what their company's official MBE management strategy is, they often say they don't have one. Business leaders must establish a formal MBE strategy before their staff starts spending money and changing processes. This should be one of the first things we do in our model-based journey.

MBD and MBE have been developed as technical disciplines. I've been involved in this space for a long time. I've helped a lot of companies, I've worked on MBD and MBE standards, I've taught many courses and workshops, and I continue to work in this space. My opinions and views continue to evolve as I learn from my experiences and convert that into action and advice for clients. Only in the last few years have I come to a clear understanding of MBD and MBE, which is presented in this book. It is a problem that MBD and MBE are treated as technical disciplines. It is not a surprise that they're considered as technical topics, as they were developed by technical professionals. Technical professionals develop technical solutions to technical problems. It's what we do. However, we don't use MBD and MBE in a laboratory or a classroom. We use MBD and MBE in a business setting. MBD and MBE must be considered as business topics. They must be understood and considered from an economic perspective. Only in this way can we transform our technical ideas into successful long-term business value.

There is a lot of hype in the MBx space. I try to cut through the noise and hype and stay focused on achieving business value from MBD and MBE. We fail if we don't think of these as business topics.

Depending on what counts as a book, this is the seventh or eighth book I've written. As usual, I learned a lot while writing the book, and I ended up with something different than I originally expected. This book is my third iteration of an MBE book. The first two attempts never got very far for various reasons, namely life getting in the way. As I mentioned above, I finally figured out what MBD and MBE are in the last two years. I hope to share that clarity with you so you can achieve the greatest value possible from MBD and MBE in your organization.

I have challenged other MBE subject-matter experts to tell me what they thought MBE was, and what we were talking about in our many discussions and debates. I could tell we were not all on the same page, not by a long shot. There are many ideas about what MBD and MBE are, and most are regressive or backward-looking. Most ideas of MBD and MBE are focused on or around CAD and 3D data. Most ideas of MBD and MBE are overly influenced by our past, where people try to shoehorn the way they work today into a completely new paradigm. Given these regressive ideas, MBD and MBE are treated as a way to keep working like we worked yesterday but using new media (e.g. replace drawings with annotated models). MBD and MBE are not about replacing drawings with annotated models.

If you want to be successful in your organization's MBE journey, you must get buy-in, support, sponsorship, and ownership at the highest levels of the organization. This is the main reason I focus so much on business value in the book. We must present our ideas in terms and metrics that matter to business leaders. There is often a big difference between the value technical people propose to business leadership and the values that are important to business leaders (e.g. technical value vs business value). I present formulas and ideas in Chapter 5 to help organizations clarify the cost of how they work today and potential savings and value that may come from transformation. Transformation takes courage. Copying what another company says they've done, what I call *Copycat MBE*, doesn't require courage. Little risk, little reward. Aim high. If you aim low, you'll get what comes with that.

I hope you enjoy the book and benefit from reading it. I spent about two years working on the book and graphics, ranging from very part-time to full-time. A lot of the new graphical content is dual use, as I use it in my GD&T training and materials. The content has evolved from ideas that started back in 2002 when I first got involved with the ASME Y 14.41 standards subcommittee. As I said, my ideas and understanding of MBD and MBE have changed a lot over the years. In 2007, I wrote an 82-page chapter on MBD in the *IHS Global Drawing Requirements Manual, 11th Edition*. That material was similar to ASME Y14.41 in scope and purpose. This text, on the other hand, is much more about obtaining business value and understanding the often hidden costs of how we work today, what I call the *status quo*.

The status quo feels comfortable for most people, as they are used to it. It's how they work; it's what they do from 8-to-5, five days a week. People generally don't like change, and they like change at work even less. There is considerable inertia in the status quo, and we need to build a lot of momentum in our model-based transformation to overcome that inertia. I hope to help you do that with this book.

Appreciation

There are many people who have influenced my ideas over the years and what ended up in this book. Many of these folks have been involved in MBD, MBE, and related areas for a long time and have contributed to the advances made in many related areas. Note that I'm not implying that the following people agree with what I have written in this book, as they haven't read it yet.

- Thank you, Roy Whittenburg, Lothar Klein, Tom Hedberg, Rick Zuray, Narayana Prasad, Doug Cheney, Tom Thurman, Allison Barnard-Feeney, Larry Maggiano, Simon Frechette, Bob Lipman, Josh Lubell, Ed Paff, Atsuto Soma, Akihiro Ohtaka, Jochen Boy, Phil Rosche, and Martin Hardwick.
- Thank you Robert Keys and my brother, Curtis Oller.
- Thank you to my wife, Janine, for her support while I was in self-exile developing and completing this book!
- And thank you to you, the reader. I hope the book helps you understand MBD and MBE better and allows you to transform your work and strive for a better tomorrow. The future starts NOW!

Author

Bryan R. Fischer is the President of Advantage MBE Inc. and TDP360 LLC. He is a leading expert in Model-Based Definition (MBD), Model-Based Enterprise (MBE), and product definition practices with the experience, vision, and leadership skills to drive a fundamental change in how design and manufacturing industries do business. Bryan helps companies understand and adopt MBE and digital transformation to achieve meaningful success.

Bryan is an American Society of Mechanical Engineers (ASME) Fellow and an internationally recognized leader, trainer, consultant, speaker, and author with over 40 years of experience in engineering design, MBD, MBE, GD&T, GPS, tolerance analysis, manufacturing, and quality. In his early career, Bryan worked as a drafter, designer, checker, GD&T expert, and tolerance analyst.

Bryan has worked and consulted with companies in many manufacturing-related industries. He has worked in low- and high-volume environments and from one-of-a-kind to mass production. He has developed many standards used in drawing- and model-based environments. Bryan also works as an expert witness, helping companies in disputes related to the topics listed above.

Bryan is a lifetime member of ASME, an ASME Certified Senior Level GD&T Professional, and a longtime member of ASME and ISO standards activities. He has taught thousands of people about MBD, MBE, GD&T, ISO GPS, tolerance analysis, and engineering drawings over the last 25 years. Bryan has authored many books, articles, and papers on MBE, MBD, GD&T, drawing practices, and tolerance analysis.

Introduction

This book is about MBD and MBE with a strong emphasis on achieving business value from digital transformation. The material is technical and technical details are covered in depth, but significant discussions about the cost and value of information and its effects on productivity are included. MBD and MBE are primarily about information. Here's a description of the major topics in each chapter. I use mostly new terms in the book, as many existing terms are stale, carry a lot of baggage, and serve to reinforce misunderstandings and regressive ideas about MBD and MBE. I am trying to provide a fresh start to achieving greater understanding and value from MBD and MBE.

CHAPTER 1: OPPORTUNITIES AT OUR FINGERTIPS

Chapter 1 introduces the purpose of the book, how we work today and why we work that way, how we create and use information, and the role of documents and manual activities in the workflow. The pervasive ever-expanding state of documents and how people act as intermediaries between information and the purpose of the information are presented. Current ideas about MBD and MBE, why many current MBD and MBE implementations do not provide adequate value, the actual purpose of MBD and MBE, and industries that may benefit from MBD and MBE are discussed. The history of how we got to our current state is also included. This chapter provides a comparison of MBE and Industry 4.0.

CHAPTER 2: THE CURRENT STATE OF INDUSTRY

Chapter 2 introduces the concept of business information as an asset, that product definition datasets and other business information sets are valuable information assets. Information and information assets are considered from a business-value perspective, and the implications of doing this are presented. The concept of the information lifecycle is introduced and explained. Document-based workflows, manual activities, and their costs and shortcomings are discussed at length. The workflow required to use manual-use information (Read-Interpret-Determine Implications-Convert-Transcribe (RIDICT)) is explained. RIDICT is an important topic in this text and eliminating the need for RIDICT is a core goal for MBE. Restructuring of the design engineering workforce resulting from widespread adoption of CAD software and its effects on drawing quality are also discussed. The premise that the purpose of MBE is automation is introduced. The origins of MBD as they relate to complex product geometry are discussed. The chapter discusses the ASME Y14.41 and ISO

16792 standards. Cultural issues and the status quo in digital transformation are also discussed.

CHAPTER 3: INFORMATION TYPE, STRUCTURE, REPRESENTATION, AND PRESENTATION

Chapter 3 discusses types of information relevant to MBD and MBE, including digitally-defined information, direct-use information, dual-use information, manual-use information, presented information, and represented information. The chapter discusses several different ways information is used, including direct use (in software), manual use, and dual use. The chapter presents an argument that information should be structured and optimized for its intended use and thus makes a case why supporting dual-use is not recommended. How we create/author information and why we do it that way are discussed. The importance of interoperability and data formats in MBD and MBE are introduced in this chapter, as well as different workflows for translating from one data format to another. Requirements for MBD and MBE information are introduced.

CHAPTER 4: FUNDAMENTALS OF MBD

Chapter 4 provides a detailed description and a deep dive into MBD, including what it is, what it isn't, what it should be, and what it must be. The chapter recommends taking a holistic approach to MBD and pursuing it from a business-value perspective. Many new definitions and a new information taxonomy are provided as core components of this chapter. The definitions and taxonomy have been tailored to help readers understand the business implications of MBD, to see MBD and related topics from a business-value point of view, and to help readers, who are likely to be technical professionals, present MBD concepts and value propositions to their management in terms that matter to upper-level managers. New terms and concepts are provided to bring the discipline of MBD up to date, to increase and elevate its relevance, and to provide a laser focus on obtaining business value from MBD. Dataset classifications are introduced, including a new classification, DCC 6, which is unique to this text. The concept of MBD organization structure or schema is introduced. Perhaps most importantly, in this chapter, a new concept of Dataset Use Codes is introduced, along with a system to use the new codes. This concept is unique to this text. The chapter concludes with 15 examples of model-based product definition datasets.

Note: While the title of this chapter is "Fundamentals of MBD," the material represents new and more useful fundamentals of MBD and advanced content, based on a comprehensive top-down holistic systematic approach to MBD and

MBE. If you have learned about MBD, from someone else or even from me, the fundamentals presented here are different than what you learned before. I hope the ideas presented here provide a significant improvement in business value and return on investment for your organization.

While the content in this chapter is technical in nature, these concepts must be understood from a business perspective. Our goal must be to increase value for the enterprise and its workforce, not just to make technical changes. As I have stated elsewhere in the text, most implementations of MBD and MBE are inadequate – the implementations changed some of how we work without carefully dissecting how we work, the reasons we work that way, and the cost associated with each step we perform, which merely leads to minor incremental benefits from adopting MBD and MBE.

CHAPTER 5: FUNDAMENTALS OF MBE

Chapter 5 provides a deep dive into MBE, including fundamentals and advanced content. Many new definitions are provided. Maximizing the lifetime value of information while minimizing its cost are emphasized in this chapter. The chapter contains many formulas, including formulas to determine the lifetime value of information, the value obtained from a task or activity, the lifecycle cost of information, change in lifecycle value, and lifecycle savings from information. Value-driven MBE is emphasized throughout the chapter. The effects of business environment and culture on MBE are introduced along with ways to overcome challenges. The role of documents, annotation, PMI, and visual information (DAPVI) in modern workflows is explained. Understanding information cost and value are discussed in the context of obtaining value from MBE. This chapter clarifies the different ways information flows in manual-use and direct-use workflows. The chapter discusses process definition datasets and their classification and use codes. Process definition dataset use codes (PrDUCs) are unique to this text and a valuable new addition to MBE. The concepts of nominal product model, specification model, and consumption model are introduced. Parts lists and bills of materials in an MBE are examined. Detailed examples of the many places MBD is created and used in an organization are included.

CHAPTER 6: BECOMING AN MBE

Chapter 6 begins with a frank discussion about what the author has encountered in industry, how some MBE implementations have failed or stalled, and thus, what not to do. The importance of having a clear understanding of MBE and its business value proposition are emphasized. Cultural inertia against MBE, its sources, its prevalence, how it has evolved over time, and how to overcome it are included, including how flawed implementations increase inertia and create additional barriers to succeeding with subsequent attempts. The chapter focuses on becoming an MBE, digital transformation, and how to get there. How to establish an MBE transformation project, the different phases, and the components and goals of each phase are explained, including the importance of understanding and mapping the current state, planning for the future state, the transformation process, and pre- and post-transformation activities. The chapter addresses the question of how many steps are needed to transform. The role and effect of standards on MBE is presented from an insider's perspective. The chapter concludes with important questions needed to start our MBE journey, a synopsis of the goals of this text, and a few motivational thoughts.

1 Opportunities at Our Fingertips

TERMS AND DEFINITIONS

Notes:

I define many new terms in this book. While I discuss familiar Model-Based Definition (MBD), Model-Based Enterprise (MBE), and related topics, I use new terms and definitions to clarify the topics, avoid the traps, pitfalls, and baggage that come with existing terms, to highlight the reasons for MBD and MBE, and to elevate our understanding. For many years, we have wasted a lot of time discussing and advocating methods that do not add value. I hope to change that with this book. I am looking at MBD and MBE from economic and business-value perspectives.

The unattributed definitions used in this book are my own (the author's). If they are from another source, that source is attributed as indicated. While I helped develop many definitions in various standards related to product definition data, Model-Based Definition (MBD), and Model-Based Enterprise (MBE), I find many of those definitions are inadequate. Many of the existing terms have been designed-by-committee (group activities), which usually yield the least objectionable alternative rather than the best definition. The definitions I provide in this text are intended to be more useful, to better represent the subject, to provide more value, to help us understand that our goal in MBE is to achieve value, and to facilitate a more meaningful discussion about MBE.

There have been a lot of missteps in MBD and MBE, where companies tried and failed to achieve lasting value from MBD and MBE. Most of the problems were because people thought about MBD and MBE too narrowly, and their goals were purely technical and not focused on the right things, such as ensuring that implementing MBD and MBE would yield adequate business value or ensuring that the organization was prepared to embrace MBD and MBE. I believe some of those problems are attributable to incompletely considered ideas upon which the work was based. Most explanations about MBD and MBE are about nuts and bolts, technical aspects, the rules for model-based product definition datasets, and how to do this or that in software. That stuff matters, but it does not guarantee that the benefits of MBD and MBE implementations will be greater than the cost of implementation. As mentioned above, I provide definitions in this text that also consider topics from economic and business points of view, because successfully becoming an MBE requires understanding the economic and business aspects.

DOI: 10.1201/9781003203797-1

ASSET

"Something having value, such as a possession or property, which is owned by a person, business, or organization"

Cambridge Dictionary (online) [Definition of asset from the Cambridge Academic Content Dictionary © Cambridge University Press]

"A useful or valuable thing, person, or quality. Property owned by a person or company, regarded as having value and available to meet debts, commitments, or legacies."

Google English Dictionary/Oxford Languages (online) (from Google search)

"An asset is a resource with economic value that an individual, corporation, or country owns or controls with the expectation that it will provide a future benefit."

Investopedia (online) ("What Is an Asset? Definition, Types, and Examples")

An asset is something that is perceived to be useful, has value, and thus is relevant to an organization. Assets usually cost the organization currency, time, and/or labor.

DISINTERMEDIATE

Economics (Verb)

"Reduce or eliminate the use of intermediaries between producers and consumers. 'with digital distribution, you can disintermediate and ship directly to your audience'."

Google English Dictionary/Oxford Languages (online) (from Google search)

"Reduce or eliminate the role of (an intermediary)

'mobile-based services threaten to disintermediate banks from consumer spending transactions'."

Google English Dictionary/Oxford Languages (online) (from Google search)

My definition:

Reduce or eliminate the need for an intermediary or middleman.
For MBE, that means

- Reducing or eliminating documents and information that must be used manually from the workflow.
- Reducing or eliminating the need to use information manually in the workflow.
- Reducing or eliminating the need for a person to intervene and actively participate each time information is used in the workflow.

INFORMATION (BUSINESS INFORMATION)

Data that is adequately structured and has context such that it is useful for some purpose.

Information that is relevant to a business, its products, processes, activities, etc. Business information has value for a business and can be classified as intellectual property.

INTANGIBLE ASSET

"An intangible asset is one that is not physical in nature. Since intangible assets have no shape or form, they cannot be held or manipulated. Common types of intangible assets include brands, goodwill, and intellectual property."

Investopedia (online) ("What Are Intangible Assets? Examples and How to Value")

Note that the information in some intangible assets may be presented visually, such as printed documents, rendered CAD models, results of analyses, weather maps, etc.

In the context of MBE, digital datasets are an important type of intangible asset.

INFORMATION ASSET

An intangible asset that contains information. In the context of business (e.g. original equipment manufacturers (OEMs) and similar organizations), information assets contain information that is relevant to their products, processes, activities, inventions, copyrights, trademarks, how they do business, etc.

Examples of information assets include documents, drawings (a type of document), information models, the digital information represented in information model files, such as native CAD, STEP, JT, or QIF files, model-based systems engineering models (SysML models), analytical models, simulation models, digital twins (a type of simulation model), manufacturing, assembly, or inspection work instructions, CNC, CMM, and robot activity programs, additive manufacturing datasets, machine programs and instructions, etc.

MODEL-BASED DEFINITION (MBD)

The system and methods used to define digital business information so that it is optimized to be directly used in software. MBD replaces document-based methods, where information is optimized to be presented in documents and used manually.

Note: Historically, MBD has had a limited definition, in which MBD was used in a mechanical design context to describe cases where 3D models are used in place of 2D drawings. This definition leads us to situations where little value is realized by the organization that implements MBD and MBE.

MODEL-BASED ENTERPRISE (MBE)

Simple Definition

An organization focused on obtaining maximum value from its information throughout the organization, the supply chain, the product lifecycle, and the information lifecycle by creating and using MBD as much as possible.

THE REVEAL

Unlike my other books, in which I slowly built a case for and the logic behind the premise of the book, expecting the reader to make their way through the material, build their understanding, and finally for the light bulb to come on in an "Aha!" moment, I lay it out clearly right up front in this book.

MBD and MBE are not just about CAD and are not just about replacing 2D drawings with 3D CAD data. Yes, that is where this discipline started in the late 1990s or even earlier, and that is where many companies are stuck today, struggling to obtain value from their investment. However, we've had a lot of time to test the ideas and methods, and to understand the value provided by testing and implementing the original and subsequent versions of MBD and MBE. There's a lot more to it than 3D models versus drawings, and if we do it right, there's a lot more value that we can derive from adopting model-based business processes. This book presents a broader vision of MBD and MBE that will drive business value. This book is written for technical professionals and for business managers.

MBD and MBE should be treated as business topics, and considered as ways to enhance the value of a company's information and intellectual property. Yes, MBD and MBE are technical topics, sets of technical methods and tools developed by technical professionals. But, if MBD and MBE are implemented primarily for technical reasons, the potential business value will not be achieved. A lot of value will be left on the table. And MBE may fade into the background as yet another failed business-hype-of-the-month initiative.

The concept of what constitutes a *model* is critical to succeed in MBE. To provide meaningful business value, the term *model* in MBD and MBE must have a broad definition. *Model* includes all sorts of model types, including SysML models, CAD models, G-code files, CMM toolpaths, costing models, analytical models (e.g. FEA, CFD), etc. Generally and broadly speaking, a *model* in the context of MBD and MBE is *a set of digital information that represents something of interest.*

MBD and MBE are about information. The concepts of MBD and MBE extend far beyond CAD, models, and drawings. MBD and MBE occur in a business context; thus, we are talking about information that matters to a business, such as OEMs and organizations that procure, design, manufacture, inspect, assemble, and service machinery and equipment and similar goods.

MBD and MBE are about how we create information and how we create information assets (e.g. digital datasets, models, documents (e.g. drawings), procedural information, etc.).

MBD and MBE are about how we use information and information assets, including how we intend to use them and how we actually use them. (This concept is critical to obtaining value from MBD and MBE, and I provide new ideas, methods, and tools related to using information in this text.)

MBD and MBE are about understanding why we use information, what information is actually needed, what information and information assets we use today, and whether we really need that information and those information assets or not.

MBD and MBE are about understanding how we work today and why we work that way, recognizing the inefficiencies and hidden costs of current work methods, and how we can do things more efficiently.

MBD and MBE are Lean concepts. MBD and MBE are about recognizing that the way we currently work drives costly and wasteful hierarchies of documents, documents that lead to other documents, which spawn more documents on and on. Every document is expensive to create and maintain throughout its lifecycle. Documents drive unnecessary manual activities, and manual activities drive the need for documents. Document-based processes and manual activities are an ever-increasing inefficient self-reinforcing cycle. Its documents all the way down (Footnote 1.1) (see Figure 1.1).

FOOTNOTE

1.1 *"Documents all the way down" is a play on words from an old philosophical concept of "turtles all the way down". Turtles all the way down represent the idea of infinite regression, often in the context of how the Earth is supported in space. The idea was that the Earth rested on a turtle's shell. Which led to the question of "what supported the turtle?", which of course was another turtle beneath it. Taken to an absurd extreme, it was an endless stack of "turtles all the way down". I use this as a metaphor of how much the discrete manufacturing industry relies on documents, and documents beget more documents, which beget more documents, which ...*

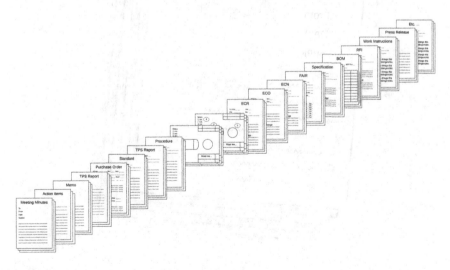

FIGURE 1.1 Documents all the way down.

MBD and MBE are not just about CAD. In fact, we should not focus on CAD as we attempt to transform our business operations and adopt MBD and MBE. A common failure mode is where companies choose their starting point by focusing on technical details, such as replacing engineering drawings with 3D CAD models. Most companies start the transformation from the bottom up and encounter a lot of problems scaling the new methods, as many issues come to light as the transformation is rolled out. We should start with the overall business goals of increasing productivity, quality, throughput, value of IP, and automation, and decreasing waste, cycle time, the number of steps in processes, time to market, scrap, rework, and the cognitive load required to produce products.

Note that in this book we will address the CAD-related aspects of MBD in detail, but this will be presented alongside the more important goals of increasing business value. Too many MBD implementations have failed to meet their potential. Many have stalled. We will address the reasons in this book.

The other day I was thinking about the role of documents in the classic movie "Office Space". The filmmaker highlights how documents and related fluff can become perceived as the most important thing, how documents and related fluff can be perceived as our goal, and how a culture and environment can be constructed to pursue pointless goals. Yes, I am speaking of the TPS Report (and its cover sheet) (see Figure 1.2).

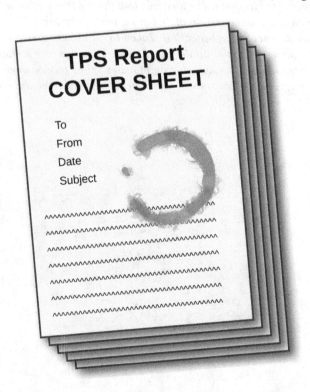

FIGURE 1.2   Did you see the memo about the TPS Report cover sheet?

I mention the film jokingly, but it presents relevant points about modern business, and in the context of this book, how the influence of inefficient past practices persists long after it should, and how easy it is to develop a culture that is focused on the wrong things. The status quo and how we work today are impediments to how we could improve tomorrow.

MBD and MBE are about recognizing that a lot of how we do business today is based on how it was done circa 1900 or earlier. While we have built tremendous computing infrastructure, the cloud, artificial intelligence (AI), information technology (IT), operational technology (OT), software, and platforms (e.g. PLM, ERP, etc.) during the Third Industrial Revolution, we largely built our systems and work methods to facilitate the same workflow we had before computers. We use our computers, networks, clouds, and technology largely to make visual information and to facilitate document-based workflows. We still use documents and visual information as the primary mover of business and as the primary record of business activities. Documents are still king. Creating, using, revising, and managing documents have become major goals. This must change. (To clarify, documents are not the goal. Documents are an inefficient means to an end (e.g. a way to get to the real goal, whatever it may be.))

MBE is a business model that wrings the maximum amount of value from information, business activities, and resources used by creating and using MBD as much as possible. Simply, MBD and MBE are about eliminating or reducing documents and the need for them throughout the workflow. This is accomplished by a series of carefully considered steps through which we scrape away misconceptions and historical baggage and inertia against progress.

MBD and MBE are about automation, about automating away the need for documents, tasks, and activities that we do today because that was how we did whatever we were doing before computers, back before we had no choice but to use information printed on paper. Now, we have information in a PDF, onscreen, in a dashboard, or other electronic digital method, but that information is still used the same way. It is read, interpreted, converted into a relevant form, written/transcribed, and used in another process or another document. MBD and MBE should start with the questions:

1. Should we be doing things the way we are currently doing them?
2. Should we be doing these things at all?
3. Are these things really necessary?
4. Does it make sense to convert our current methods and information into MBD and MBE?
5. Will doing so make things better for the organization?
6. If we could start with a clean slate, what would we actually need to do? What information would we actually need?
7. How much can we change our workflows and information to achieve our goals?
8. Do we have the right goals? Are our goals truly important, or are they intermediate milestones related to archaic practices and beliefs? (Such as creating a report for people to read and act on when we could easily automate the process in PLM. Yes, a TPS report... With a cover sheet...)

Unfortunately, most of the time, these questions are not asked, as we are expected to do our jobs, be productive, and achieve the metrics for whatever activity we are performing. And those metrics were built around an older way of thinking about what we do, why we do it, and how we should be doing it. In companies that have migrated to MBD and MBE, they've asked questions, but they haven't asked the right questions. They decided to go model-based, but they didn't figure out the real reason to do so.

We do a lot of things in our companies and in our jobs that are regressive, unnecessary, a waste of time, and a huge drain on resources, time, budget, and labor. As stated above, we likely don't have a choice, as we are being measured and evaluated relative to metrics designed around inefficient workflows. Many companies have implemented quality and process improvement methods, such as Six Sigma, Lean, Right the First Time, etc., but still overlook one of the most critical and wasteful parts of how they work – the role of information, how they use it, how they expect it to be used, how they create it, and how it is transmitted and communicated. We need to consider the cost of ownership of information and whether the information has a net positive or negative value to the organization. Surely, we need a scorecard for our information and information assets. How else can we ascertain the value and the cost of information and information assets? We can increase the value obtained from all our information by asking these questions and carefully analyzing the items listed above. And we should do so.

MBD

The main reason MBE implementations fail is that people are focused on MBD and they have a limited understanding or idea of what MBD is. There are several layers to this problem.

Layer 1:   Many people think MBD only pertains to CAD and how it is used in design engineering to define a product definition dataset. Some people think MBD is synonymous with a product definition dataset, as if MBD is a thing that you could attach to an email. This is a mistake. In this book, I provide a more complete definition of MBD. MBD is jargon. Like most jargon, initially, MBD was used before being defined, and people came up with their own interpretations. Fast forward to today, and we still need a clear and simple definition of MBD. Hence my offering in this text.

MBD is defining something (useful to a business or organization) using digital data, preferably as a model that is in a predicable format that can be used directly by software. (Note that the recent proliferation of large language models in AI based on relatively unstructured data and their ability to provide value from that data means that all MBD data does not have to be structured to be useful. However, structured relevant datasets are much more useful and trustworthy for most use cases.) The methods used to create the digital information and the resulting information can be described using MBD as an adjective, such as MBD dataset or MBD product definition practices. MBD can be applied broadly to many types of digital data. Examples of MBD applied to other datasets and uses besides product definition data include model-based systems engineering (MBSE), model-based process definition (e.g. CAE, CAI, CAM, and CMM datasets), model-based costing and procurement, model-based activity datasets (e.g. model-based work instructions, A/R, V/R, etc.), and others.

Note: See information in Chapter 4 covering *built for use*. For MBD, *built for use* is a more granular and useful concept than *structured* or *unstructured*, as it correlates how information is structured and optimized with how it is used.

Layer 2: People focus on MBD rather than MBE. If, as stated in layer 1, we think MBD is only the CAD stuff that design engineers create and release, and if we think that should be our focus, we end up omitting most of the potential value of model-based digital transformation. While we might realize benefits and savings from using MBD in design, the vast majority of benefits occur elsewhere in an organization and the product lifecycle. MBE is a way of doing business. MBE is based on maximizing the business value of MBD throughout the enterprise and product lifecycle to the organization. MBE is about maximizing the value of an organization's information and intellectual property. MBE focuses on the information lifecycle and the lifecycle cost of information versus the value obtained from information. If MBD is limited to the CAD stuff that design engineers do, we have lost the race before we even start it.

Our focus must be on MBE first, recognizing that MBD can and should occur everywhere possible in the enterprise and across the product lifecycle and that our focus is not limited to CAD and related software that enable this process. Many people think MBD and MBE are IT, OT, or software disciplines. Many software companies certainly want us to think about it that way. Their goal is selling software and generating recurring revenue. We need software to do MBD and to become an MBE, but software is merely one of the tools we need, a tool in the toolbox. From the point of view of an OEM and other organizations pursuing model-based methodologies, software is not and should not be our focus. For an OEM, MBD and MBE software is a means to an end – it is not the goal.

Layer 3: While many people interested in MBE recognize the importance of MBD and creating model-based information, their understanding of what MBD is and *how it should be used* is limited. People tend to focus on creating model-based information to replace drawing-based information, with the goal of using the model-based information in a similar manner to how they use drawing-based information. Their goal is to use 3D MBD information manually, like they use drawings and documents today.

One of the main purposes of this text is to clarify that we need to focus on using information directly in software rather than using it manually, and thus we need to create information that is optimized and intended to be used directly. I will spend a lot of time discussing manual use and direct use throughout the text. Our goal is to automate tasks in our workflows. We already do this today; you already do it today, although you may take it for granted and not even recognize that some of the tasks you commonly perform have elements that have been automated or semiautomated.

There is a concept in economics called *disintermediation*. See the definition of *disintermediate* at the beginning of this chapter. As stated in my definition, disintermediate means to remove or eliminate the need for a middleman or intermediary.

In the context of MBE, the middleman is the person who must read, interpret, and determine the meaning and implications of presented information, and convert and transcribe that information in document-based manual processes, such as an inspector reading a drawing or annotated model to determine what the product requirements are so they can inspect a product for conformance to its requirements. In such cases, we (people) act as intermediaries between the information and the result of the use case. If the information is provided in a computer-useable form (e.g. as digital information that software can understand), we could bypass the steps of reading/interpreting/determining the implications of/converting/transcribing the information and use the information directly in a target process. This book will help you understand these ideas and the opportunities that MBD and MBE facilitate.

In our discussion of becoming an MBE and digital transformation, it is helpful to understand the ten steps of transformation:

1. Understanding
2. Accepting
3. Adopting
4. Evaluating
5. Planning
6. Developing
7. Testing
8. Optimizing
9. Implementing
10. Continuously improving

We will also cover traditional MBD and MBE in detail as it relates to drawings and models. However, as we want to understand MBD and MBE and maximize the value we can achieve from them, we will consider each topic in the broader context of business value.

INCREASING BUSINESS VALUE

Increasing business value is the point of any business process improvement activity – increasing value for the business and its stakeholders. Hopefully, our outcomes also increase value for society. MBD and MBE were developed by technical professionals like me. As I stated in the front matter, I am an expert in MBD, MBE, GD&T, GPS, product definition data practices, tolerance analysis, engineering standards, and related disciplines. I've been deeply involved in the development of model-based methods since 2002. All the people developing model-based methods are technical professionals. In a professional sense, we are focused on technical topics with technical benefits and on doing technical things better for technical reasons. For example, the reason to use GD&T is that it is the only way to unambiguously define the geometric requirements and limits of acceptability for manufactured parts. However, when this true statement is shared with a nontechnical upper manager whose primary concern is business operations, schedule, profitability, and other business criteria, the statement is likely in the wrong context to make a difference to the manager. A realistic response would be something like, "Okay, so what does that mean to me? What's the bottom line?" A long discussion could follow about the business consequences of using or not using GD&T. That long discussion would start as a technical discussion in a technical context, and by necessity, would end as a business discussion in a business context.

MBD and MBE are about business. MBD and MBE are business topics. MBD and MBE are about increasing the value obtained from information by the business. If MBD and MBE are to truly make a difference and provide their potential value, they must be understood and embraced at all levels of an organization, from the lowest-level worker to C-Suite executives. Too many MBD and MBE projects have been driven from the bottom up, by technical teams trying to solve technical problems, motivated by technical reasons. While I recognize that MBD and MBE are considered as technical topics, they must be primarily considered and adopted from the point of view of enhancing business value. Only in this manner will MBD and MBE provide meaningful long-lasting benefits to an organization. Many times I've seen MBD and MBE approached for the wrong reason. We will discuss these at length in the book. Only when core staff and upper management understand why the company is pursuing MBD and trying to become an MBE will the concepts gain enough momentum to overcome the inertia of the status quo and persist beyond the first hiccup or speed bump encountered after rollout.

The biggest challenge to successfully become an MBE is cultural. To achieve significant value from MBE, we must understand our organization's culture, our expectations, priorities, and values, what is expected of us and what we are rewarded for, how we work today, why we work that way, and the inertia that must be overcome to succeed. These may be the most important statements in this book.

The biggest challenge to successfully become an MBE is cultural.

The power of the status quo and the inertia against progress or change are tremendous. Technically, we have most of the tools in place today to completely embrace MBD and become an MBE. That said, the situation in each organization is different. How much, how fast, where, when, and for which products or processes to adopt MBD are business decisions. Organizations must carefully answer those questions, and build the momentum needed to overcome the inertia. I read a lot of books over the last few years about many areas related to digital transformation, the Fourth Industrial Revolution, 10X transformation, Lean this and that, and related topics. I don't take credit for this idea, but several authors have described the often hostile response to change and process improvement as an attack and the reason is that the change is viewed as a threat. Change is often seen as an existential threat, a threat to productivity in the context of the status quo, in the context of doing what we do and what we are rewarded for doing. We need to clearly and completely understand our goals, what we are doing, why we are doing it, and what we truly need to do. Thus, a big part of a successful MBE digital transformation is to create an environment in which change can occur and flourish, and the ten steps of transformation listed above can be achieved. We'll discuss this at length in Chapter 6.

Let's make a difference. The future is here. No more TPS Report cover sheets.

INTRODUCTION

It's the twenty-first century. Companies and organizations have invested tremendously in IT, OT, software, and customization of those systems. Indeed, since the start of my career in the 1980s, computers, networks, the Internet, smartphones, the cloud, AI, and other technologies have become ubiquitous and are now commodities. Our world, our jobs, and industry run on technology. Amazing changes have been enabled by these changes, changes that many of us now take for granted.

I started my career in the mid-1980s as a design drafter with a mechanical pencil in my hand. My skills and tools focused on defining products and their requirements manually on engineering drawings, on paper, vellum, mylar, …. The way we worked was design engineering would design products, document the designs on drawings, procurement, manufacturing, assembly, quality, inspection, service, suppliers, and other groups would receive and read the drawings and use them to order materials, manufacture parts, create plans, estimates, work instructions, and manuals, evaluate whether items conformed to their requirements, etc. Everyone using the drawing had to read it, interpret its contents as best they could, convert that information into parameters that mattered to them, to their role in the product lifecycle, and to how the information was to be used (its use case), and enter that information into their system, in their terms. It was a slow and laborious process. Everyone had to be good at dealing with drawings, be it creating and reading them in design engineering, or reading them and converting what they read everywhere else. Most engineering drawings used orthographic projection, where views of a product are shown at right angles to one another (e.g. top view, front view, side view). Specifications were presented on engineering drawings using specialized methods and languages, many of which required specialized training to understand. These include GD&T, surface texture, welding symbols, assembly nomenclature, notes and notation systems, parts lists, etc. All these methods were used on engineering drawings, which are a type of a document.

There were other documents used in this process. Parts lists were often converted into bills of materials (BOMs). Lists, specifications, instructions, and requirements were defined on documents. Change requests, orders, and notifications were defined on documents. Change documents defined changes required for or made to an item or a product, and the change was first implemented on engineering drawings. A document (engineering change order) was created to drive changes in other documents (drawings and other documents). Document A explained changes shown on document B. We used documents to document changes to other documents....

In the rest of the company, across the organization and throughout the product lifecycle, the information created and maintained by design engineering was used to create documents relating to each group's processes, activities, goals, and results.

Everyone used documents to explain what was needed, what that meant to their group and activity, what they were doing and how they would do it, and to represent their progress and success at satisfying the constraints placed on them by the documents.

Documents, documents, documents. Documents everywhere.

In the 1990s, the move away from paper and printed documents accelerated. Blueprints (actually Diazo prints), like I used in my early career, were largely gone by this time. As the use of CAD increased, blueprints were replaced by drawings created on plotters. Most knowledge work in an OEM was done digitally, using computers. However, the move to computers didn't make documents go away. No, we made documents on the computer. We entered information into documents using a keyboard rather than writing manually on paper or using a typewriter.

Recently, information dashboards and other visualization aids have become popular. Many companies have spent a lot of time and money developing dashboards to enhance understanding of visual information for various processes and activities. A lot of people are excited about dashboards. Except me. I see dashboards as more of the same, more visual information that people read, interpret, transcribe, transpose, and convert to something else in a manual process. I assume that we don't need to be using information manually in manual processes that could easily be automated.

Today, many of the aforementioned workflows are still being used. Documents are the official information assets in organizations. (*Document of record*, anyone?)

Making, understanding, maintaining, and managing documents is a huge part of how organizations operate. Documents and their role are deeply embedded in organizational culture.

Question: Why do we use documents in business today?

Answer: We use documents in business today because we used documents in business yesterday.

If we use the five whys method of root cause analysis to find why we use documents today, if we ask enough questions, the answer is because we used documents hundreds of years ago.

We have drastically changed how we create, manage, share, store, and maintain documents, but in the end, we use digital documents the same way we used them in the past. We read them, interpret their contents, convert our interpretation into information that is useful to us, and transcribe that information onto other documents or into a process. This is an inefficient and costly way to work. And it requires skills that are expensive to acquire, skills that are less commonly found in recent graduates and new hires, and skills that are often seen as archaic and uninteresting.

There is a cost to using documents. A large cost. It is a hidden cost, as we take documents and the need for them as a given. We quibble about what information should be included on or excluded from documents, or how that information should be depicted, if the format is correct or in the right place, but the question of whether the document is actually needed is usually not part of the discussion.

It doesn't have to be that way.

Today, with the ubiquitous power of the cloud, IT, OT, software, platforms, smartphones, workstations, and AI at our fingertips, we should be working differently. We should not be relying upon documents and the visual information they contain to drive our business.

We should be working to replace documents and the information they contain with information that can be used directly in software. We should reduce and eliminate the role of documents and the need for documents in the workflow. This book is about doing that.

DRAWINGS AND DOCUMENTS

I have mainly been discussing documents and how they are used up to this point. I haven't written as much about drawings as you may have expected. A drawing is a type of document, and drawings are not the only documents used in design and manufacturing environments. While drawings and how to replace them with digital model data is a major topic of this book, we will also focus on the broader issue of the pervasive use and role of documents and the visual information they contain in industry. I believe that broadening our focus from drawings to documents is a critical component of ensuring that MBD and MBE implementations provide the maximum value possible for an organization.

I am not sure of the exact dates, but the term *MBD* originated circa 1990–1995, and the term *MBE* came later, circa 2000–2005. MBD was originally used to describe processes where 3D CAD models are used to replace some, or all uses of 2D engineering drawings in design and manufacturing environments. Many people think that MBD only describes the information created by design engineering (e.g. CAD models, drawings from CAD models). This is an outdated and limited idea of MBD. As I explain in this book, MBD is the approach used to model any business information digitally, preferably with the intention of using that information directly in software. Indeed, if MBD is only used in design to define a product and its requirements in a product definition dataset, and no one else uses MBD methods, the benefits of creating a model-based product definition dataset are squandered.

MBD is and should also be used in systems engineering, analysis, procurement, manufacturing, quality, inspection, assembly, service, technical publications, inventory management, shipping, and wherever doing so allows business value and savings to be achieved.

Many MBD and MBE projects have been initiated, but few have lived up to the hype and promise, and most have failed to deliver their full potential. There are several reasons for this, first and mainly that people had a limited idea of what MBD means and what being an MBE entails. Their limited view has limited their opportunity space.

The reason most MBD and MBE implementations are insufficient and have not yielded adequate return on investment is that people's ideas of MBD led them to pursue the wrong goals, for the wrong reasons, and in the wrong way.

Most MBD and MBE implementations have been based on the idea of replacing 2D drawings with annotated 3D models as the main representation of product definition. The goal of the activity was to replace 2D drawings as the driver for engineering, manufacturing, and other downstream activities with annotated 3D models. People believed that moving the information typically represented on a 2D drawing to a 3D model and using that model elsewhere in the enterprise would drive savings. This is a valid belief – there are savings that can come from this transformation. However, it is insufficient by itself, as more change is needed to really drive business value.

Most MBD and MBE implementations use geometry-only models and drawing annotations, or annotated models. The model-based information created and released by design engineering is similar to the information provided on the 2D drawings it replaces. Models are annotated. Models views are created to facilitate improved visualization and viewing. The annotation is presented to facilitate visualization. A well-executed drawing or other document is optimized for visualization – it should be optimized to allow the viewer/reader to quickly understand what they are viewing, what the product looks like, the requirements imposed by the specifications, etc. The same criteria apply to MBD models intended for viewing.

Most MBD and MBE implementations have merely changed what is used for visualization – people read or view models instead of reading drawings. *The issue is that people are still reading. They are still using visual information to drive their work. They are still working manually.* Visualization requires several steps outlined above: viewing/reading what is presented, interpreting that information, extracting the relevant information, converting that information into the necessary form or context, and transcribing that information into another document or system.

The prevailing MBD approach today is where annotated models are used as documents (3D documents). Their primary use case is for people to view the model and read the information and specifications they contain. Given the proper hardware and software (e.g. a tablet or monitor and PDF viewing software), this version of MBD provides improved visualization over 2D drawings. This method mainly improves the method of visualization. The savings it offers are mainly from improved visualization, which comes from quicker uptake and understanding of the product and its requirements – less time to understand the information and fewer mistakes from misunderstandings.

However, MBD is much more than improved visualization. MBD is much more than replacing 2D drawings with annotated 3D models. MBD is defining and optimizing business information so it can be used directly in software, potentially without human intervention. Note that we already do this in many places today, and in some cases, we have been doing it that way for decades.

An MBE is much more than an organization that has replaced 2D drawings with annotated 3D models. An MBE is an organization that uses MBD (as defined in this text) as much as possible, throughout the enterprise and the product lifecycle. An MBE is optimized to use digital information directly in software such that processes are automated, high-skill low-value and low-value activities are eliminated or automated, and people are able to focus on interesting high-value activities that are worth doing.

So much of our time and effort is wasted on trivial activities. We spend an inordinate amount of time creating and working with documents. Workers leave our companies because they are tired of working on documents, on being mired in the trivialities and nuances of document-based workflows.

As a contextual reminder, I've spent a good part of my career working on, developing, writing, and teaching people about engineering drawings and engineering drawing standards. I have decades of experience and expertise in these topics. I have also been working in and around MBD and information modeling standards since 2002. As part of the team implementing PMI during the development of ISO 10303-242 (STEP 242), I ran face first into the paradoxes and issues between drafting methods and goals and information modeling methods and goals. The engineering workflow, engineering drawing practices, and engineering drawing standards are optimized to present visual information that people will read, interpret, and hopefully understand. I am a leading expert in GD&T and GPS. These are extremely complex systems and languages, so complicated that even most of the experts don't quite understand them. Trying to break these complex systems down into information models made it clear that we have serious problems.

As a side note, GD&T was developed in the middle of the twentieth century. As drawings were the main method of representing and communicating product and engineering requirements, a symbolic language was invented to represent geometry and geometric requirements on engineering drawings. GD&T was born. Over time, GD&T, and, to a much greater extent, GPS, its international ISO-based cousin, expanded into complex graphical languages. GD&T and GPS are hard to understand and almost impossible to master. The reason is simple – GD&T and GPS were developed to define 3D geometry and geometric requirements on engineering drawings, engineering drawings are 2D media, and that we're trying to define 3D geometry and its constraints symbolically. It's very abstract.

Fast forward to the ASME Y14.41-2003 Standard. We migrated 2D engineering drawing practices into 3D space. It was a great idea. However, for many reasons, and none really good, we retained a lot of the 2D thinking and adapted 2D methods for 3D. We made a standard that explains the rules for making annotated 3D models optimized for visualization and manual use. We made a standard for 3D drawings.... There is a simple fix, and that is to make sure that the information in our annotated 3D models is optimized for direct use, that the information can be used directly in software instead of having to be read, interpreted, converted, and transcribed manually into another document or system. We need to recognize that a model is more than a pretty document that can be rotated or zoomed into onscreen.

Note: I participated for many years in the development of the ASME Y14.41 Standard. I am an advocate of what it is and what we accomplished. However, the methods in the ASME Y14.41 Standard are not enough – they do not lead to appreciable business value, as most of the content in ASME Y14.41 is focused on manual use of 3D information. Having digital data opens the door to using that data directly in software, to automating steps in processes, and eliminating error-prone and time-consuming low-value manual activities. ASME Y14.41 was a good place to start this transformation, but we must do more and expand our ideas of MBD to obtain greater benefit.

Engineering standards and practices that have been optimized for engineering drawings pose many challenges for MBD and MBE. We will discuss these challenges and how to mitigate them in later chapters. Keep in mind, there is a better way. And we can do it now.

PARALLEL WORLDS

Twenty years from now, we'll look back and think of our current workflow and methods as archaic. We will chuckle about how our work methods were inefficient and wasteful. The way we work today will join the abacus, slide rule, mechanical calculator, and even electronic calculators that I used throughout school and much of my career. We will use generative software in all aspects of engineering and production, such that the minutia and nuances that consume much of our time will be overcome by new methods. There is much precedent for this in other areas of our lives.

ELECTRONIC CALCULATOR STORY

The first electronic calculator I saw was an HP-45. My stepfather had been a land surveyor since WWII, and he used slide rules and mechanical calculators for many years. In 1973, he got an HP-45 (see Figure 1.3). The calculator was relatively expensive, but it improved his productivity tremendously and quickly made up for its cost. For the first time, he could use logarithmic and trigonometric functions directly in his calculations rather than looking up multidigit approximations of logarithm and trigonometric values from reference material he carried. Fast forward to 2008 or so. I was teaching MBE workshops at a US aerospace company. A senior engineer was attending the workshop. He was older than me and had at least 20 years more experience. As was common at that time, someone in the class was complaining about changes that would come with MBD. The senior engineer asked if he could respond to the complaints, and he explained how electronic calculators were viewed when they were first introduced in the workplace. He reminded us that at that time, everyone in engineering had learned to do calculations longhand and to use slide rules. They developed proofs manually and using slide rules. He explained that many engineers and scientists were initially skeptical of electronic calculators. He said that when some engineers were presented with a result from a calculator, they would do the math manually or use a slide rule to determine if it was correct. Eventually, after comparing the calculator's results with manual and slide rule results enough times, they acquiesced and accepted calculators.

At the time of the workshop, in 2008, electronic calculators were already on their way to obsolescence. I say obsolete loosely, as I still have my HP 50G and several other calculators, including calculators designed for architectural and facilities work. But my calculators rest in a desk drawer. Today, I can use a Google

FIGURE 1.3   HP-45 calculator (MoHPC – Museum of HP Calculators).

search to perform many mathematical functions in a search query. In fact, I can do this verbally. A takeaway here is that at the time of their introduction, electronic calculators were new, adopted by some, and distrusted and rejected by others. Calculators were disruptive. At some point, electronic calculators became THE way we did calculations, and they were used everywhere they could be. Eventually, calculators became commodities because of their widespread adoption. Finally, they were replaced by computer software, more recently by calculator apps on smartphones, and finally by cloud-enabled AI that will solve verbally presented math problems. What was once new and scary becomes accepted and ubiquitous and is later replaced by new technology once again.

Someday we will look back at our ubiquitous use of documents, drawings, and visual information as the official drivers and records of business, and we will think it is quaint and outdated. If someone suggests that we should return to the old way of working (the way we work today), it will be perceived as a joke. "You've got to be kidding!"

However, for many people, thinking about how to get to a future state is stressful and a mystery. Many people don't like change and definitely don't like change at work. I assume one of the reasons for this is that people like to be in control, and at work, most of us are not in control. The organization sets the rules, and we must operate according to its rules and structure. There are generational aspects to this, and there are many books and articles addressing the psychology and potential reasons. Suffice it to say, that same psychology engages when we look back. It is difficult to imagine what it was like to operate in a less technologically advanced state, such as horse and buggy, pre-factory putting out systems for manufacturing, before the telephone, radio, television, Internet, smart phones, CAD and solid modeling, PLM, and ERP to name a few.

It's strange how we think about how we work today or do things today with respect to the past and the future. We often have a baked-in preference for where we are, what we are doing, and how we are doing it. Looking back to older work methods, methods that are less productive because of the state of technology at the time, we may chuckle about their inefficiency and how much better it is now; and looking ahead to work methods that use technology we don't understand, accept, or haven't even been invented yet, we think it is ludicrous or even impossible. I remember a conversation with a group of standards developers, people who, like me, have been working with drawings and documents our whole career. As I explained automation use cases for engineering information, and how we should develop processes so we don't need to see some or all of the information currently shown on engineering drawings, one of the responses was "You can't do that".

We can't do that? We can't provide information to users in another form, such as replacing visual information with digital data? Of course, we can! We do it today and have been doing it for decades. In fact, the person who made that comment worked for a large aerospace company that uses that practice. Today, most, if not all companies that design products with complex-shaped geometry use 3D CAD or neutral model geometry to define the complex product geometry. As I mentioned above, the first time I saw complex product geometry represented by 3D CAD model data was in 1999, in Detroit. However, I think my aerospace friend's response was emotional. He knew his company routinely used 3D CAD model data as the main representation of complex geometry. However, maybe the idea of completely replacing an engineering drawing or annotated model with an information model to be used directly by software was too much of a stretch.

Years ago, I had a conversation with a friend who is an expert in information modeling about how we could improve the slow uptake of model-based methods and the acceptance and use of information models in OEMs. His response was, "We need to retire for this to happen". I said something else, and he repeated his comment. After I calmed down, he explained that we needed to get out of the way so younger people who had not been polluted with our current document-based manual-first approach to business, could move industry forward. His comment was that even us, two of the leading thinkers in model-based methods, were biased because of our past.

If we consider our smartphone-enabled connection to the cloud, AI models, and applications, we can do many things almost immediately and anywhere. These are things that used to take a lot of time, which had to be done in a specific place and were thus expensive in terms of time and/or money. Remember the term "banker's hours"? Today, on the rare occasions I get a physical check, I usually don't have to go to the bank to deposit it. I can use the banking app on my phone and take photos of the check for automated validation. I remember the first time I saw an ATM when I was a kid in the 1970s in California, where I grew up. When ATMs first appeared, many people were very skeptical of them.

Think of all the other changes enabled by our smartphones connected to the cloud, and how these affect our lives, hopefully in a good way. Obtaining information, communicating, shopping, dating, navigating, managing finances, fitness tracking, managing our vehicles, smart homes, security, …. There are numerous ways these have improved our lives. In fact, for younger people, much of this is taken for granted. Apps and what they enable have become commodities. Many of us expect the ease, speed, and ubiquity that technology brings to our lives. So, why do people who have accepted change in their personal lives via technology and their smartphones resist similar changes at work? What makes work so special? Why can we upend something as critical to our lives as banking and our bank accounts, but even small changes at work are greeted with dread? These are rhetorical questions – I will not try to answer them here. Human psychology and our behavior at work are complicated to say the least.

In 2018, I attended the twenty-fifth anniversary of Wired Magazine, the Wired25 Summit, in San Francisco. The summit celebrated and looked back at the previous 25 years in Silicon Valley. It also looked ahead to the next 25 years. It was fun, entertaining, and thought-provoking. At one point, I was asking myself why I was at the event. I wasn't thinking this way because I was dissatisfied with the event, I was feeling perhaps that I was not their target audience. Indeed, most of the attendees were from tech, Silicon Valley, and related industries from around the world. There I was, a technical guy, a mechanical engineer, focused on improving engineering, our methods, our output, our products, and how we interact with the rest of the world. Many presentations were about how some new technology or app that did something completely new, replaced and made something currently done much cheaper, faster, less expensive, less exclusive, or obsolete. I asked myself, "why I am here? How do I fit into this?" All these discussions were about disruption from disruptive technologies. As one of my friends liked to say, "there ain't no such thing as a free lunch". Change, even breathtakingly beneficial change, comes with a price. At the least, the price is disruption, the cost of which is the apparent pain from people's inherent aversion to change. In the last few years, I've thought a lot about change, the disruption it brings, and how to help companies understand, accept, embrace, adopt, and implement change for the better. Many of my ideas in this area are in this book.

To merge the ideas of the last several paragraphs, at some point, most people have embraced disruption to our personal lives from tech via our smartphones as we learned it really was beneficial and that we could trust it. That's an oversimplification, but it works in this case. The usual aversion to change at work is partially attributable to corporate improvements and programs that didn't lead to things being better for the worker; thus, workers become distrustful of whatever Important-Thing-of-the-Month that the company rolls out. I run into this sometimes teaching GD&T, where people sometimes expect something like a course about how to be safer when using the stairs....

RECOMMENDATION TO THE WORKFORCE

I strongly recommend accepting that the way you work, and the quality of your work can be improved by properly implementing MBD and MBE. Start with an open mind. Read this book and consider how much of the way you work today, how much of your time and effort are spent on trivial activities, or to obtain the skills to perform the trivial activities. MBD and MBE can help eliminate much of that tedium. Consider how many skills and how long they took to develop to create and work with drawings and other documents.

Consider the cognitive load that working in a document-based environment requires. We need a lot of skills, and we spend a lot of time working on trivial activities related to documents. Our skills and intellect are too valuable to waste. In this book, I am using the term *"cognitive load"* to mean the burden imposed to understand and use information in an activity and the skills required to be able to do so effectively. To illustrate, we can ask questions like: "How much of my mental resources are required to understand and use this information to do that activity? How much time does it take? What is the risk of misunderstanding or missing something? Is that a good use of my resources and time?" Consider the training required to develop and maintain those skills. Consider the time it takes to do the activities, to create and/or use documents. For many of us who have already developed the skills and are used to working in document-based workflows, we take these skills for granted. We accept the burden they posed as we have already developed those skills. We may even forget that acquiring those skills took a long time, had a cost, and was burdensome. Keep this in mind when thinking about what it takes for people with no work experience to join our companies. Think about the cost of getting them up to speed, in terms of time, money, and efficiency.

**RECOMMENDATION TO LEADERSHIP
(C-LEVEL, VPS, UPPER MANAGEMENT)**

You can drive significant improvements in the business value you obtain from your intellectual property, workforce, CAPEX, and OPEX by adopting MBD and MBE as described in this text. Productivity, schedule, cost, time to market, labor requirements, churn, market position, etc. can be significantly improved with MBE. As presented in this book, MBE is a Lean initiative. Properly implemented, MBE is a necessary competitive advantage, one that your competition is considering or has adopted. MBE will be the de facto way we do business. We've been using it in piecemeal fashion for years – it is time to embrace it fully and realize much greater benefit.

TERMS – MBD, MBE, MODEL

Many of the acronyms, initialisms, and terms in MBE are used in other disciplines and have other meanings. This is common with any jargon.

In this book, *MBD* represents *Model-Based Definition* and *MBE* represents *Model-Based Enterprise*.

Note about Terms

The initialisms MBD and MBE have other meanings and uses. Such as

MBD: Model-based definition (the use in this book and relevant industry and academia)
Model-based design (software engineering)
Motor belt drive (from ASME Y14.38-2007 Abbreviations)
Minimal brain dysfunction (medical condition)
Metabolic bone disease (medical condition)

MBE: Model-based enterprise (the use in this book and relevant industry and academia)
Model based engineering (department of defense)
Multistate bar examination (law)
Minority business enterprise (business classification)
Member of the Most Excellent Order of the British Empire (sounds like Bill and Ted, dude)

The term "model" has many meanings, so many that its use in MBD and MBE can be ambiguous and confusing. No wonder early definitions of *model* in MBD were limited to meaning a 3D CAD model. However, as we have learned over the years, even that usage has caused confusion, as people have different ideas of what a CAD model is. Some believe that a model in MBD is only geometry, some believe that it is geometry and annotation or PMI, and some believe that it is everything in a CAD file, and there are other ideas as well. A wise friend suggested to me years ago that we should not describe this discipline as *model-based*, for there are too many uses and ideas of what *model* means. These issues illustrate why definitions are critical for any technical discipline. I will define major MBE terms in later chapters. Below is an excerpt of the definition of model returned from a Google search for "definition of model". Note that the definitions below do not contain all technical uses of the term. Also, note that none of the definitions found in the dictionary searches below are quite right for our needs. The definition of model appropriate for our use is provided after Google, Oxford Languages' definition.

DEFINITION OF MODEL

MODEL (FROM GOOGLE, OXFORD LANGUAGES, FROM GOOGLE SEARCH)

Noun

1. a three-dimensional representation of a person or thing or of a proposed structure, typically on a smaller scale than the original.
   'a model of St. Paul's Cathedral'
   (in sculpture) a figure or object made in clay or wax, to be reproduced in another more durable material.
   'wax models were used by sculptors in the lost wax method of bronze casting'
2. a system or thing used as an example to follow or imitate.
   'the law became a model for dozens of laws banning nondegradable plastic products'
   a person or thing regarded as an excellent example of a specified quality.
   'as she grew older, she became a model of self-control'
   an actual person or place on which a specified fictional character or location is based.
   'the author denied that Marilyn was the model for his tragic heroine'
3. a simplified description, especially a mathematical one, of a system or process to assist calculations and predictions.
   'a statistical model used for predicting the survival rates of endangered species'
4. a person employed to display clothes by wearing them.
   'a fashion model'
5. a particular design or version of a product.
   'trading your car in for a newer model'

Verb

1. fashion or shape (a three-dimensional figure or object) in a malleable material such as clay or wax.
   'use the icing to model a house'
   (in drawing or painting) represent so as to appear three-dimensional.
   'the body of the woman to the right is modeled in softer, riper forms'

2. use (a system, procedure, etc.) as an example to follow or imitate.
    'the research method will be modeled on previous work'
    take (someone admired or respected) as an example to copy.
    'he models himself on rock legend Elvis Presley'
3. devise a simplified description, especially a mathematical one, of
    (a system or process) to assist calculations and predictions.
    'a computer program that can model how smoke behaves'
4. display (clothes) by wearing them.
    'the clothes were modeled by celebrities'
    work as a model by displaying clothes or posing for an artist,
    photographer, or sculptor.
    'he's been modeling for just two weeks'

DEFINITION OF MODEL (APPLICABLE FOR MBD AND MBE)

The optimal use of the term "model" in MBD, MBE, used alone, or in other
relevant contexts is close to the definition of an information model as defined
for software engineering. As MBD and MBE are not about software or software
engineering, I have modified the definition for our needs.

MODEL (INFORMATION MODEL, DIGITAL)

Noun

1. A set of digital information that represents something of interest.
    A model contains and defines digital objects and their relation-
    ships, constraints, and other criteria as applicable. Preferably, a
    model is structured in an expected, predictable, and consistent man-
    ner that is appropriate for its intended use.
    In a business context, models should contain information that is
    relevant and useful to the organization.

ON AI AND MBE

Note that in the context of AI, big data, and some models used in that
domain, the information may not be as structured as information created
for a specific use. For example, consider information scraped from the
Internet versus information obtained by mining a company's product definition datasets. The information from the Internet has some structure, but
it can be considered as less structured, as it is from multiple sources using
multiple formats, it is intended for different uses, and presented primarily

for being viewed on screens (note that information used to train large language models is often described as *unstructured*). Generally speaking, information in a company's product definition datasets will be much more structured and consistent, as companies typically follow standards for preparing, entering, and structuring information in product definition datasets. Each dataset will have a lot in common with other product definition datasets prepared by the company.

A model may represent a product, a process, a system, an activity, an actor, a group or team, a discipline, a division, an enterprise or organization, an object, or another entity. A model may be general in nature, such as a model of an ideal product (such as an ideal chair) or a model of a typical OEM, it may represent instances of something, such as a particular chair (chair model XYZ_01), or it may represent a singular instance of something, such as the chair XYZ_01 that is in my office (an imperfect chair that is no longer new, etc.). As I am not in the business of information modeling, and I am not a data scientist, I am not attempting to completely define this topic in a manner that would please information modelers and data scientists. I am not building a theoretical framework that in the end might not be particularly useful. I want the concepts in this book to be pragmatic, useful, and to significantly increase our understanding of MBD and MBE and the value obtained from them.

Prior discussion of structured versus unstructured information notwithstanding, a model in the context of MBD and MBE is not merely structured information, but information that is also predictably structured and conforms to a desired relevant data format. I say this because having properly and predictably structured information is important for business. Information that is structured, but the structure is unknown, unrecognized, unexpected, or inapplicable for the intended use case means that the information is likely of little or no use. Our goal is to be able to use information optimally, with the least time, cost, and effort possible. Our goal is to be as efficient and productive as possible, which necessitates a definition of model that has broad applicability across many of the different disciplines that are relevant to business.

ON AI AND MBE

Today, the need for information to be rigorously structured is less important than in the past. With capabilities like AI, machine learning, data mining, and other powerful technologies, we can obtain useful information and insight from unstructured data. Given the right expertise and enough computing horsepower (and enough resources and money to cover the cost of training the model), we can obtain meaning from unstructured or loosely structured information. However, this is not the best scenario for business processes that rely upon predictably-structured information. Less structured information used in conjunction with LLMs and AI will play a big role in industry, alongside more directly applicable use cases targeted by MBD and MBE.

IN WHICH INDUSTRIES AND WITH WHICH PRODUCTS ARE MBD AND MBE RELEVANT?

Some of the more interesting discussions we've had during the development of MBE and MBE standards have been about our scope. What does MBE cover, what areas of industry, which sectors, which disciplines, how much of the life-cycle…? Our answers and ideas led to a long list. For now, I will focus on industry and sectors. We will address other questions in later chapters.

I have found it challenging to succinctly define what to call the industrial sector I work in, the one I assume most readers work in. My work and experience is mostly with OEMs and their supply chains, software companies, and it also encompasses organizations that do not manufacture items but rely upon others to do so, such as government procurement activities. If we consider the domain of OEMs, the E in the term is critical. Of course, equipment is a broad category in itself. We are not tied to one type of equipment, one discipline, one region, or even parts, assemblies, systems, or finished products. That said, the ideas in this book and the business models I describe apply to more than OEMs, manufacturers, government agencies procuring equipment, and other support services in their supply and execution chains.

If we look at where MBD and MBE originated, our original ideas of MBD and MBE were limited by the people who drove their early development. MBD and MBE were developed in the US by people working at large OEMs such as Boeing, General Motors, and Lockheed Martin, smaller OEMs, and consultants and suppliers working to assist these companies. The original ideas for MBD were driven by many factors, but the most compelling idea was related to complex product geometry and the challenges of trying to define complex shapes using traditional dimensioning techniques. While prismatic and other simple geometric shapes were easy to fully dimension, many complex shapes could not be fully dimensioned on a drawing. At best, we used dimensions to define what we hoped were enough points or lines on a complex surface to represent the surface for manufacturing, inspection, and other uses. This technique relies heavily upon interpolation. People who worked at companies in industries where it was common for products to have complex shapes faced this dilemma – how to adequately define a complex shape using traditional drafting methods. Methods in dimensioning and tolerancing standards such as ASME Y14.5 have inadequately addressed these challenges for many years.

As 3D CAD became more commonly used, and as software and hardware that could use 3D CAD geometry directly became available, people realized that there was a better way to define complex geometry. If we can send a digital, high-resolution, geometrically-accurate, mathematically-defined representation of a complex surface to processing machinery and use that data directly in the process, maybe we don't need to try to dimension the surface. We can avoid the problems that come from dimensioning points on a surface and requiring manufacturing and inspection to interpolate (and guess) what the intermediate geometry is – we can send them a complete definition of the surface as a 3D model (Footnote 1.2). We will discuss this transition and use of 3D geometry later in the book. For now, this introduction provides background and context for the scope of MBD and MBE.

The original ideas for MBD and MBE were created by people who worked in companies that designed and manufactured equipment and products in the aerospace, automotive, shipbuilding, and similar industries. Many other industries faced similar challenges and were interested in our work, such as medical devices, consumer products, mostly industries or products that used cast or injection molded surfaces. Early in my career, my boss had worked at an OEM that made hot tubs. The company faced the same issues of dimensioning complex geometry faced by OEMs as listed above.

The idea for MBD preceded MBE. People were motivated to address the issue of not being able to fully define complex shapes by dimensions on engineering drawings. Once people could define complex shapes in 3D CAD and use the CAD model geometry or a neutral version of it, such as an IGES, JT, or STEP file, they realized that they didn't need to use dimensioning to represent the surface.

I believe that the difficulty or impossibility of fully defining complex surfaces using traditional dimensioning practices led to MBD. There are other reasons, some of which are less technically relevant, such as many people just don't like drawings, and many engineers were not trained to create drawings as part of their degrees. The earliest MBD attempt I saw was in 1990 or so. A coworker decided not to make a drawing of a plate with complex-curved edges because he knew an IGES file would be used to represent the geometry on a plasma cutting machine in our shop. He didn't like to spend time making drawings, as he liked to spend his time solving design problems and preferred not to spend the sometimes inordinately long time required to make a drawing and get it through the checking process. He felt he had better things to do....

To return to the scope of MBD and MBE, these topics apply to OEMs and organizations that design, manufacture, inspect, assemble, service, and use equipment of all sorts. We have tried to put some limitations on where MBD and MBE apply. For example, years ago, in a meeting at the National Institute for Standards and Technology (NIST), we discussed whether MBE should apply to bulk processes for products such as laundry detergent, breakfast cereal, and pharmaceuticals. At the time, we decided that these product types were not what we were trying to cover with MBE standards. However, given my clearer and broader definition of MBE, we can apply MBD and MBE methodologies to such products. That said, I do not address or include examples of this broader application in this book.

The core area and focus of MBD and MBE relate to OEMs, their products, and the processes and activities that are used and occur as part of their business. The industries most interested in MBD and MBE are aerospace, automotive, shipbuilding, medical, consumer products, and industries that include products with complex geometry. Most of the examples in this book are drawn from these industries, but keep in mind that I am not limiting which industries or product types that MBD and MBE would benefit. Note, too, that MBD and MBE are internationally applicable. Members of Japan Electronics and IT Association (JEITA) producing consumer electronics (e.g. digital cameras, printers, scanners, etc.) and Japanese Automotive Manufacturers Association (JAMA) are or are becoming strong proponents of MBD. European organizations are also becoming more proficient with MBD.

FOOTNOTE

1.2 *According to experts, the definition of a complex surface provided by CAD software may differ from a translated version of that surface or a representation of the surface in processing software. In some cases, the difference between the versions is minor and may be classified as noise. The methods or algorithms used to generate the surface may be different in different software. So, while I consider using model-based methods to define complex surfaces to be critically-important, I recommend defining a consistent process and formats that yield adequately equivalent results. We achieve this through SOPs, official formats, versioning, and verification and validation of results.*

I find that thinking of MBE as a Lean discipline is useful and appropriate. MBE focuses on information, how information is created, why it is created, how it is used, what information is necessary, and how to obtain the maximum business value from information and its use. In MBE, we are concerned about the value we obtain from information over the entire information lifecycle, and we are concerned about the information lifecycle. Obviously, with a definition like this, there is no limit to where these principles could be applied. However, the primary focus of MBD and MBE remains related to OEMs and organizations, as described in the previous paragraphs.

COMPARISON OF MBE AND INDUSTRY 4.0

This section includes simplified descriptions, origin stories, and evolution of MBE and Industry 4.0.

The concepts of MBE and Industry 4.0 were developed separately and considered as different disciplines affecting different parts of organizations. (Note that Smart Manufacturing is a similar concept to Industry 4.0). MBE was developed from a design engineering point of view and Industry 4.0 was developed from a factory or production point of view.

MBD and MBE started in the US as design engineering methods to partially or completely replace engineering drawings of mechanical components with 3D CAD models, and to use the models in manufacturing, inspection, and other activities that traditionally use engineering drawings.

Industry 4.0 (Industrie 4.0) started in Germany in 2011. The focus was on industrial production and the digitalization of industrial production. Industry 4.0 is a logical extension and component of The Fourth Industrial Revolution. Process and production-related equipment, sensors, information, products, etc. are digitally connected to facilitate communication between process elements and between product and process. Smart manufacturing originated in the US circa 2006.

MBE and Industry 4.0 were initially focused on opposite ends of the design-production spectrum. Over time, people developing MBE and Industry 4.0 began to realize that to be successful and derive greater organizational value, the initiatives needed to be broader in scope. MBE, being based on MBD, must apply to production, as the digital information created in design activities should be used in production activities. Likewise, Industry 4.0 must apply to design activities, as smart products and smart product definition data are required to drive factory automation and optimization.

Today, MBE and Industry 4.0 overlap and both aim to accomplish very similar objectives: digitalization and optimization of design-production-lifecycle workflows and getting the most from the product and process information. MBE also focuses on production information creation and use, and Industry 4.0 also focuses on design and product information – these are logical extensions that address input, output, and use cases. From an MBE perspective, creating digitally optimized design data without considering how that information is used in production provides little or no value. From an Industry 4.0 perspective, developing digitalized production systems provides less value if product definition data from design and systems engineering is not optimized to work with their systems.

I see MBE and Industry 4.0 as different ways to look at the same thing, to look at optimization of the product, process, and information lifecycles. MBE and Industry 4.0 are developed by different teams of experts, using different

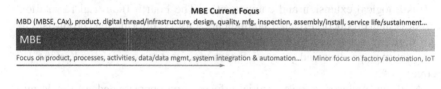

MBD (CAD, MBSE), design engineering, product data		Manufacturing, process data, smart factory, service life/sustainment
MBE		**I4.0**
Design, product-related activities, product	Systems integration and automation	Manufacturing activities and data from field

MBE Original Focus

MBD (CAD, MBSE), design engineering, product data

MBE
Focus on design, product-related activities, product visualization   Did not focus on factory

I4.0 Original Focus

Manufacturing, data management, smart factory

I4.0
Did not focus on design and product-related activities   Focus on factory and data from field

FIGURE 1.4 MBE and Industry 4.0 – Current Focus, Overlap, and Original Focus.

MBE Current Focus

MBD (MBSE, CAx), product, digital thread/infrastructure, design, quality, mfg, inspection, assembly/install, service life/sustainment…

MBE
Focus on product, processes, activities, data/data mgmt, system integration & automation…   Minor focus on factory automation, IoT

I4.0 Current Focus

Data mgmt, product data, digital thread/infrastructure, quality, mfg, inspection, assembly, install, service life/sustainment, …

I4.0
Less focus on design and product   Focus on factory, processes, data mgmt, system integration & automation, IoT, data from field

FIGURE 1.5 MBE and Industry 4.0 – Current Focus.

terminology, with different emphases. As with all technical development activities, those of us working in these disciplines are down in the weeds, focused on details, and may not see the similarities between our goals. Some of us might not want to believe that what we are doing is similar to what the other group is doing, or that we have the same goals. See Figures 1.4 and 1.5 for historical comparisons of MBE and Industry 4.0.

2 The Current State of Industry

TERMS AND DEFINITIONS 1

Note about Terms and Definitions:

Some of the terms and definitions in this text are new and different from terms we've used in the past or are used by others for this discipline. Many of the terms currently used elsewhere are limited in intent, and in many cases, the limitation inhibits clear understanding of our goals, what's possible, and the scope of MBD and MBE. Many existing terms evolved from different sources and in different contexts, often to satisfy narrow views of MBD. Most of the existing terms are based on the simple limiting concept that a model is 3D CAD geometry and MBD is using 3D CAD to define products. While these ideas are part of MBD and what I present in this text, there is a lot more to it. New ideas and a more systematic top-down approach to terms are needed, and I offer that in this text. This will allow us to extend MBD and MBE and to focus on obtaining value from what we do. Much of the current activity focuses solely on MBD, with little recognition of how to obtain value for the organization.

ASSET

"A useful or valuable thing, person, or quality. Property owned by a person or company, regarded as having value and available to meet debts, commitments, or legacies."

Google, Oxford Languages (online) (from Google search)

See definitions of Tangible asset, Intangible asset, and Information asset.

AUTOMATION (NOUN)

Using machines, hardware, software, or a combination of these to do something that was previously done manually by people.

Notes: Software includes AI, machine learning, etc.

Automation may be used to completely replace a manual activity, or it may be used to replace some of the manual tasks that make up a manual activity. Automation does not necessarily mean complete removal of human activity, interaction, or oversight. I assume we can also automate a process that has already been automated, say, with new technology, but I don't get into that level of detail in this text.

DOI: 10.1201/9781003203797-2

In the context of MBE, automation primarily applies to information, information use cases, and information-related events that occur throughout an extended organization and the information lifecycle. Examples of information use cases include creating, distributing, securing, evaluating, validating, scoring, using, expanding upon, learning from, benefiting from, etc. information and information assets.

For MBE to provide maximum benefit to an organization, automation must be considered at the lowest level, at the level of individual tasks that combine into activities, which combine into processes.

Automation may completely eliminate the need for an activity; it may reduce the number of tasks needed to perform an activity; and it may require new tasks and activities to support automation. The value proposition of automation is that the end result is more efficient, quicker, safer, less risky, requires less time, less labor, less cost, less materials, less floorspace, less burden, less training, and increases productivity, quality, utilization of CAPEX and OPEX, staff utilization, staff satisfaction, staff value, staff retention, and value of all expenditures.

EXTENDED ORGANIZATION (EXTENDED ENTERPRISE)

An organization or enterprise, its supply chain, its customer base (if they use, purchase, require oversight of the organization's information), regulators (as applicable), other organizations (such as contractors, partners, etc.), and end users (if they use or add to the information lifecycle). In the context of MBE, the concept of extended organization is intended to capture all potential entities and actors that create, use, or contribute to information in the information lifecycle.

INFORMATION (BUSINESS INFORMATION)

Data that is adequately structured and has context such that it is useful for some purpose. Information that is relevant to a business, its products, processes, activities, etc. Business information has value to a business and can be classified as intellectual property.

INFORMATION ASSET

An intangible asset that contains information. In the context of business (e.g. OEMs and similar organizations), information assets contain information that is relevant to their products, processes, activities, inventions, copyrights, trademarks, how they do business, etc.

An information model, such as the digital information represented in an information model file, such as a native CAD file STEP, JT, or QIF file, model-based systems engineering model (SysML model), analytical model, or manufacturing, assembly, or inspection programs are also information assets.

Note: The uses listed above of model-based product definition, SysML, analytical models, or code-defining manufacturing, assembly, or inspection toolpaths and operations are forms of MBD. MBD is not limited to the output or methods of design engineering.

INFORMATION FOR MANUAL USE

Information that is defined and intended to be used manually (e.g. information depicted in documents, on drawings, in lists, visualization data (e.g. FEA deformation diagram, flowchart, schematic, logic diagram, wiring diagram, P&ID, weather map)).

INFORMATION LIFECYCLE

The timespan that information, information elements, or information assets are used by an organization, affect an organization, or remain relevant to an organization.

The concept of information lifecycle is based on the concept of product lifecycle. Initially, my discussions about MBD and MBE were mainly in the context of the product lifecycle. However, as the ideas evolved, I realized several things:

1. A product and information about the product may have different life-cycles. The information about a product, leading to, derived from, or affected by a product and its use may exist before the product is developed or after it is retired.

   The product lifecycle and information lifecycle are distinct (they are not the same thing).

   In the best case, for organizations that develop similar products or new products based on old products or information (e.g. lessons learned, customer feedback) over time, the information will continue to be relevant and valuable to the organization long after a product is no longer in production.
2. It is useful to discuss the information lifecycle directly and separately from the product lifecycle in the context of MBD and MBE. My intent is that the approaches in this text increase understanding and the value of information to the organization throughout the entire information lifecycle.
3. Information may not be related to a product. Information may be related to a process, an activity, a task, a service, an event, a disruption, etc.

INTANGIBLE ASSET

"An intangible asset is an asset that is not physical in nature. Goodwill, brand recognition, and intellectual property, such as patents, trademarks, and copyrights, are all intangible assets."

Investopedia (online) ("What Are Intangible Assets? Examples and How to Value")

See definition of Tangible asset for comparison.

INTELLECTUAL PROPERTY

"A work or invention that is the result of creativity, such as a manuscript or a design, to which one has rights and for which one may apply for a patent, copyright, trademark, etc."

Google, Oxford Languages (online) (from Google search)

"Intellectual property is a broad categorical description for the set of intangible assets owned and legally protected by a company or individual from outside use or implementation without consent. An intangible asset is a non-physical asset that a company or person owns."

Investopedia (online) ("What Is Intellectual Property, and What Are Some Types?")

DOCUMENT

Information asset that contains information for manual use.

> **Note:** For this text, documents are information assets, and information assets are intangible assets. A printed or hardcopy document is obviously tangible, as it can be touched, held, etc. However, today, in many corporate environments, printed copies of documents are marked as copies and not considered as the official binding version of that information.

DRAWING (ENGINEERING DRAWING)

A specialized category of documents that convey engineering and related information, usually describing a company's products but also pertain to its processes and activities.

TANGIBLE ASSET

"Tangible assets are physical and measurable assets that are used in a company's operations. Assets like property, plant, and equipment, are tangible assets."

Investopedia (online) (What Is a Tangible Asset? Comparison to Non-Tangible Assets)

See definition of Intangible asset for comparison.

HOW DO WE WORK TODAY?

Most organizations targeted by this book primarily use document-based workflows. OEMs and similar organizations or organizational structures (e.g. DoD) use documents as the formal definition and record of most business products, processes, and activities. Documents are used manually, or at least they have been since their initial use thousands of years ago. Before the advent of digital systems, documents were tangible assets. Cave paintings may have told stories and cuneiform inscriptions on clay tablets may have delineated inventories and

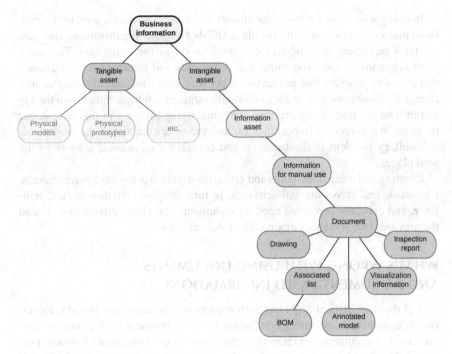

FIGURE 2.1 Information to document hierarchy.

transactions, but in a general sense, they functioned as documents. I define a document as a type of information asset. Information assets are things that contain information, generally considered valuable intellectual property, for the purpose of retaining and conveying that information. See Figure 2.1 for a hierarchical map of definitions related to information relevant to OEMs and similar organizations.

DOCUMENT-BASED WORKFLOWS AND MANUAL ACTIVITIES

It is a safe assumption to say that all companies operating the industries I described in Chapter 1 as being in scope for MBD and MBE can be classified as being document-based. I use the term "document-based" a lot in this book. It is critical in order to understand what MBD and MBE are (or rather, what they are not) and why MBD and MBE are important. Documents, their purpose, their intended use, their actual use, and the skills and effort needed to create, optimize, release, manage, and maintain them are probably the third most important thing at most companies. Corporate mottoes include statements about safety, society, lifestyle, etc., but after financial performance, which is their primary focus, and meeting production targets and schedules, which are their secondary focus, companies are focused on documents and the manual activities they drive. However, emphasis on documents and the inefficient work methods they drive is not part of companies' mission statements.

In many ways, this focus on documents and related activities goes unnoticed. From many conversations with people at OEMs and other organizations, they just see their inefficient document-centric work world as the status quo. That said, there are many people who think that documents and drawings are a nuisance that impedes progress and productivity, but they either assume that they cannot change their environment or focus on small, targeted solutions rather than the big picture and the underlying problem. The underlying problem in this case is our focus on, the perceived importance of, and use of documents in the workplace. A corollary problem is the focus on and prevalence of manual activity in the workplace.

Creating and using documents and visual information in the workplace embeds, reinforces, and drives manual activities. In turn, manual activities embed, reinforce, and drive the perceived need for documents and visual information. (Read the previous two sentences again.) It is a vicious circle.

WHAT'S WRONG WITH USING DOCUMENTS AND DOCUMENT-BASED INFORMATION?

One of the most critical problems with documents and document-based information is that there is no direct connection between business critical information presented in documents and how the information is used to execute business processes. This problem comes from how the information in documents is structured and how it is used. The information in documents is presented visually, to be read or viewed, interpreted, and used. There are several steps in this process with several corresponding failure modes.

Using visual information as the driver for and record of business assets and activities is inefficient, costly, and error-prone. The main problems in document-based workflows are that information is intended to be used manually and it is used manually. Thus, information must be read or viewed, understood, converted to some useful form, and manually transcribed into some other process or document. Note that *understand* is actually *interpret* – we *interpret* what we view or read – thus, we may miss something, misinterpret something, add something that isn't there (read between the lines) ... There are several failure modes in this workflow. See Figure 2.2 for a simple model of the Read-Interpret-Determine Implications-Convert-Transcribe (RIDICT) workflow for using presented information.

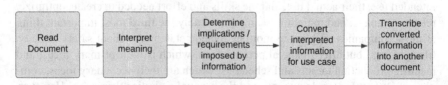

FIGURE 2.2   The read-interpret-determine implications-convert-transcribe (RIDICT) sequence for using visually presented information.

Forward-thinking value-centric MBE implementations seek to forestall or eliminate these problems by linking information in the information architecture. With document-based information, information is presented visually, and the workflow does not facilitate such linkage.

In an MBE workflow, carefully defined digital information is provided to a process that is designed to use that information. The source information is directly linked to the process and its outcome or product. Nominal values, acceptable limits, requirements, and conformance criteria are defined for a product, product step, or outcome, and they are defined for the process used to achieve that product or outcome. This is one of the underlying premises of Industry 4.0, that information about and from processes and the tools and services that perform them is collected and tied to the information that defines the values of their settings and limits.

The individual components or fields within the information are directly related to components or fields within a process; the process parameters are monitored; their values and outputs are compared to and reported against the input parameters that relate to the process; the parameters and values of the process outputs are compared to and reported against the input parameters that relate to the output; and the process parameters are adjusted to bring the process and product into control. See Figure 2.3 for a diagram of the relationships and potential linkage between product, process, and inspection information. Note that there are many more feedback loops possible than the one shown in the figure. Information for each step could/should be linked to its inputs and outputs, thereby enabling an order of magnitude increase in value for the information. Adjustments to all values can be optimized to ensure the most efficient, lowest cost, and fastest outcome that satisfies the requirements. This scenario enables an organization to validate and optimize the requirements and their limits, the process and its settings, and the outcome. Considering that most tolerances specified on drawings

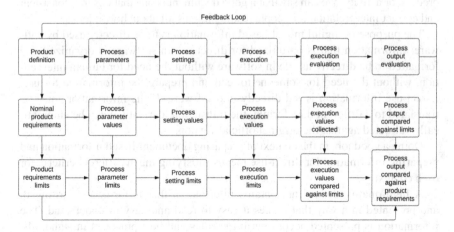

FIGURE 2.3 The relationship and linkage between product, process, and inspection information.

are educated guesses or barely-considered defaults, such feedback and optimization allow some reality to creep in and business value to drive each step, including
the requirements and their limits.

> **Note:** Information cannot be directly linked, as shown in Figure 2.3, in a docu-
> ment-based workflow. Any "links" between information elements in a doc-
> ument-based workflow are manual, subjective, informal, and non-existent.
>
> The overarching problem with manual document-based processes
> is that the information in the source material and the information
> in each step is not and cannot be directly linked. There are no logi-
> cal (digital) relationships between data elements in such a document-
> based workflow. The information representing something in one step
> is not linked or logically related to the representation of that informa-
> tion in corresponding steps. There are no links between information
> in each step. All relationships are managed manually, in the mind of
> someone trying to manage the information and processes. Everything
> hinges on someone trying to understand and manage the informa-
> tion. From product definition to specifications to process definition to
> operations to validation, in manual document-based processes, there
> are no links between the step and the source document for each step.
> The inputs to each step are manually interpreted from visual material
> and transcribed into the information driving that step.

The purpose of a document and visual information is to be read and/or viewed
by a person. So, how do we define our goal for documents, or what constitutes document quality? An optimized document is one in which information is organized and
presented in a manner that is easy for someone to correctly understand as quickly as
possible for its intended use. A document is read and/or viewed by someone, its contents are interpreted, and that interpretation is transcribed into another document,
process, or activity. We can say that a good document is one that ensures consistent
and correct interpretation for people reading it – it's all about how it looks.

The purpose of digital model-based information is to be directly used by software. Information that is optimized for direct use in software is organized and
formatted to be directly usable in software without the need for human intervention, without the need for someone to read and prepare the information for use.
Direct-use information is fed directly into software. If the information must be
translated to a different data format, the translation (derivative) can be automatically validated against the original official version.

Doing a good job, in the context of preparing document-based information and
preparing information for direct-use means satisfying much different criteria for
each context.

A high-quality document contains information that is formatted, arranged,
and presented in a way that makes it easy to read and easy to understand. The
information is presented according to the rules and best practices in standards,
style guides, and corporate documents. Adhering to formatting, arrangement, and
presentation rules are critical components of making a high-quality document.

The way a document looks is critical to its purpose. How well the information presented visually relates to other elements is subjective; it is difficult to validate and must be performed manually by people with a lot of experience.

Doing a good job when creating a document means optimizing the presentation so people can read, view, and understand the information presented quickly and easily.

In a document-based workflow, doing a good job means presenting visual information in a well-organized predictable manner that makes it easy to use the information manually.

We are trying to ensure that the information is easy to read and understand.

A high-quality direct-use information model contains information that conforms to a consistent, recognized, logical structure (e.g. data format), contains the right information, contains the correct type of information in each field, and contains the correct logical relationships between information and data elements. The way direct information looks is irrelevant. How well elements of direct-use information relate to other elements is logically encoded and can be validated using software.

Doing a good job when creating information for direct use means optimizing the information content and structure so it can be easily used (consumed) by software.

In an MBE workflow, doing a good job means structuring the right digital information in a well-organized predictable manner that conforms to the applicable information model and data format.

We are trying to ensure that the information can be directly used in software.

The goal in both cases is to fully represent the required information in a manner that is best for its intended use. For a document-based workflow, the goal is to create visual information for people to read or view. For an MBE workflow, the goal is to directly use the information in software.

For visual use cases, the underlying digital and logical structure of the information is irrelevant and how the information looks is critical.

For direct use cases, the underlying digital and logical structure of the information is critical and how the information looks is irrelevant.

The preceding paragraphs explain a few fundamental differences between document-based and model-based workflows. It's about what's important, what matters, what isn't important, and what doesn't matter. The only overlap between the requirements for these workflows is to completely and correctly represent the required information for the intended use case. From there, the methods diverge. A primary concern for document-based information is how it is presented and whether it looks nice. A primary concern for direct-use information is the structure of the information – how the information looks is completely irrelevant – software doesn't have eyeballs. In later chapters, I will expand on these differences and the relationship between the intended use of information, which use it is optimized for, which type of information is officially released, and how the information is actually used. Misunderstanding and/or mismanaging these aspects of business information can cause an MBE implementation to underperform and miss its ROI targets. I will also cover the cost of focusing on document-based information and compare that to the cost of focusing on model-based information. Note that there are significant business-affecting costs related to focusing on documents and document-based information.

THE CHANGING EMPHASIS ON QUALITY IN DOCUMENT-BASED WORKFLOWS – THE DEMISE OF DRAWING QUALITY

As I described at the beginning of Chapter 1, I started my career in the 1980s. I did manual drafting and design for the first two years of my career. I learned drafting in Junior High School in the 1970s and saw civil engineering drawings that my stepfather used in his work – these were rolled-up blueprints (or more accurately, prints created using the diazo process) – I took a drafting class again in High School, and eventually got a design drafting degree in college. All that time, drafting was done manually. There was a large workforce of drafters and designers in the US and around the world. Companies of all sizes had drafters on their staff – some large companies employed hundreds of drafters.

In addition to technical, scientific, engineering, mathematics, and other relevant subjects, drafters were trained in communication. They were trained to communicate technical information. Design drafters and designers were also trained to solve technical problems, similar to engineers, but usually with less mathematical and scientific rigor. Being trained to communicate technical information meant that drafters had specific training and skills to create engineering drawings and to represent products and specifications graphically. Much of the workday was spent with a pen or pencil in hand, working on a drawing on a drafting board. I am describing a simplified hierarchy of drafters, designers, and engineers before the drafting activity was absorbed by/assigned to engineers. See Figure 2.4 for some of the historical roles, skills, and responsibilities in engineering departments.

The prevalence and roles of engineers, designers, and drafters have changed since the 1980s. With the introduction of 2D CAD, drafting and design moved from the drafting board and manual drafting to the computer. In the 1990s, as 3D solid modeling CAD software became more powerful and widely adopted, the role of drafters and checkers began to diminish. As companies adopted CAD software, which was often expensive, difficult to learn, required sometimes expensive training, and required specialists to manage the company's installation of the software, the emphasis of companies' engineering shifted from high-quality

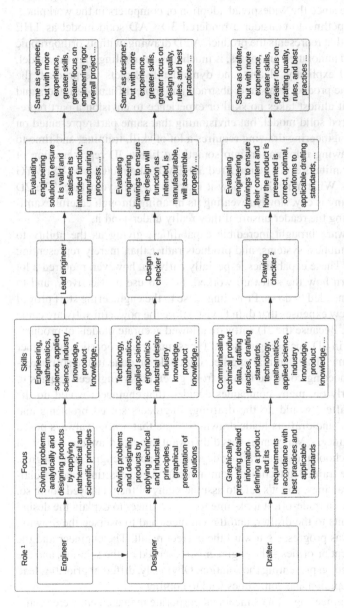

FIGURE 2.4 Historical roles, skills, and responsibilities in a design engineering department.

Notes for Figure 2.4:

1. The roles shown are idealized. Often, there was overlap, and the distinction between the roles was blurred and varied by organization. Individual skills and experience, predilection, specialization, company size, and past practices also affected how work was assigned. For example, some small companies only had one person in their engineering department who fulfilled all roles.

2. Typically, a checker served both roles, checking the design (Will it work? Is it manufacturable? Is it economical? Does it use standard components?) and the drafting (How is the design information presented on the drawing? Does it follow applicable standards and best practices? Does it clearly and completely convey the design and its requirements?). The checker's sole focus was checking. Note that all drawings were reviewed by a checker, regardless of who created the drawing (engineer, designer, or drafter).

drawings to 3D CAD solid models as their output. The thinking changed with regard to what represented the product and what was most important.

Figure 2.5 shows estimates of the changes that have occurred in the design engineering workforce since the widespread adoption of computers in the workplace.

It is very compelling to consider a rendered 3D CAD solid model as THE information asset that represents the product. A 2D drawing, with its orthographic views and stylized annotation and notes, is much less compelling than the model, for while a model explicitly, clearly, and dynamically depicts the product, the drawing depicts the product in a static, abstract, and more difficult-to-understand manner. That is, it almost takes no skill or experience to envision a part represented as a rendered solid model, but envisioning that same part represented on a traditional 2D engineering drawing requires a lot of skill, training, and time to assimilate the drawing information into something like a 3D representation of the product in our mind's eye. The rendered 3D CAD solid model is WYSIWYG (What You See Is What You Get) – how it looks is right in your face. The 2D engineering drawing is coy, slowly revealing its meaning through careful examination, often leaving the reader unsure if they really understand it.

3D CAD software brought incredible capabilities, giving us the ability to model parts, assemblies, systems, and products rather than merely representing them graphically. These capabilities, especially early on, however, required a lot of training to learn how the software worked, how to use it effectively, and to acquire the skills needed to do so. Providing these to the engineering staff proved to be expensive, new expenses that didn't exist before the powerful software came into the workflow. Because the 3D software allowed accurate models of products to be represented in 3D, including material types, material properties, and other attributes, it became clear that real engineering could be done in the software – much of the engineering could be moved from green paper engineering pads to the computer. Early on, many companies decided that engineers would do the modeling and drafters would do the drafting. Engineers solved problems and designed products, and drafters made drawings. Note that drafters always posed challenges for engineers. Many engineers didn't want to make drawings – after all they didn't go to school and get a drafting degree, they got an engineering degree. There were people trained in technical communication and drafting who specialized in making drawings, and it's safe to assume many of them enjoyed doing so. However, there was a trade-off. It took time for the engineer to explain the design and its requirements to the drafter, and the engineer had to oversee the drawing to ensure that it was progressing toward the correct result. The engineer and the drafter had different priorities – the engineer was focused on the solution, and the drafter was focused on presenting the solution. Obviously, different priorities lead to differences of opinion and sometimes lead to conflict.

Meanwhile, engineering management and corporate management were seeing their operating costs increase with the costly new CAD software, its management, the training it required, and the loss of production time for training, gaining skills, culture change, and getting the staff to accept the new tools and to migrate from the board to the computer.

Circa	CAD Operators	Drafters	Designers	Engineers	Checkers	Toolset	Focus	Focus	Response
1970	0 nil	10	10	10	8	Manual drafting	Drafting, design, engineering	Execution, product	Status quo
1980	0.5	10	10	10	7	Manual, 2D CAD	Drafting, design, engineering, CAD	Execution, product	Reliance on software begins. Emphasis on CAD skills begins.
1985	2	10	10	10	7	Manual, 2D CAD	Drafting, design, engineering, CAD	Execution, product	Increasing emphasis on CAD skills and capabilities enabled by CAD.
1990	5	9	9	10	6	Manual, 2D CAD	CAD, drafting design, engineering	Execution, product	Increasing emphasis on CAD skills and capabilities enabled by CAD. Decreasing focus on drawing quality.
1995	7	6	7	10	5	2D CAD, 3D CAD	CAD, drafting design, engineering	Profitability, stakeholder returns, short term goals	Deemphasize drafters and checkers; Move their work to engineers
2000	10	3	6	8	3	3D solid modeling CAD	CAD 1st, engineering, 2nd	Profitability, stakeholder returns, short term goals	Drafters and checkers are gone. Designers are deemphasized. Their work is moved to engineers.
2010	10	1	3	7	2	3D solid modeling CAD	CAD 1st, engineering 2nd	Profitability, stakeholder returns, short term goals	Engineering, design, drafting and checking are engineer's responsibility
2020	10	0 nil	2	6.5	0.2 nil	3D solid modeling CAD	CAD 1st, engineering 2nd	Profitability, stakeholder returns, short term goals	Drafting, checking, and drawing quality are seen as impediments to schedule.

Legend [staffing]
Increasing
Static
Decreasing

Notes
Numeric values shown for CAD Operators, Drafters, Designers, Engineers and Checkers represent relative amount of staff in industry
Scale of 0 - 10: 0 = none present, 10 = present everywhere (ubiquity)

Core engineering tasks have remained static, but labor has decreased due to computerization and digitalization

Engineering workload has increased from inheriting drafting, design, and checking activities

Labor required for drafting, design, and checking activities has decreased due to computerization and digitalization

All values shown in the table and the changes over time are approximations based on author's observation and experience

FIGURE 2.5 Change in design engineering staff over time: 1970 – Present.

The last part of the story is that the 3D solid modeling software included a drafting module.

The software made it easy to create orthographic views of a solid model. As easy as drag and drop … With the right training, a CAD user could make a drawing quickly, creating the needed views in just a few minutes. The annotation and other information could be added to the drawing views.

Managers realized that maybe they didn't need all these people anymore. Maybe they didn't need engineers, designers, drafters, design checkers, and drawing checkers. Managers realized that their 3D CAD software had the much of the analytical capabilities needed by their engineers, that the product could be modeled and rendered so people could easily understand what they were looking at, and the software made it easy to make a drawing. As the software facilitated engineering and drafting, and only the engineers had the training to do rigorous engineering and analysis, maybe managers could increase the engineer's workload by requiring them to make drawings and phase out the designers and drafters. Fast forward to today, and drafters are a rare commodity in the US. Designers still exist but in far lower numbers than in the 1980s. Also note that today job titles are more creations of HR departments than actual descriptions of people's skills and careers.

Thus, in many companies, making drawings became the engineer's responsibility and drafters were history. Engineers were making drawings. Engineers already had a lot on their plate. In many cases, they were responsible for the entire design from start to finish. Most engineers didn't have much drafting training. They didn't learn much about drafting practices and drafting standards as part of their engineering degrees. Engineers had to learn a lot after they left school and joined a company. A lot of training and practice using CAD software was required. Note that modeling and drafting in a 3D CAD system are different activities, and the thinking, approach, and skills required for these activities are different.

One thing remained constant throughout these changes, however – there were still deadlines and a schedule to be met. With the drafters gone, it was clear that something needed to change, as the engineers were doing everything. The checkers were still there. The checkers were still focused on design and drafting. They were still ensuring that the engineering department's drawings were high quality and suited for use by people in other departments who had been trained to use engineering drawings. Checkers were often relentless in their pursuit of quality. Checkers pushed the engineers to do what the checkers considered was important. The engineers, however, didn't have the same emphasis on drafting as the checkers. They felt that the checkers, often older, more experienced drafters and designers, were impeding their productivity and ability to meet their metrics (e.g. schedule, drawings released per week) – checkers were often considered as bottlenecks to productivity. The engineers, their managers, and corporate management looked at the drawings the engineers were creating and thought the drawings looked fine.

Slowly, sometimes quickly, checkers were no longer part of the workflow. Formal checking became peer review, getting another engineer (who was already busy with his or her own work) to review the drawing. Checking went from a rigorous drawing quality control activity to someone who sat next to you quickly glancing at a drawing, or worse, a friend who tended not to give a critique. Now the design engineering staff was increasing production. However, drawing quality was decreasing, abysmal in some cases. The metric was *drawings per week* – it wasn't *drawings that were easy to understand per week* or *drawings that completely and correctly represent the product and its requirements per week...*

In the end, these changes in the workforce have had long-lasting consequences. Drawing quality, and by extension, annotated model quality have suffered. The effects of this change have been to transfer the labor, time, and cost spent making and checking drawings to downstream operations. People downstream have to complete the design. Requests for Information (RFIs) and mistakes generally increase in response to poor drawing quality. Costs and schedules have shifted from early in the product lifecycle, during the design process, to later in the lifecycle, during procurement and production planning and execution. Rather than finding and solving all the problems during the design process, many problems are found during the prototyping, manufacturing, assembly, and installation processes.

I have spent a lot of time since 2000 consulting and training in GD&T, MBD and MBE, drawing standards, and product geometry management. While I don't have hard data to back up the statements in the preceding paragraphs, I have seen and continue to see their effect today. One time I was teaching a course to a roomful of manufacturing engineers at a large aerospace company. I asked them what the main thing they did each day as a manufacturing engineer at that company. To my surprise, they said their job was to fix design's mistakes. I was dismayed because design engineers are supposed to do our best not to make mistakes and make sure that the design and product requirements are defined completely, correctly, and unambiguously. In another case, I was applying GD&T to a new vehicle design. We were developing the GD&T scheme for a large assembly, which we would complete in a week or so. I found out in a meeting that a team in manufacturing had designed and was building an inspection fixture for this large assembly. I was dumbfounded, as we hadn't finished developing the specifications for the assembly, which should be what an inspection fixture would verify conformance to. This group hadn't even seen what we were doing yet. I told my client that developing the inspection tooling before the specifications were released didn't make sense and that it was very risky, and my client's cynical response was "They think they make vehicles in spite of the drawing." In retrospect, I am convinced that these events are not only a function of bad management and bad execution but also a result of consistently poor drawing quality. Why would workers trust the drawing if their experience taught them otherwise if the drawings they received usually had problems and required workarounds?

A side effect of the changes listed above is that product definition has changed. Some aspects of products and their requirements have increased in usefulness and fidelity, while other aspects of products and their requirements have decreased in usefulness and fidelity. Much of the information that used to be conveyed is now incomplete, incorrect, or missing altogether. Tolerancing is a good example of this. With the larger and more specialized workforce of the past, all the different aspects, disciplines, and specialties that were part of inventing, designing, and defining products and representing that information were shared across different specialists. Now, the responsibility for these aspects often falls on a single person or role, the design engineer. A specialist understands their chosen area of expertise more in depth than a generalist. With that understanding, the specialist understands the relative importance of the different aspects of their domain. If that level of understanding of the domain and its details are lost, then the reasons for some of what the specialist did are also lost. A good example is drawing checkers. Drafters and drawing checkers focused on drawing quality. Drawing quality (to correctly, completely, and optimally define a product and its requirements in a predictable and standardized manner) was their primary emphasis. To a large degree, understanding of the importance of drawing quality disappeared along with the drafters and drawing checkers. Today, I routinely see drawings or annotated models that are incomplete, inconsistent, disorganized, and incorrect. I see problems waiting to happen. These problems will happen at a much more expensive part of the product lifecycle than if they were prevented during the design process.

Today, a lot of the specifications I see on drawings and annotated models are reactive rather than proactive. The cost consequences of this in terms of profitability and schedule are huge.

Back to the topic of this book, MBD and MBE, which provide a way to solve some of these problems. Reducing the role and importance of documents (such as engineering drawings) and visually presented information mitigates the problems caused by poor drawing quality. MBD obviates the drawing, and MBE obviates the need for visual information as the official representation of products, processes, and activities. To say this differently, model-based methods provide an opportunity to reduce or eliminate the problems related to poor drawing quality.

To recap, I am not saying that visual information is no longer important or will no longer be available, I am saying that it should not be the primary official definition of products and processes; visual information should be included such that it is available on demand, and that products, processes, and activities should officially be defined by information optimized to be used directly by software. Define your products, processes, and activities with digital information that can be used directly in software and release that digital information as the official representation of the product, process, or activity. Whatever visual information is included in the official release should be designated as reference only. Visual information for manual use should be secondary to the digital information for direct use, and the visual information should not be used to formally define products, processes, or activities.

MBD, MBE, AND AUTOMATION

Note that in this book, my approach to MBD and MBE is strongly influenced by business value and economics. While MBD and MBE are typically defined as technical disciplines, MBD and MBE must be considered in a business sense rather than a purely technical sense. Implementations of MBE focused on technical details and technical value propositions are not focused on providing value to the organization, and thus leave a lot of potential business value on the table (i.e. implementations that create MBD for manual use don't provide much value to the business). The approaches defined in this book aim to provide as much value as possible from MBD and MBE.

Looking ahead, the primary goals of MBD and MBE are to increase the value of information and the value obtained from it, and to streamline the use of information and activities. We accomplish these goals by creating information for direct use (which increases accessibility of the information), by using the information directly in software, and by automating tasks and activities, thereby freeing staff to work on more interesting, rewarding, and value-driving activities. We focus on information, its lifecycle, the value obtained from it, and the cost of creating and maintaining it through its lifecycle. There are many benefits that come from doing these things.

This is not a new idea. I have been thinking about MBD and MBE for many years, thinking about what we are doing versus what we should be doing. Many people seem uncomfortable when I first mention automation. I think they envision human-shaped robots, lights-out factories, AI bots, or a combination of these. They think I mean total automation. That may be the ultimate goal for some industries (see Sidebar/Info Box), but I am not explicitly talking about automation at that level. I am thinking of automation at much lower level. Automation can and should be considered from the ground up, at a detailed level. We need to review all workflows in our organizations to increase efficiency by streamlining activities and optimizing the value obtained from information during its lifecycle. We need to focus relentlessly on how we create and use information, whether we need information in its current form (i.e. visual, for manual use), and whether we need that information at all.

The goal I describe for MBE, *to automate tasks and activities, thereby freeing a company's valuable staff to work on more interesting and rewarding activities,* is value-driven. The prevailing idea in industry today is that the goal of MBE *is to replace engineering drawings with annotated 3D models.* While I agree that MBD and MBE include replacing drawings with models, the purposes of MBD and MBE are much broader and provide much more value than drawings versus annotated CAD models. In fact, thinking that MBD and MBE are just ways to migrate from drawings to annotated CAD models significantly limits the potential value that can be obtained from MBD and MBE. Migrating from drawings to annotated CAD models in most cases doesn't fundamentally change the way we work and may increase cost and decrease productivity. The key is how information is created, its intended use case, and whether it is used as intended. Many companies migrate to MBD only to work almost exactly as they did with drawings. They use model-based information manually as a visualization tool. Often, some experts use a derogatory term and call these *3D drawings*. Unfortunately, that is much of what is being done today.

Thinking that MBD and MBE are just ways to migrate from drawings to annotated CAD models significantly limits the potential value that can be obtained from MBD and MBE.

Figure 2.6 breaks down business operations by organization, role, activity, task, subtask, and action, where subtask and action represent the most detailed and discrete activities. In the figure, the selected organization is an assembly team, the role is assembler, the activity is assembling parts, the task is fastening parts together, the subtask is tightening specific fasteners, and the action is using specific tools to tighten those fasteners to specified torque values. Business value can be increased at every level of this hierarchy by replacing document-based methods and manual activities with model-based methods and automation. As we envision and design new processes to replace and automate existing processes, we should start at the activity, task, and action levels.

We must decide how much or how little of a process should be changed, how much document-based and/or manual activity will be replaced with direct use of digital information and automation. In many cases, careful examination of the value, basis, and actual need for a process step or its output will yield the understanding that the activity or task and its output are not required at all. We will discover that the activity, its constituent tasks, and their output are historical artifacts that are part of how we do business and considered important or necessary because no one has asked or challenged whether they are actually needed. We do a lot of work today simply because that's the way we did it yesterday. That must change.

We must focus on replacing document-based workflows and the manual activities and tasks that are driven by them with digital information used in automated digital processes. We may be able to automate and replace entire processes and benefit from consistency, productivity, and concomitant savings, but we do not need to start at that level of automation to obtain significant business value from MBD and MBE.

Returning to Figure 2.6, the automation might only occur in the last action to validate that the applied torque is within limits. Perhaps tightening the nuts and applying torque will be partially automated, changing from a simple torque wrench to one with digital collection and output, or to a multi-headed hydraulic manual tool with preset torque limits. Perhaps we will automate interaction with the assembler or use augmented reality to show the assembler animations of sequential tasks overlaid on an actual in-process assembly. Perhaps the activity will be fully-automated, where the parts are automatically loaded onto a fixture by pick-and-place robots, a robotic arm inserts the fasteners and applies torque using a multi-headed end effector fitted with sensors that validate that the torque meets requirements, and that information is automatically recorded in process management software, thereby enabling the assembly to proceed to the next step in the process. The level of automation is a business decision, and how much or little to automate will vary company-by-company and department-by-department. However, companies should follow consistent and clear goals laid out and supported by the highest levels of management.

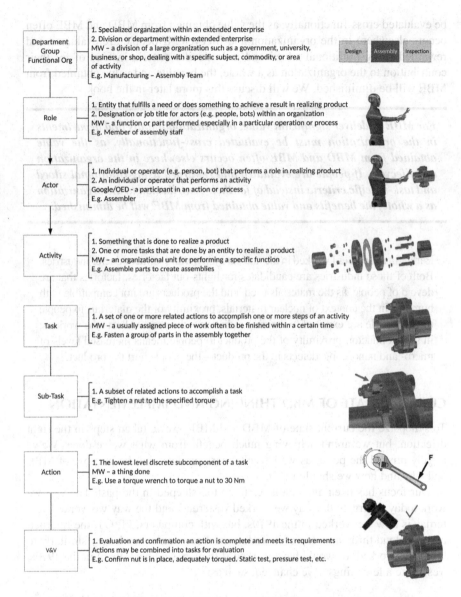

FIGURE 2.6 Department, role, activity, task, and subtask action taxonomy.

MBE must be visible to, owned by, and supported and championed by upper management.

Middle managers and workers must also support MBE, and it is critical that their metrics, KPIs, whatever you want to call the criteria upon which they are evaluated and rewarded, are aligned with how MBE works. For MBE to deliver significant value, organizational units and departments in the organization must be evaluated cross-functionally, as the value obtained from MBD and MBE often occurs elsewhere in the organization and lifecycle. If people are evaluated and rewarded using traditional siloed and task-specific criteria instead of for their contribution to the organization as a whole, the benefits and value obtained from MBE will be diminished. We will discuss this more later in the book.

For MBE to deliver significant value, organizational units and departments in the organization must be evaluated cross-functionally, as the value obtained from MBD and MBE often occurs elsewhere in the organization and lifecycle. If people are evaluated and rewarded using traditional siloed and task-specific criteria instead of for their contribution to the organization as a whole, the benefits and value obtained from MBE will be diminished.

Earlier in my career I worked in the nuclear systems and semiconductor industries. Both of those industries are candidates for lights-out factories, factories that are devoid of people, as the materials used, and the products are not compatible with humans. In the context of nuclear materials, proximity of the materials to people means people are exposed to ionizing radiation – the process hurts the workers; in semiconductor, proximity of the product to people means increased levels of micro- and nanoscopic defects in the product – the people hurt the product.

CURRENT STATE OF MBD THINKING AND IMPLEMENTATION

To synopsize the current state of MBD and MBE, we've taken steps in the right direction, but we aren't achieving much benefit from what we've done. We've largely missed the point, as we haven't clearly understood the purpose of MBE and why and how we should use it.

Our focus has been and continues to be too steeped in the past. The way we work today is largely the way we worked yesterday, and the way we worked yesterday is how we worked in the 1970s, but with computers, IT, OT, the Internet, and the cloud thrown on top of it. Today, most companies work very similarly to how they worked (or would have worked if they existed back then) in the 1970s. Yes, quite a few things have changed, such as:

- Typewriters, dictation machines, and typists have been replaced by computers and speech recognition technology.
- Mimeograph machines have been replaced by copy machines, which have been partially replaced by digital versions of documents (e.g. PDF).
- Drafting boards, drafting equipment, and drafters have been replaced by engineers using 3D CAD software displayed on monitors.
- Slide rules and manual calculations have been replaced by calculators, which have been replaced by software.
- Hard copy interoffice correspondence, which was typically sent in ruled envelopes with handwritten date, sender, and recipient name and

department (see Figure 2.7), have been replaced by email, chat, texts, e-message boards, and other productivity enhancing- applications (along with the social malaise that came with them).

- Financial, accounting, inventory, and other records management in physical ledgers were replaced by spreadsheets and later cloud-based enterprise software.
- Operations and materials management in physical ledgers were replaced by spreadsheets, databases, and later cloud-based enterprise software, such as PLM and ERP.
- Staff training performed in person in a training facility on specialized equipment has been replaced by self-paced online training viewed at one's desk, or in their home, on a laptop, or anywhere on their smartphone.
- Standalone, manually operated factory equipment by highly trained and skilled operators has been replaced by programmable, automated, and semiautomated multi-function digitally-controlled machinery that is linked to and communicates with factory oversight software and other machinery in the factory.
- And more recently, artificial general intelligence is on the rise, on its way to replacing many manual activities.

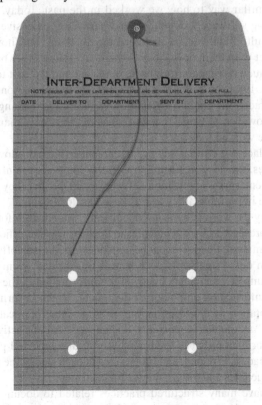

FIGURE 2.7 Interoffice envelope – How we used to communicate at work (while we communicate digitally today, we still use information pretty much the same way).

There have been many changes, and I only listed a few above. The changes are everywhere across an organization. A lot of the changes were driven by the widespread adoption of computers and related systems in the workplace. With computers came software, and with these came networks, networks inside the enterprise, and later the Internet, intranet, the cloud, and ... Along with the capabilities and power these technologies provided came the need to manage and optimize them and manage how people used them. With them came the Information Technology (IT) department.

I remember the first time I heard of IT. My first two design jobs were for small companies that didn't have IT departments. In 1988, at my third design job, I heard people talking about the IT department. Most people in the office and the shop used computers that were connected to local area networks, and much of the business was managed using early enterprise software on IBM AS/400 computers. My employer, similar to most companies, realized that they were creating and using a lot of information, and that information was valuable to the company. Very soon thereafter, in the early 1990s, it was apparent that the now ubiquitous IT department at companies had a lot of power and big responsibility.

I describe all this to set the stage for where we are today. As I said earlier, we work in a similar way to how we worked in the past. Today, we have powerful and capable software, computers, and hardware, extensive networks and connectivity, multimodal means to interact with them, and how, where, and when we interact with them have vastly expanded. Taken as a whole, it is mind-boggling how much our capabilities have increased and the cost to use them has decreased since 1980. As discussed in Chapter 1, in our personal lives, many of us have embraced the enhanced capabilities offered and changed how we do many things. However, at work, much of what we do is only marginally different than how we worked long ago.

In the workplace, we still largely use document-based workflows. Documents are still king. Yes, the documents are now on the computer or online, such as on GUI screens in enterprise application software or PDFs, but they are still documents, and more importantly, they are still used as documents.

The main way we work is by using documents. We create information that is intended to be presented in a document, the document is the official representation of the information, the document is signed/stamped/made official, the document is stored in our systems, the document is transmitted, recipients read and/or view the documents, recipients interpret the information in the document and transcribe that information onto another document, which in turn is used as the official representation of the information for some other application. To repeat from Chapter 1, "*documents, documents, documents*". All this activity is to create and manage information that is used manually in disconnected processes – the information is read, interpreted, and transcribed at each step of the process. There are many inefficiencies and failure modes for each step.

Companies have many structured practices related to documents: for creating them, for using them, for managing and maintaining them, for storing them, and for all aspects of their lifecycle. When we work with clients, it is clear that much of what they do is driven by their document-based culture and workflows. It often seems like documents have a life of their own and are perceived to be more important than the products and results they are intended to foster.

When we work with clients, it is clear that much of what they do is driven by their document-based culture and workflows. It often seems like documents have a life of their own and are perceived to be more important than the products and results they are intended to foster.

MAIN PROBLEMS OF USING DOCUMENT-BASED WORKFLOWS

- The information stored in documents is intended to be used manually and thus is disconnected from how it is used in the enterprise.
- Documents are often embedded in the workflow and treated as one of the most important assets in the enterprise. This is a distraction – documents should not be the goal of any milestone. Documents should be a means to an end, not the goal.
- Information in documents should be optimized for manual use.
- Information optimized for manual use often follows specialized and abstract rules for visual format and best practices.
- Information optimized for manual use is very difficult to use directly in software and in model-based workflows.
- New hires and recent college graduates (RCGs) lack the skills for creating and understanding documents and knowledge of these specialized rules. The very idea that documents and their arcane rules and skills are so important conflicts with the expectations of digital natives entering the workforce. The use of document-based workflows is a disincentive and turn-off for young people entering the workforce.
- Using documents decreases productivity and increases cost, labor, and time required to achieve goals that are actually important to the organization.
- Using documents distracts everyone from the actual organizational goals.

SYNOPSIS OF PROBLEMS OF DOCUMENT-BASED WORKFLOWS

- Document-based processes and the manual activities they drive are inefficient and expensive.
- Information in documents must be viewed, interpreted, and manually converted into another form before it can be used.

- Document-based processes are error-prone and require a lot of skills in the workforce, skills that are becoming less and less prevalent.
- Information in documents is optimized for manual use and thus is poorly suited for direct use in software as part of model-based workflows.
- Document-based workflows distract everyone from what really matters to an organization. People think the documents are the goal in many cases.

Note that we have the tools, technology, and technological framework to work much more effectively today. We need to change how we work, how we use technology, what we consider as important, critical, or necessary, and most importantly, we need to change our culture at work. Of the changes needed, cultural change is the most important and most challenging.

TERMS AND DEFINITIONS 2

PRODUCT DEFINITION DATA (PRODUCT DEFINITION INFORMATION)

1. Information that defines a product and its requirements. Product definition data defines a nominal (perfect) product, its requirements, and its allowable limits of variation.

   Product definition data is included in engineering drawings, models, associated lists, BOMs, databases, digital files, and other information assets to represent a product. Product definition data is typically created and owned by a design organization. Note that product definition data is also included in PDM, PLM, and ERP systems, which may be owned by a different organization.

   Note that in this text, I retain and predominantly use the terms product definition data and product definition dataset rather than product definition information and product information model. I do this because of the pervasive use of the "*data*" terms and their use in ASME Y14 and other standards. However, as the data and datasets are provided with context, they contain information and not just data.

   The idea that we are dealing with information and not just data is a core idea in this text.
2. "Denotes the totality of product definition elements required to completely define a product. Product definition data includes geometry, topology, relationships, tolerances, attributes, and features necessary to completely define a component part or an assembly of parts for the purpose of design, analysis, manufacture, test, and inspection."

   (Reprinted from ASME Y14.41-2019 by permission of
   The American Society of Mechanical Engineers. All rights reserved.)

EXPLICIT DIMENSION

(On Drawings) Dimensions that are shown on the drawing that formally represent the product and/or its requirements. Reference dimensions are not explicit dimensions.

(On Annotated Models) Persistently-displayed dimensions on the model that formally represent the product and/or its requirements. Reference dimensions are not explicit dimensions.

Explicit dimensions may be presented using dimension lines, extension lines, and leaders, they may be presented in tabular form or represented by formulas or other means depicted in the product definition dataset.

If explicit dimensions are used, they represent the official definition of the geometry they represent. Using explicit dimensions in MBD is problematic, as there is potential conflict between explicit dimensions and model geometry and confusion as to whether the dimensions or the model geometry are the official representation of product geometry. In such cases, there are two definitions of product geometry, and people can choose which definition to use. While the ASME Y14.41 Standard states that the model value and dimension value shall be in agreement, in practice the model value and dimension value often differ, and in many cases ensuring that the values agree is not possible. The best practice is to only provide one definition for each geometric feature.

(This is a fundamental issue that must be addressed in MBD and MBE implementations. See later chapters for more information.)

EARLY REASONS FOR DIRECTLY USING MODEL GEOMETRY

The idea of using geometric CAD data directly in manufacturing and inspection processes has been around for many years. As I mentioned earlier, the first time I saw this in action was in 1990, as 2D IGES data translated from VersaCAD software, and again in 1999, in Detroit, where 3D solid model data from Unigraphics was being used to define complex product geometry (automobile interiors and exteriors). CAD software capabilities increased dramatically in the 1980s and 1990s. The need to model complex geometry, such as automobile body components, aerospace structures, consumer products, and medical devices, to name a few, drove increased capability to model these shapes. Once we were able to model these shapes, we realized that it was a good idea to use the model geometry directly in analysis, manufacturing, inspection, assembly, and for other processes. Design engineering could benefit from these new capabilities, and other activities, such as prototyping, analysis, manufacturing, tooling, inspection, and assembly could also benefit.

Complex geometry and the need to define it in product definition data have been challenging for as long as complex geometry has been included in designs. Defining complex geometry in product definition data was a major driver in the development of early MBD. See Figure 2.8 for an example of a part with complex geometry.

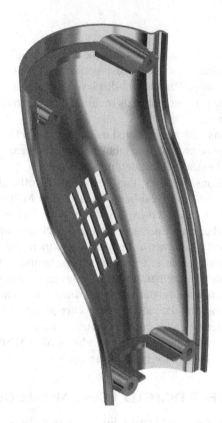

FIGURE 2.8   Part with complex geometry.

CHALLENGES OF DEFINING COMPLEX GEOMETRY ON DRAWINGS

1. How to design complex geometry.
2. How to depict or visually represent complex geometry.
3. How to completely define complex geometry using annotation (dimensions).

Note that 3D CAD is powerful geometry and data modeling software (e.g. a powerful database optimized for graphical and related data) that has a graphical user interface (GUI), graphical display of the information, and graphical output of that information.

Also note that most CAD software is modular, with modeling and annotation functions provided separately, such as in separate software modules, and using different menus and commands.

Geometric modeling functionality is used to create a geometric model that represents a desired product or outcome. Annotation functionality is used to create specifications that define the product, its requirements, and metadata information about the product (e.g. part number, revision) and the information asset in which it resides (e.g. drawing, model, annotated model). See Figure 2.1 for more information about information assets.

Keep in mind that creating a geometric model and defining the specifications to apply to the product are different activities that require different skills and different training. Modeling is like sculpting, while defining annotation such as dimensioning, GD&T, welding symbols, and notes are something else entirely, more like being a writer.

CAD solid modeling provided powerful capabilities to design complex geometry (item 1 in the list above) and to represent it visually for visualization purposes (item 2 in the list above). Today, creating and visually representing complex geometry are usually accomplished simultaneously during the modeling process. Defining complex geometry using annotation (item 3 in the list above), defining complex geometry using dimensions, has always been difficult and imprecise and continues to be so today. The reason is simple: most complex geometry cannot be completely defined by explicit dimensions. In my GD&T work, the taxonomy I developed that defines different types of geometry, from simple to complex, is based on the number of dimensions and rules it takes to define each class of shapes. Complex shapes, such as the ergonomic shape of a computer mouse, require an infinite number of dimensions to define the shape (if we consider the idea of limits and that a complex shape can be broken down into infinitely small divisions). Obviously, no one can do this, and no one should. At best, we've used simplified methods to define complex shapes with dimensions on drawings for many years. Dimensions are used to define a grid or pattern of points with X, Y, and Z coordinates, and a magical note is added to the drawing that says something like "The entire complex surface must be smooth and continuous and not just conform to the dimensioned points." Well, the note doesn't usually mention conformance to the dimensioned points, but it is expected. The problem is that only a few points are selected on a surface to represent the surface; all other points on the surface are implicitly defined by interpolation. It is a very messy and imprecise way to define design requirements. This method has worked because of craftsmanship, informal methods, and agreements between stakeholders (workers). Note that informally-invoked craftsmanship does not work with MBD and MBE. Obtaining value from MBD and MBE requires formal methods.

As CAD matured and we gained the ability to model complex geometry precisely in CAD, and later as we gained the ability to apply dimensions to models,

the problem of dimensioning complex shapes persisted. We still have more surface geometry and points on the surface than space or time to dimension them.
Luckily, someone realized that we could bypass the dimensions and use the CAD
model geometry instead.

To clarify, we have a few options for how to represent product geometry in
product definition datasets:

1. Explicit dimensions officially represent product geometry.
2. Model geometry officially represents product geometry.
3. Explicit dimensions officially represent some product geometry and
   model geometry officially represents other product geometry.
4. Include model geometry and explicit dimensions in the product definition data, and indicate whether the model geometry or the explicit
   dimensions officially represent the product geometry. (This is messy,
   and it is a common problem in MBD implementations.)

For parts with complex geometry, option 2 is best. Option 3 can also be used,
with the caveat that we must clearly define where the model geometry officially
represents the product geometry and where explicit dimensions officially represent the product geometry. To leap ahead, I recommend not using methods where
people have to guess or use their judgment as to whether a dimension or the model
geometry represents product geometry. Make it clear. Keep it simple.

The industries that I first saw model geometry being used to represent product geometry made products that included complex geometry. People working at
these companies realized that if they had the complex shape defined mathemati-
cally in their CAD software, they could bypass the problems caused by traditional X, Y, and Z grid dimensioning methods and interpolation, and use the CAD
model geometry (or a neutral derivative like STEP) directly in their analytical and
processing software. Early on, however, their analytical and processing software
didn't accommodate CAD or neutral 3D data input. Over time, the ability to use
3D CAD and neutral data directly was added to analytical and processing software, and the idea of MBE started to grow.

People realized that they could use complex and even simple model geometry
as the basis for many activities. Today, model geometry and other model-based
information are used directly in engineering analysis, procurement, estimating,
tooling and fixture design, manufacturing (e.g. machining, fabrication, assembly, inspection), toolmaking, die, mold, and pattern design, defining toolpaths
(e.g. CNC machining, CMMs), defining point clouds for inspection, as building,
reverse engineering, training (e.g. rendered animations of models, A/R, and V/R),
shipping container design, and many other uses.

From my experience, the desire to use CAD model geometry directly and
the business case to do so initially were driven by complex product geometry
and the need to define, understand, work with, produce, and inspect it in the
workflow. Companies that had such products seemed to get there first when it
came to MBD.

THE ASME Y14.41 STANDARD: DIGITAL PRODUCT DEFINITION DATA PRACTICES

According to ASME, the initial meeting that led to the ASME Y14.41 Standard occurred in 1997. ASME Y14.41-2003 was the first major standard to cover model-based definition in the context of traditional engineering documentation and drawing practices. Here's an excerpt from the Foreword of ASME Y14.41-2003:

> "The development of this Standard was initiated at the request of industry and the government. A meeting was held to determine the interest in this subject in January 1997 in Wichita, Kansas, hosted by The Boeing Company in their facility. A subsequent meeting was held during the spring ASME meeting in 1997 to enlist membership of those who would be interested in working this project."
>
> *(Reprinted from ASME Y14.41-2003 by permission of*
> *The American Society of Mechanical Engineers. All rights reserved.)*

During the development of ASME Y14.41-2003, it became apparent that many things that were important for engineering drawings were not as important for annotated models, and annotated models required many things that did not apply to engineering drawings. There was a lot of debate around the new rules and recommendations surrounding this work, with some old timers who lived their careers around drawings objecting or feeling discomfort with the new material. The ASME Y14.41-2003 Standard set many rules and requirements for MBD. ASME Y14.41-2003 was a groundbreaking standard, further improved by later editions released in 2012 and 2019. Work on the next edition continues today. Keep in mind that most of the people who developed this standard were fluent in drawing practices, and their experience occurred in environments where drawings were the official record of products. That is, official product definition data resided on drawings and related documents (e.g. parts lists, associated lists). A few people were from the CAD companies, whose experience included engineering drawings from the point of view of how the CAD software supported making acceptable drawings.

We were trying to develop digital product definition data practices, modeling practices, and MBD, and many of us used 3D CAD solid modeling software. But we had a lot of experience with drawings, and thus we had a lot of baggage. Some of the most influential people on the committee, the old timers, were not CAD users. They represented other areas germane to stakeholders, such as corporate standards, configuration management, and engineering management. Remember, back in Chapter 1, I described the discussion with my friend about how we were going to finally get people to start accepting MBD, and his response that *"We have to retire"* (and thus get out of the way)? His comment also described limitations in the ASME Y14.41 Standard. And those limitations persist today.

ASME Y14.41 contains a mixture of documentation/presentation rules and practices and modeling rules and practices. While we tried to clearly separate the documentation/presentation rules from the modeling rules, our experiential bias of working in, on, and around drawings for so many years led to overlap, confusion, and contradiction in the final release. Over the years, we've rearranged and fixed some of these issues, and we've added more coverage for related topics. However, the overall intent of the standard, which was driven by the mindset of the people making it and their early ideas of MBD, left us with a slightly flawed standard. Please understand that I am not criticizing the standard or the people who made it, as I participated in its development for more than 10 years. I am criticizing the workflow we thought the standard should support and the way we intended people to use models and model-based information (sp. *a mano, manualmente*).

ASME Y14.41 emphasizes manual use, using annotated models manually. The version of MBD described in the standard is where an annotated 3D model replaces a 2D drawing as the product definition dataset, and the official information asset for a product is an annotated 3D model. MBD is not merely an annotated model, as additional information (ancillary information) may also be required, information that doesn't reside in a drawing or model.

For years, I felt we did a great job and hit it out of the park with our work. Annotated 3D models are better than drawings in many ways for presenting product definition data. While 2D drawings are static and all we can do is read them, an annotated model can be interrogated, there can be links between annotation and model geometry, so we can highlight geometry that a specification applies to, we can zoom in and out and rotate the model, and we can suppress or make certain geometry transparent to increase and speed up how well we understand the information. Annotated models offer many improvements over 2D drawings as product definition datasets. However, we made (and continue to make) one fundamental error. We are still focused on using model data manually (e.g. for visualization). All the benefits I listed above are improvements in visualization, and there's only so much value we can obtain by improving visualization.

The best use of MBD datasets is to use them directly in software. The only benefit we get by using them manually, e.g. by viewing them and manually interrogating them, is improved visualization. While improving visualization is important if we are using visual information, the benefits provided by improving visualization are low. The benefits are decreased time to understand what we see, potentially less errors, fewer questions and RFIs that come from the difficulty of understanding engineering drawings, and possibly less scrap and rework from misunderstandings. Much greater benefits can be realized if we use MBD datasets directly in software.

Our goal for MBD and MBE must be to use model-based information directly in software wherever possible. We can make annotated models for manual use, and we can make them for direct use. Directly using MBD information leads to much greater savings and benefits than manually using MBD information, likely an order of magnitude greater savings and benefits. As we will discuss later in the book, by itself, MBD used in design to create and release a model-based product definition dataset is of limited value if that is the only MBD done within the enterprise. The goal of MBE is to create and use MBD in as many places and in as many activities across the enterprise as possible. The real benefit of an MBD product definition dataset is when we use that information directly in software. For example, using 3D CAD model geometry and PMI directly in CAM or CMM software. The savings and business value gained from these activities in terms of cost, time, and labor are an order of magnitude more than from using MBD product definition data manually.

THE ISO 16792 STANDARD: TECHNICAL PRODUCT DOCUMENTATION – DIGITAL PRODUCT DEFINITION DATA PRACTICES

Soon after the release of ASME Y14.41-2003, members of International Organization for Standardization (ISO) asked if they could use ASME Y14.41-2003 as the basis for a new ISO standard on the same topic. ASME agreed and three years later ISO 16792:2006 was released. The standards were similar, with the main changes relating to the differences in terminology used for drawing practices between ISO and ASME, and the way that engineering drawing and related practices were structured in ISO and ASME. Here's an excerpt from the Introduction of ISO 16792:2006:

> "Every effort was made during the preparation of this International Standard — adapted from ASME Y14.41:2003 — to apply existing requirements developed for two-dimensional (2D) presentation equally to the output from three-dimensional (3D) models. Where new Geometrical Product Specification (GPS) rules have proved essential, these have been drafted with a view to their being equally applicable to both 2D and 3D. Therefore, in order to maintain the integrity of a single system, these new rules are being incorporated in the relevant existing ISO standards for cross-reference. Application examples have been included where, due to the specific requirements of 3D modelling, additional guidance was deemed beneficial."

> *(©ISO. This material is reproduced from ISO 16792:2006*
> *with permission of the American National Standards Institute*
> *(ANSI) on behalf of the International Organization for*
> *Standardization. All rights reserved.)*

The content, how it was developed, and the pros and cons of the content of ISO 16792 are almost the same as described above for ASME Y14.41-2003.

TOOLS IN PLACE

Today, we have the tools and technology in place to achieve greater productivity. Our IT and OT systems are ready, with a few additions and tweaks, to support model-based workflows. The technical aspect of transforming from our current document-based workflows to model-based workflows is the easy part. Buy some software and hardware, customize and integrate them into our workflows, do some training, and set a few new rules, and the technical part is taken care of. While the tools are the easy part, tools alone will not get us anywhere near where we need to get to.

CULTURAL ISSUES AND THE STATUS QUO IN DIGITAL TRANSFORMATION

The cultural part of transformation presents a challenge. Developing the right vision, recognizing our current costs and inefficiencies, understanding what we need to do, and getting people to accept, adopt, and implement it and follow it consistently are the real challenges. The primary challenge of becoming an MBE is cultural, not technical.

At the very core of this is that people have the wrong vision and the wrong motivations. Most people think we are trying to replace 2D drawings with 3D models in our workflows, that this is our main goal. We can replace 2D drawings with 3D models in our workflows but end up with very few benefits. There's a lot more to MBD and MBE than replacing 2D drawings with 3D models in MBD and MBE. It is not always easy to convince people that they could be working smarter, being more productive, and enjoying a greater sense of self-worth in their job if they were willing to adopt a new method. The status quo, how we do things today, is a potent and real danger to progress.

Most of us wake up each workday with the understanding that we will spend the day doing our job. Our job, loosely defined, is doing what we did yesterday and the day before, in the environment our employer constructed for us to operate in. Our goals, our methods, our tools, our systems, our metrics, and the expectations placed on us are in the context of the status quo, how the business operates, and what we are expected to do within that framework. Our careers, the metrics we are evaluated against, and our livelihoods are based on the status quo and how our performance is measured against it.

There is always a better way to do things. This is an oversimplified statement, but it is true. Change should be cautiously welcomed. The idea that change may improve our personal work situation, that it may improve organizational efficiency, should be cautiously accepted. Often, the response to change is worry, suspicion, and resistance. Change is often viewed in a negative context, sometimes met with quiet groans and grumbling, sometimes by overt action to challenge or block the change.

As a very generalized statement, a lot of people don't like change, especially at work. People settle into routines, and they adapt to organizational politics. They learn the written and unwritten rules and expectations of their jobs and take comfort from knowing these things. They know what's expected and how to perform adequately in that environment. If a new method such as MBE comes along, a method that seems like a radical change, people's comfort may be undermined. In some cases, change comes from one department or organization and another feels it is their *turf.* Turf wars are real and unfortunate. I've seen them with regard to MBD and MBE.

A more beneficial way to think about how people in an organization react to change is to consider it in the context of a system. The organization and its constituents operate as a system, or a system of systems, and the system and its constituents are productive. The system operates, produces, and meets its goals. In this environment, the goal of most people is to maintain the status quo. The goal is to keep things running smoothly to ensure productivity. The metrics for operations and people performing operations are based on maintaining or incrementally increasing productivity. In many organizations, people are expected to fight fires (deal with emergencies and unexpected problems) and are rewarded for doing so. Firefighting at work is a sign of poor or inadequate planning. Nonetheless, we shouldn't have to be firefighters at work, unless our jobs are truly as firefighters (e.g. wearing red helmets and flame-resistant PPE and putting out real fires...).

A common response to change and process improvement is to attack it, as the change is viewed as a threat. And, while that threat might not be seen as a personal threat, it is seen as a threat to productivity in the context of the status quo, in the context of doing what we do and what we are rewarded for doing. Thus, a big part of a successful MBE digital transformation is to create the environment in which change can occur and flourish.

Returning to the ten steps of transformation, we see that step 1 is technical and cultural, steps 2 and 3 are cultural, and steps 4–10 are technical and cultural (Table 2.1). This is a simplification, but the point is that cultural considerations are important in every phase of digital transformation.

The core concept of MBD and MBE is that we are trying to reduce or eliminate manual document-based processes and workflows. Using documents drives

TABLE 2.1

The Ten Steps of Transformation (from Chapter 1)

	Steps	Technical	Cultural
1.	Understanding	x	x
2.	Accepting		x
3.	Adopting		x
4.	Evaluating	x	x
5.	Planning	x	x
6.	Developing	x	x
7.	Testing	x	x
8.	Optimizing	x	x
9.	Implementing	x	x
10.	Continuously improving	x	x

manual workflows, and manual workflows drive the need for documents, documents as input, documents as output, and documents as records of activity and milestones. Eliminating or reducing this cycle of documents and associated manual activity is how we can achieve significant benefits from MBD and becoming an MBE.

I've found a few simple ideas to help us understand what we are trying to accomplish with MBE, what stands in our way, and how to achieve our goals. One idea relates to the status quo and resistance to change. We can think of the status quo, people's understanding of it, their expectations, their comfort with it, and resistance to changing it, as inertia. Cultural inertia. Our job in an MBE digital transformation is to develop enough momentum to overcome the inertia of the status quo. I like using these physics concepts to describe what we are up against and how change can be accomplished. It would be easy if all we had to do is plug numbers into a formula. That isn't how organizational change works, but as we will discuss later in the book, we will establish an organizational baseline and plan for digital transformation to a future state that embraces MBE. In those activities, along with the phase where we estimate return on investment (ROI) from transformation, we will develop the numbers that matter to business, and to real success. The time, money, and labor savings and the improvements in people's jobs from properly implementing MBD and MBE will create the momentum needed to overcome inertia.

Critically, an environment must be established and maintained that allows a team to develop, test, and optimize model-based workflows that is outside the influence of production. The organization must embrace, own, and support this work if it is to succeed.

STATUS QUO SYNOPSIS

In many ways, business culture at OEMs and similar environments is backward looking. What we do today is influenced by the past. The longer a company has been around, the further they look back and the deeper their values, processes, and goals are rooted. Our business methods are primarily manual document-based processes and workflows. We are organized in siloes, and our metrics, rewards, and the expectations placed upon us reinforce the siloes, creating insular thinking, values, and behavior. Because we are using manual document-based processes and workflows, much of what we do is tangentially related to our real goals (e.g. we spend a lot of time and effort working on documents to drive or record the results of manual processes rather than on adding value to the product). We spend an inordinate amount of time, effort, and money fighting fires. We are distracted because a lot of our otherwise productive time and brainpower is spent working on non-value added tasks (NVATs), like creating and managing documents. Our skills and effectiveness at work revolve around documents and manual activities. Because of the emphasis we place on manual document-based processes and workflows, it takes a long time to bring new hires up to speed, especially RCGs. There are underlying morale issues, as people are drained chasing documents around all day – RCGs must wonder why what they learned in school didn't prepare them to spend all day chasing documents. The metrics and rewards in our firms are siloed, they are based on manual document-based processes and workflows, they (explicitly or implicitly) reward firefighting, and they are backward-looking.

When we think about process improvement and change, we often think in very incremental terms. Most MBD and MBE implementations represent small incremental change, change that may or may not really be an improvement. This is largely because we are stuck thinking in the context of manual document-based processes and workflows. We may decide to use MBD in design, but we release product definition data that is essentially the same as what we had on drawings, information that is intended to be used manually in document-based workflows. We don't really change how we work; we dot an "*i*" or cross a "*t*" and think we made a big change.

Most MBD implementations create annotated models that include the same information presented essentially the same way as on drawings – most of the information in the models is used the same way as the information on drawings – it is read or viewed and used manually – people work the same way with annotated models as they did with drawings. The media changed, but the work methods didn't. The official representation of the product is still visual information, and that visual information still drives all other work. The costs of changing from a document-based/drawing-based workflow to a model-based workflow are about the same if we make only a very small change in how we work as described above (e.g. changing the media we read from a drawing to an annotated 3D model) or if we make a radical change that eliminates 50% of the steps and most of the NVATs in our workflows.

This is an important point: The cost in terms of culture shock, money, time, labor, and resources is about the same whether we make a trivial change that provides no real value to the organization or a meaningful change that provides significant value to the organization. We should opt to make a meaningful change that significantly increases value for the organization. It's going to cost about the same amount whether we do window dressing or if we change how we create and use information to obtain maximum value for the organization.

The cost in terms of culture shock, money, time, labor, and resources is about the same whether we make a trivial change that provides no real value to the organization or a meaningful change that provides significant value to the organization.

MBD AND MBE INDUSTRY BREAKDOWN

If we start with the traditional idea of MBD, defining product geometry with 3D CAD model geometry instead of by dimensions on engineering drawings, then MBD was initiated at companies that worked with complex product geometry. Some geometry is too complex to be completely defined by dimensions on drawings (explicit dimensions).

Complex geometry may be driven by function, such as shapes that interact with fluids (e.g. airfoils, pump impellers, propellers, and turbine blades), medical shapes optimized for interface with the human body, (e.g. anatomically shaped medical implants and structural components), shapes optimized for certain manufacturing processes, (e.g. die casting and injection molding), shapes optimized for ergonomics, (e.g. a computer mouse), and shapes optimized for aesthetic purposes, (e.g. freeform shapes of styling surfaces on consumer electronics, watercraft, automobiles, hot tubs, and appliances). All these examples include complex geometry. One thing that separates simple geometry and complex geometry is whether the geometry can be completely defined by explicit dimensions. Simple geometry like planes, cylinders, cones, and spheres can be easily dimensioned. Complex shapes, not so much. Back in the days before 3D CAD modeling, if a complex surface was designed, it was approximated in design, it was approximated in manufacturing, and it was approximated in inspection. We used tools like French curves to design such shapes before CAD. Specifying explicit dimensions on engineering drawings is a relatively recent phenomenon – prior to that drawings for complex shapes were undimensioned templates of cross-sections. This was common in automotive, aerospace, shipbuilding, and architecture. In fact, there are still a few holdouts using this technique in some environments.

3D CAD software enabled us to create complex 3D CAD model geometry, which in turn drove manufacturing and inspection software capable of directly using complex 3D CAD and neutral data model geometry. 3D CAD also allowed us to easily make drawings, although it didn't solve the problem of not being able to completely dimension complex shapes. Industries that dealt with complex geometry were pioneers in using 3D CAD and neutral model geometry to represent product geometry.

Thus, early leaders in MBD included industries such as automotive, aerospace, medical, and consumer products. In my experience, products manufactured using processes requiring dies, such as die casting, injection molding, and transfer die sheet metal stamping processes, were early candidates for MBD.

More recently, an entire ecosystem has developed that is model-based – additive manufacturing or 3D printing. I like the example of additive manufacturing because a model is required. The model is the basis for the product definition. The process requires a model, thus, the process demands a model – I like to think that the process has *pull*, it *pulls* (demands) a model from design, as the model is the basis for the manufacturing process.

Today, companies in every industry are becoming interested in MBD and MBE. As people realize that there is a better and more productive way to work, they are turning to MBD and MBE. There seems to be a lot more interest in MBD, which I assume is because more has been written about it and there are more MBD software products being promoted online. Unfortunately, most software products being touted are focused on the traditional idea of MBD, replacing 2D drawings with 3D CAD and neutral annotated models. I am not saying that the only MBD software available is focused on replacing drawings with models – I am saying that most of the noise seems to be for such products. There are many other products that facilitate MBD and MBE available that are not necessarily marketed as MBD and MBE products. PLM software is a good example. PLM is one of the most critical parts of the MBE ecosystem, as it is the backbone of product and related information throughout the organization and the product lifecycle.

I use the terms *push* and *pull* throughout this book. I use the terms in slightly different ways, but in general *push* means something is forced or pushed onto others and *pull* means something is requested or pulled by others. In the context of technological change in a business setting, *pull* is better than *push*. As a comparison, *push* can be equated with throwing something (like a drawing) over the wall, which is often thrown back over the wall broken and abused, whereas *pull* is requested by the recipients and lands in welcoming arms. Most early MBD/MBE implementations were *push*, where design engineering decided that it was a good idea to use 2D or 3D CAD geometry instead of a drawing with explicit dimensions to define product geometry (without telling anybody outside design). I am a staunch supporter of using CAD model geometry to define product geometry. However, it is best done in a controlled manner with careful planning, risk assessment and mitigation, coordination with and buy-in from other stakeholders. This is achieved with carefully thought-out value propositions and statements of roles, goals, and benefits for all involved. Most of the early MBD initiatives I witnessed were like guerilla strikes, where, without notice, drawings were partially or wholly replaced by CAD models.

3 Information Type, Structure, Representation, and Presentation

TERMS AND DEFINITIONS

DIGITALLY-DEFINED INFORMATION

Information defined digitally that represents something of interest. In the context of MBE today, digitally-defined information includes all computer files, digital datasets, information models, and other digital objects of interest to an organization. Examples include digital documents (including drawings), models, datasets, lists, instructions, code, etc., that represent information about a product, its production, its status (e.g. twin), etc.

Note that digitally-defined information may or may not be structured to be understood and used directly by software. Direct-use information is a special type of digitally-defined information that is optimized to be used directly by software.

DIRECT-USE INFORMATION (INFORMATION FOR DIRECT USE) (INFORMATION FOR AUTONOMOUS USE AND AUTONOMOUS OPERATIONS)

Information that can be read/parsed by software, understood by software, and used directly in software without the need for human intervention (a person does not have to act as an intermediary to use the information). The information is intended to be used autonomously. In most MBE use cases, direct-use information is intended to be used directly by software – it is not intended to be used by people. Thus, direct-use information does not need to be presented.

In the case of recent AI models in which large sets of unstructured data are parsed and learned from to train a model, the data wasn't created with the intention of being used directly by software. In many cases, it was just information that was collected from web-based sources and from records of web-based transactions. While I contend that an MBE should use information intended to be used directly, methods such as general AI allow businesses to obtain value from information intended for other uses.

A similar case has evolved with data mining of corporate information. MBD opens up a company's intellectual property (IP) and makes it visible to and

DOI: 10.1201/9781003203797-3

available for other uses. Thus, model-based IP is structured to facilitate other uses. As MBD information is structured data supported by metadata, it can easily be parsed and understood, which allows the company to use that information as it wishes. This is the promise of big data in an OEM and its supply chains. While unstructured information or information that is not structured for a new use case may be used for the new use case if it is enabled by programming, information structured for its intended use is preferred.

As I stated above, direct-use information does not need to be presented. This may make many readers uncomfortable. In many cases, we will still want to see or read visualization information. In those cases, direct-use information should be used as the basis for presented information. For example, a CAD model visible onscreen is a visual depiction of the underlying digital representation of product geometry, material, visual attributes, etc. in the model. We use the digital representation directly in our CAM and CMM software, but we view the rendered image onscreen. In all cases, however, it is critical that we designate which version of information is the official version – I recommend designating the direct-use version of information as the official version as much as possible.

Direct-use information may be used directly in software:

- As the basis for presenting manual-use information (e.g. direct-use information is used to present visualization, audio, or other manual-use information).
- To automate and directly drive a process, such as estimating, procurement, manufacturing, inspection, etc. Note that the entire process or portions of the process may be driven by direct-use information. This is an autonomous use case.

Information is commonly used in more than one way or for more than one purpose. See the following examples:

- Example 1: 3D CAD model data. CAD model geometry that represents a part's geometry adequately to directly drive a manufacturing, inspection, or other process is direct-use information. The portion of the CAD file that defines how the model appears visually on screen (view/display state/scene/...) is directly used by the CAD software to present manual-use information. The information defining visual presentation is not required by manufacturing, inspection, or other processes. In this example, both types of information exist in the same file. Also, the manual-use information (the visualization) should be based on the direct-use information.
- Example 2: Weather data. Large datasets are used to model and forecast/predict the weather. Digital data are fed to models (computer programs) that process the data to represent current weather conditions and forecast future weather conditions. Note that the current and future weather models exist in software as digital data, not necessarily for human use. This data is a form of direct-use information. The data is used directly in the weather modeling software to populate a weather model. However, this

is not how most of us humans interact with this information. We don't interact with the information model, the 1s and 0s. We interact with its presented form – we interact with visualizations of static and dynamic weather maps, charts, graphics, or histograms in a variety of media. As in example 1 above, the underlying digital data and its presentation satisfy different use cases: digital data constitutes the underlying weather model, and the underlying digital weather model is processed and used to drive visualizations of the model in ways that people can understand.

In an MBE context, direct-use information adheres to an applicable data format (e.g. native CAD, STEP, JT, QIF, etc.), is properly structured, and elements and objects in the information are properly associated with other elements and objects in the dataset.

DUAL-USE INFORMATION

Information that is intended to be used directly and manually.

It is difficult to optimize information to be used manually and directly. That is, it is difficult to optimize the same information for manual use and for direct use. In many cases, released information has more than one form, with one form derived from the other. Preferably, the direct-use information is designated as the official information and the manual-use derivative is designated as reference. The direct-use information is used to drive manual-use information.

INFORMATION (BUSINESS INFORMATION, FROM CHAPTER 1)

Data that is adequately structured and has context such that it is useful for some purpose.

Information that is relevant to a business, its products, processes, activities, etc. Business information has value for a business and can be classified as intellectual property.

INFORMATION ASSET (FROM CHAPTER 1)

An intangible asset that contains information. In the context of business (e.g. OEMs and similar organizations), information assets contain information that is relevant to their products, processes, activities, inventions, copyrights, trademarks, how they do business, etc.

Examples of information assets include documents, drawings (a type of document), information models, the digital information represented in information model files, such as native CAD, STEP, JT, or QIF files, model-based systems engineering models (SysML models), analytical models, simulation models, digital twins (a type of simulation model), manufacturing, assembly, or inspection work instructions, CNC, CMM, and robot activity programs, additive manufacturing datasets, machine programs and instructions, etc.

INTANGIBLE ASSET (FROM CHAPTER 1)

"An intangible asset is an asset that is not physical in nature. Goodwill, brand recognition and intellectual property, such as patents, trademarks, and copyrights, are all intangible assets. Intangible assets exist in opposition to tangible assets, which include land, vehicles, equipment, and inventory."

Investopedia (online) ("What Are Intangible Assets? Examples and How to Value")

Note that the information in some intangible assets may be presented visually, such as printed documents, rendered CAD models, results of analyses, weather maps, etc.

In the context of MBE, digital datasets are an important type of intangible assets.

MANUAL-USE INFORMATION (INFORMATION FOR MANUAL USE)

Information that is defined and intended to be used manually, such as information depicted in documents, on drawings, in lists, visualization data (e.g. FEA deformation diagram, flowchart, schematic, logic diagram, wiring diagram, P&ID, weather map, topographical map), recorded messages, audible instructions, haptic feedback, etc.

PRESENTED INFORMATION

Information that is presented in a manner for a human to use as sensory input. In the context of MBE today, most presented information is visual information. Visual information may be presented on hard copy media, it may be presented in 2D or 3D, it may be presented on a 2D drawing or a document, it may be a view of a rendered 3D model onscreen, it may be a color-coded FEA deflection map, it may be static or dynamic, it may allow user interaction, etc.

If information is intended to be understood by a person by interacting with the information, it is presented information. Note that some information may be visible or audible but not intended to be viewed or understood by a person, such as smudged ink on a printed drawing, malfunctioning pixels on a display, static in an audio recording, and unintentional visual artifacts in a video recording.

Note that presented information can take many forms, such as audio information, visual information, tactile information, olfactory information, and other types of information. This text primarily discusses visually presented information.

REPRESENTED INFORMATION

Digitally encoded information that can be understood and used by computational systems.

Represented information may be direct-use information (e.g. to be used directly in software or computation), it may be manual-use information represented by digital information (e.g. underlying data driving the display of onscreen graphics intended as official information), or it may be information that is represented digitally but is not useful in a particular context.

TYPES OF INFORMATION

There are several types of information to consider in MBE, some of which are presented in Figure 3.1.

INFORMATION STRUCTURE BY USE CASE

Information should be structured for its intended use. This concept is obvious to most of us, as we learn this throughout our education. Historically, we were limited to using information manually, so we focused on structuring information to be used manually. We have learned to structure and optimize information for manual use. These methods and the need to use them are embedded in most of our heads and work methods. This has biased our thinking toward using information manually and optimizing information for manual use.

If we want to be successful with MBD and MBE, we must create direct-use information and use that information directly in software. Specifically, we need to structure information intended to be used directly in software for direct use. If we want to use information manually, then we need to structure that information for manual use. To summarize, we need to structure information for its intended

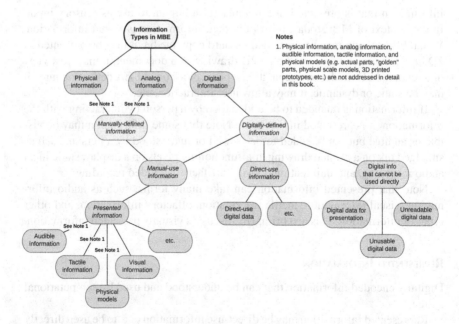

FIGURE 3.1 Types of information used in MBE.

use. This means we must clearly define the goals, work methods, and use cases for our information, *before we create the information.*

Here's a synopsis of best practice for structuring information:

- Presented information should be structured and optimized to be used manually by people.
- Direct-use information should be structured and optimized to be used directly by software.

The needs of these two use cases differ, with little overlap. Thus, we need to make sure we understand the needs for each use case and the criteria that we need to satisfy. Note that in many cases, we will want to use some information manually and other information directly, even if that information is in the same dataset, and even if the same information is used to support manual and direct use. This idea is expanded below.

PRESENTED INFORMATION

Note: For the remainder of this book, presented information is limited to visual information. Audible, tactile/haptic feedback, physical models, and other presented information are not addressed in detail.

Presented information should be structured and optimized for manual use. The information should be presented so it is easy to read, easy to view, easy to find information, easy to find related information, easy to find similar information, easy to understand, and, by contrast, difficult to misunderstand. While we cannot ensure that someone will understand the information we create, we must try to make sure that what we create is laid out and presented optimally for its purpose. Doing a good job of authoring information for manual use means trying to make it as easy as possible for someone using the information manually to get it right.

For presented information, it is critical that the information is presented in a way that is assimilable, and if possible, is presented in a familiar manner. This is why drafting manuals, drawing practices, and standards have been created and evolved over time. Drawing practices and standards provide methods to present graphical and textual information clearly and consistently. Note that a drawing or annotated model can be difficult to understand even if drawing or model annotation standards are followed. Understanding involves two parties: a sender and a receiver. A document, drawing, annotated model, or other presented information may be misunderstood if the recipient is not familiar with what is being presented. From an authoring point of view, we must try to do our best to make sure what we create is easy to use and understand.

The visual format of presented information (how it is visually arranged, how it is formatted, and how it's presented, etc.) is very important. Whether it's the sentences, paragraphs, chapters, table of contents, index, footnotes, and other written material, or the views and graphical presentation of geometry, or annotation, notes,

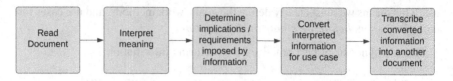

FIGURE 3.2 Read-interpret-determine implications-convert-transcribe (RIDICT) sequence for using visually presented information.

title block and information on engineering drawings, and annotation and meta-data elements presented on an annotated model, or the fields and records in a parts list or Bill of Materials (BOM), the format is critical. It is also critical that presented information is presented consistently from dataset to dataset.

Not only should presented information be easy to understand, but it must also be easy for a reader to recognize if other information is related to information being considered, and if so, that the related information is easy to find. In writing, in drafting, on engineering drawings, and on annotated models, we have developed methods to help readers recognize and use related information. However, there are many, many ways for people to miss and misunderstand presented information.

Figure 3.2 repeats the steps of using visual information manually shown in Figure 2.2.

Figure 3.3 shows a few common failure modes associated with each step in the process of manually using visually presented information. The steps, inefficiencies, potential failure modes, and risks of creating and using presented information will be addressed in detail in Chapter 6.

Digitalization has improved our ability to find related information, as evidenced by hyperlinks in digital documents and online. Metadata, tags, and hyperlinks weave the web and provide many paths down rabbit holes…

Metadata, tags, and hyperlinks in engineering and business information significantly improve finding related information. Database structures also help define relationships between information, be it the more traditional approach of standardized structures and naming of records and fields to more flexible approaches like graph databases and vector databases. Digitally defined information can be used in more ways than non-digital information, such as information printed on hard copy, analog information, etc. Digitally defined information can easily be used in more ways than originally intended. Case in point, consider this or any other of my books or articles: in printed form, or if just reading the material in digital format, the reader may traverse the material linearly, from start to finish, chapter-by-chapter, or perhaps the entire book. To circumvent this process, readers may refer to the table of contents or hope for a good index to guide them to specific topics or keywords. If the reader is using software like Acrobat Reader, they can open the Find function (ctrl-F) and search for particular strings, regardless of the table of contents or the index. If the string is in the digital representation of the text, the software will return all instances.

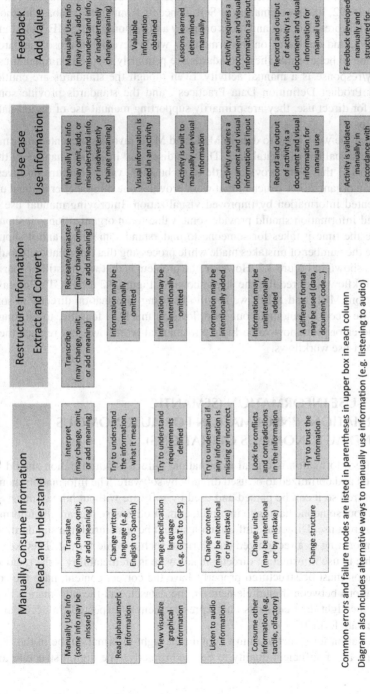

Steps/Stages and Problems of Manually Using Visual Information

Manually Consume Information		Restructure Information		Use Case	Feedback
Read and Understand		Extract and Convert		Use Information	Add Value
Manually Use Info (some info may be missed)	Translate (may change, omit, or add meaning)	Transcribe (may change, omit, or add meaning)	Recreate/remaster (may change, omit, or add meaning)	Manually Use Info (may omit, add, or misunderstand info, or inadvertently change meaning)	Manually Use Info (may omit, add, or misunderstand info, or inadvertently change meaning)
	Interpret (may change, omit, or add meaning)				
Read alphanumeric information	Change written language (e.g. English to Spanish)	Try to understand the information, what it means	Information may be intentionally omitted	Visual information is used in an activity	Valuable information obtained
View visualize graphical information	Change specification language (e.g. GD&T to GPS)	Try to understand requirements defined	Information may be unintentionally omitted	Activity is built to manually use visual information	Lessons learned determined manually
Listen to audio information	Change content (may be intentional or by mistake)	Try to understand if any information is missing or incorrect	Information may be intentionally added	Activity requires a document and visual information as input	Activity requires a document and visual information as input
Consume other information (e.g. tactile, olfactory)	Change units (may be intentional or by mistake)	Look for conflicts and contradictions in the information	Information may be unintentionally added	Record and output of activity is a document and visual information for manual use	Record and output of activity is a document and visual information for manual use
	Change structure	Try to trust the information	A different format may be used (data, document, code...)	Activity is validated manually, in accordance with visual information	Feedback developed manually and structured for manual use

Common errors and failure modes are listed in parentheses in upper box in each column
Diagram also includes alternative ways to manually use information (e.g. listening to audio)
Shortcomings and other problems are also listed

FIGURE 3.3   Steps of using information manually with common failure modes.

A lot of the material in ASME Y14.41 and ISO 16792 Standards is a variation on this theme, of user-defined searches ("queries" as defined in ASME Y14.41 and ISO 16792, likely originating in SQL concepts from early database methodologies). The prevalence of and focus on query and response in ASME Y14.41 and ISO 16792, and our focus on these methods while developing those standards, has led me to believe that these standards are primarily focused on manual use, as query/response is a manual activity. Even though the standards are entitled "Digital Product Definition Data Practices", and the standards provide some support for direct use, they are primarily supporting manual use of digital data. Baby steps.

Most of the work done to date in MBD and MBE involves presented information and manual use of digital data. Thus, as discussed in earlier chapters of this book, most of that work provides little to no business value or return on investment to an organization. Much of that work provides one thing: better manual use of presented information by improved visualization. Improving manual use of presented information should provide some value to an organization – it should decrease the time it takes for someone to understand something, and it should decrease the number of mistakes made while processing that information through the steps shown in Figure 3.2. However, the problem is that we are still primarily using workflows that require the steps shown in Figure 3.2 (RIDICT). To really benefit from MBD and MBE, we need to be using information directly in software – we need to automate our workflows as much as feasible, to eliminate mundane error-prone low-value high-skill manual workflows and replace them with direct-use workflows.

DIRECT-USE INFORMATION (SEMANTIC INFORMATION) (INFORMATION FOR AUTONOMOUS USE AND AUTONOMOUS OPERATIONS)

Direct-use information is digital information. Direct-use information should be structured and optimized to be used directly by software. The information should be structured so it can be parsed and used by software without human intervention, without someone having to interpret, edit, or otherwise prepare the information for direct use. Direct-use information should be structured in a standardized way, conforming to a recognized data format, such as a file format. They key to being able to use information directly in software is predictability – the required information must be structured properly, have the correct content, have the correct relations between digital elements in the dataset, and the information must contain the right data elements and represent them as expected (in accordance with the data format).

Information for direct use must contain the right information, be in the right form, and be of sufficient quality to be used directly. The digital structure and content of direct-use information are very important. However, what direct-use information looks like, how it is visually arranged, its graphical format, and how it's presented are irrelevant. The visible appearance of information does not matter for direct use. If text appears upside down or backward, if colors are wrong or missing, if geometry appears one way or another, none of these presentation errors matter for direct use. What matters is if the 1s and 0s are correct, understandable, and usable.

The quality criteria for manual-use information and direct-use information differ greatly.

The distinction between what constitutes high quality for direct use and what constitutes high quality for manual use is difficult for people used to working in manual-use environments to accept and adopt. We will address this in detail in Chapter 6. While humans adapt to incomplete, incorrect, or conflicting input, software is less adaptable and resilient. Humans can use messy, ambiguous, conflicting, and incomplete documents and drawings, but software will not be able to process improperly structured digital data. Thus, not only are quality criteria different for presented and direct-use information, but the level of quality required for direct use is higher than for manual use. A good example is how, when people

ON AI AND MBE

I explained that direct-use information must be unambiguous, complete, structured, high quality, etc. This is indeed very important. However, recent advances in generative AI bring several surprises. To varying degrees, generative AI:

- Is able to create relevant content from seemingly irrelevant source material (e.g. the relatively random information the large language model was trained on).
- Can deal with ambiguous, incomplete, inadequately structured, and lower-quality information.

Generative AI, by its very nature, creates information based on its source material and underlying model. The AI can do something similar to what people do when they use the RIDICT process (reading, interpreting, determining, implications, converting, and transcribing information). The AI can act as an intermediary between an organization's information and how the information is used, similar to how a person acts as an intermediary when using the RIDICT process. However, the ways that people and AI address these circumstances differ. People acting as intermediaries (e.g. RIDICT) use their judgment, refer to other materials, use their experience, consult others, and make decisions to address ambiguous, incomplete, inadequately structured, lower-quality information. People use their judgment to fill in the blanks so to speak. Generative AI chatbots are all about filling in the blanks (or more accurately, statistically determining the most likely next character and word in a sequence). However, there is a difference between how a person deals with ambiguous, incomplete, inadequately structured, lower-quality information, and how AI would deal with it. I find it interesting that computational systems can deal with messy information and achieve an acceptable result at all. Remember my stories about the aerospace manufacturing engineers and the people building the vehicle inspection fixture – both are examples of people used to working with ambiguous, incomplete, inadequately structured, lower-quality information, people who expect information to be ambiguous, incomplete, inadequately structured, and low-quality. Historically, to work in those environments required people to act as intermediaries and fill in the blanks. Now we are gaining computational tools that can perform a similar role. I stand by my contention that we shouldn't be producing ambiguous, incomplete, inadequately structured, lower-quality information in the first place. However, generative AI is rapidly changing the landscape of what we do and how we work. I look forward to seeing how much value can be derived by training LLMs with proprietary datasets within our organizations.

are reading written text, often fill in missing words. (If you didn't notice, I omitted "they" after the comma in the last sentence.)

DOES PMI HAVE TO BE VISIBLE TO BE CONSIDERED PMI?

As we were developing the first edition of ISO 10303-242 (STEP AP 242), I was explaining drafting rules and presentation rules for specific GD&T and GPS constructs. A friend on the PMI team, who is a talented longtime information modeler, asked me a question that has driven a lot of what is in this book. He asked, *"Does PMI have to be visible to be PMI?"* I had to stop and think. My background in drafting, design, engineering documentation, engineering standards, drafting standards, ASME Y14.5, ASME Y14.41, etc. affected by thinking. After a little consideration, I realized what he was getting at, as we were developing an information modeling standard, to define information models that could be used directly by software. Truly, information does not have to be visible or follow any

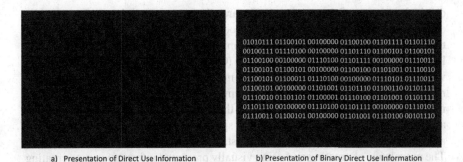

a) Presentation of Direct Use Information
   (yes, it is intentionally blank)

b) Presentation of Binary Direct Use Information

FIGURE 3.4 What digital data for direct use looks like.

presentation format to be used directly in software. The software "sees" ones and zeroes, or whatever digital content is being consumed. Some years later, illuminated by this and similar experiences, I presented a paper at the MBE Summit in Gaithersburg, MD. In the presentation, I had a series of slides that eventually showed what a digital dataset looks like (see Figure 3.4). Note that the screen went dark, with a blank background with nothing to view or read (see Figure 3.4a). I then filled the screen with binary code. Feel free to translate the code from Figure 3.4b below. People expected to see screenshots of pretty rendered 3D models decorated with GD&T, but I gave them this.

The information above is important and not just a funny story. Current methods to define specifications for direct use are inadequately addressed in engineering standards. The idea that PMI does not have to be visible to be legitimate, that specifications can be encoded digitally and not presented visually, is so important that I am proposing new rules in standards that state that digital information that defines a specification per the standard is legally equivalent to a presented specification with equivalent content. This must be true if we are to be able to use digital information models directly in software. It's possible that requirements defined in the digital information model will not be presented visually, but still must be legally binding. There are many examples where this is required, from material specifications to GD&T to welding specifications to product identification metadata, revision, and version information.

HOW WE CREATE/AUTHOR INFORMATION

As previously discussed, most activities that create or use information in OEM environments are manual-use/visualization/document-centric activities. Our focus is visualization. We focus on presenting information for people to use manually to read or view the information. We do this without a second thought, as this has been the way things have been done for centuries. This is still true today, even after the Third Industrial Revolution, in which we computerized most of what we do, deployed powerful computers, software, platforms, networks, the Internet, the cloud, robotics, AI, etc. Our authoring tools, the way we create information and the way we expect it to be used are primarily visual. Our interfaces are visual. Everything has a graphical user interface (GUI). In the case of mechanical design, we model parts visually to create visual models. The entire process is visual. 3D modeling and rendering are orders of magnitude better than the 2D drafting methods they replaced. Been there, done that. But we use modeling processes that are inherently visual. Product geometry, specifications, and metadata are entered and presented visually, so we can understand and use the information manually. The prevalence of creating and using visually presented information puts creating and using information directly in software at risk. Quite simply, our priorities, rules, and thinking are biased toward visual and manual use, and our software tools have been developed and optimized for visual and manual use. One of my goals in this book is to help people understand self-imposed limitations and help software companies develop tools that are intended to support direct-use information and truly model-based workflows.

In contrast to the visual in, visual out method for manual-use cases described above, there are some digital model-based methods that focus on direct-use workflows. I am thinking about generative design. Autodesk® and a few other companies have cloud-based tools in which design constraints are input, processed in their software, and a series of solutions are offered as results. The process is similar to our traditional mechanical design process performed by engineers, except instead of humans developing the solutions to satisfy the constrained design space, software does it, typically providing many options. In the current generative paradigm, a designer enters the requirements for the product and its operational and environmental constraints, and the software generates solutions (hence the term "generative").

Someday all design will be done this way. Today, the solution space is mainly limited to geometry. In the future, specifications (GD&T, welding symbols, surface texture, thread specifications, etc.) will also be generated. A little further into the future, we won't need specifications. We will input the requirements and constraints for a system, and the system and its constituent components will be generated, simulated, analyzed, manufactured, evaluated, and shipped.

Today, we use CAD to create information that we can visualize and use manually. That is where we've been with CAD since its inception. While there is some direct use of digital information, most use cases today are still manual, which is okay. The tricky part is that today, if we want to use MBD and become an MBE, we need to make sure that the CAD data is also directly usable. This is where a lot of companies stumble, as they don't understand the difference between the quality criteria for manual use and direct use. Trying to use digital information intended for manual use directly in software often leads to problems and yields unsatisfactory results.

Most of the time we don't start out with direct-use information. We start out with manual-use information, and either by being careful during the development process or by cleaning up our initial models, we can end up with direct-use information. In these cases, direct use is an afterthought. In fact, at some point in a model-based modeling process, the information bifurcates into manual-use information, the stuff we see and can understand (pretty pictures and text), and direct-use information, the stuff that software can use and understand (1s and 0s). Usually, we end up with direct-use and manual-use versions of the information in the same dataset.

Note that many technology companies are not focused on documents and document-based workflows. Many companies, such as Google, Meta, PayPal, Block, Stripe, Shopify, Zoom, and many others, focus on providing digital data for various uses. Google and Meta provide services to users that come with caveats that allow them to harvest the users' information, add value, and to sell that information to advertisers. PayPal, Block, Stripe, and Shopify provide financial information management services that facilitate web-based commerce and Zoom provides information management services that facilitate online communication. While these companies may use document-based workflows internally, their customer-facing offerings are not about documents. Of course, there are many software companies that cater to and support document-based workflows.

With OEMs and most other non-tech companies, document-based workflows have been and continue to be how business is done. What is different about newer tech-oriented companies is that many started as greenfield companies, without history and its commensurate inertia. I've spent a lot of time and had a lot of discussions with people in the MBE space about how to crack the nut and get OEMs to see the light and pursue meaningful MBE (or even to get them to accept the old models versus drawings idea of MBE). As I searched for companies that didn't have historical bias, new tech companies seem to be different. I assume this distinction will fade as time goes on and more companies adopt optimized digital workflows and information management.

Sometimes I am surprised when I see what comes out of traditional OEMs as they move toward digitalization and digital transformation. Their document-based manual-use status quo is evident in what they see as progressive and the next big thing. A lot of their digital transformations are trivial and will not yield big payoffs, as they are not improving how they work in a meaningful way.

In 2019, I attended a major CAD/CAM/PLM software conference in Las Vegas. The event was mostly focused on CAM and related software, although the sponsor offered a full suite of impressive CAD, CAM, PLM, process management, and related software. As with most events of this type, many of their internal experts made presentations about what was possible, and their OEM customers made presentations about what they had done and implemented at their companies. I was surprised to see that most of the presentations were about presenting information for manual use. Many people were focused on information dashboards where information for a facility, process, or activity was displayed on a single screen with pretty graphics. Such dashboards are great tools to help people understand a process and act on what they see. But that's a manual activity, a manual-use case. Common to many such events, people took baby steps but thought they took giant leaps. I've seen the same thing at MBE conferences. People are excited about switching from 2D drawings to 3D PDFs that ended up being used almost exactly the same way as 2D drawings, in manual-use cases. Such activities are often snake holes into which companies pour money only to eventually realize the payback isn't there or it is insufficient to justify the work and expense of the activity.

The moral of the story above is that the goal of these activities is to provide presented information that people will and must use manually. Our goal should be to reduce manual-use cases, not to spend more money on them or to provide prettier pictures to drive them.

The challenge of having two versions of the information is that we need two quality processes and two versions of quality criteria to manage and ensure the information is suitable for each use case. What is important and constitutes quality for presented information is different from what is important and constitutes quality for direct-use information. We will expand on this later.

Thus, current practices mostly prioritize and focus on manual-use cases. We create information to use manually, we expect it to be used manually, and it is used manually. This approach has worked well enough throughout the time we were constrained to work manually. However, with the Third Industrial Revolution and computers, the cloud, powerful software and computer-based machinery, and more recently MBE and Industry 4.0/smart factory, people are trying to use information directly in software. We are at an inflection point where the manual methods and information assets of the past and their current digital incarnations are clashing with direct-use workflows. People are trying to use MBD and MBE principles, they are trying to adopt Industry 4.0, but their inherent manual-workflow bias makes it difficult to adapt to direct-use workflows. To put it simply, the way we are used to working and what we are used to creating are inadequate for direct use. Authoring to suit manual use makes it more difficult to create and optimize information for direct use. Because we are used to manual-use workflows, these issues are overlooked, as manual-use information assets seem well suited to our needs, as most people judge their needs in the current context, not the context of a future based on direct use.

INTEROPERABILITY AND DATA FORMATS

Most software saves information in a native format. Some native formats are unique to the software, and some native formats are proprietary to (owned, limited, and controlled by) the company that develops the software. Thus, options to save a digital data file in a desired format may not be available in software, even software that is similar to or serves a similar purpose (e.g. different CAD software), or in software designed to use data from that type of software directly (e.g. using native CAD data in CAM software). Also, the different programs used in a digital workflow (e.g. from CAD to CAE, CAD to CAM, CAD to CAI) might not read or write the same format. So, ways to convert data from one digital format to another are needed.

Note that different programs that serve a similar purpose (e.g. CAD software) may use different data structures and data formats, may have different ways of representing information, and may even represent different information. See the following examples:

Example 1   Software A and software B both serve the same purpose. Software A allows for and supports 500 data elements (fields) in a record, while software B allows for and supports 450 data elements in a record.

Example 2   Software A and software B support the same number of data elements in a record, but each software supports different data elements.

Example 3   Software A and software B structure data in a data file very differently, and thus converting a data file from software A into a format used by software B requires restructuring the data.

The examples above are incomplete and present just a few of the challenges of ensuring interoperable data. While off-the-shelf data translation software is available for MBD and MBE, often the software only converts a portion of the source data. Sometimes you need a more powerful customized tool to convert the rest of the data. In some cases, it is easy to convert data from a source format to a target format so that the data can be used as intended. In other cases, converting data from a source format to a target format in a usable form is difficult.

Interoperability of data formats and information is critical to model-based business activities. Information created in one activity must be understood to be used in other activities. Organizations have many different software applications that use different data formats. Some data formats are proprietary, some are standardized, and some are somewhere in the middle. Regardless of the format, if digital data created in one system is intended to be understood and used by a different system, the systems must understand and use the same format, or there must be a way to translate the data from its source format to a target format.

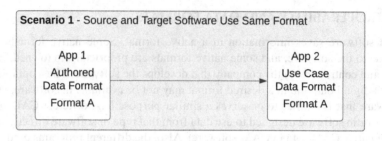

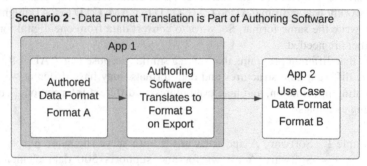

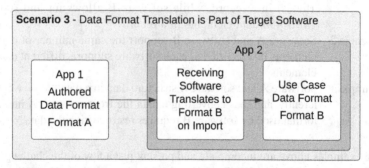

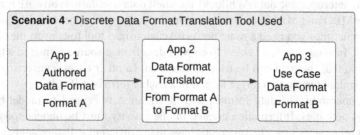

FIGURE 3.5   Data format translation workflow options.

If sending and receiving software do not share a common format, the data file that contains the relevant information must be converted to a suitable format. In such cases, there are several ways to convert from a source format (e.g. Format A) to a target format (e.g. Format B). Most authoring software includes tools to export to other formats than the native format. See Figure 3.5 for several scenarios depicting how to translate from a source format to a target format.

INFORMATION REQUIREMENTS

In an MBE, we need information that can be used without human interpretation or intervention. A target application needs to be able to use information from a source application, preferably without additional steps beyond those shown in Figure 3.5. (Note that in addition to the steps above, it is a good idea to use data validation software, software that compares the contents of a source file to a target file for equivalence.)

In manual processes, people spend time (often a significant amount of time) to review the source information, find problems, address problems, fill the gaps, guess as to what is needed, ask the source questions about the information (e.g. RFIs), and try to ensure that the information is sufficient for its intended use. In manual processes, there is a lot of interpretation and guessing, and a lot of back

For example, consider a manufacturer has a high-volume order to produce machined parts for a client, the client's drawing does not use GD&T, ± title block tolerances are used, and the drawing doesn't include surface texture specifications. The manufacturer interprets the drawing and the specifications as best they can. They recognize some information is vague, ambiguous, or missing, but they think they know what the client wants. The manufacturer does a small trial run and makes five parts, inspects them (outgoing inspection), and sends them to the client. The client receives the parts and inspects them (incoming inspection). The first issue is differences in the data in outgoing and incoming inspection. This occurs because using directly-toleranced dimensions to locate and orient features is ambiguous, so both sets of inspectors have to guess what the specs really mean and what the conformance criteria are. After some back-and-forth negotiation, the sets of inspectors agree on inspection methods and align their inspection results. Let's say that the inspection results indicate that the parts pass inspection. Note that this is informal, as the client has something else in mind to decide if they accept these parts or not.

The client sends the parts to manufacturing for prototype assembly, but the parts don't fit with the mating parts. Holes are misaligned, fasteners do not fit through clearance holes into threaded holes, edges are misaligned, etc. The client decides that the problem is that the parts are out of specification, regardless of the negotiations that occurred earlier. The clients send photos and markups indicating where interference occurs, where holes had to be enlarged, surfaces ground, etc. Wanting to continue with this contract, the vendor takes this information and adjusts their process to where they think it will produce parts that do not cause the assembly problems found earlier. The steps above are repeated until the client stops rejecting parts. Why did the client stop rejecting parts? Because they were able to use the parts in their assembly. Did the client know why they could use the parts in their assembly? (Not really.) They might answer, "Because the supplier finally made acceptable parts". After many adjustments to their manufacturing process, and many discussions and negotiations with the client, does the supplier have a clear idea of exactly why the parts are now working and what they did to achieve that state? Do these parties understand the role of the mating parts, if a mating part caused the problem, if it was the ambiguous specifications, if it was changes made to the assembly process, environmental conditions, or other factors? The answer is usually "Of course not". In many cases, the reason is poorly designed parts, parts with inadequate analysis during the design process and inadequate specifications, the parts end up working because:

1. Manufacturing does a better job than the specifications require. Manufacturing does not use the entire tolerance range or may even target nominal values that are outside the design limits.
2. Parts are flexible, they bend as forces are applied during the assembly or installation process.

It is unfortunate that design methods are sometimes insufficient and therefore require the informal reasons above to obtain acceptable parts. Design engineers are supposed to design parts that will work, will assemble, etc. We are supposed to have thought things through, done analysis and built simulations, developed rigorous and necessary specifications, etc. Well, we don't know how to do that in all situations, and unfortunately, given that design engineers now also do the work of designers and drafters, engineers don't have time to make sure every feature on every part will function as required. The whole design, manufacture, and build process is often a crapshoot that is cleaned up at or after physical prototyping. This is expensive, and the process does not translate well into MBD and MBE. We should not use a problems-in-problems-out-fix-it-with-informal-methods-later approach with MBD. We should do the work up front. We should spend the time to use GD&T, do tolerance analysis, build twins, simulations, and analytical models during the design process. An hour and a dollar well spent during design will save orders of magnitude in time and money later over the life of the product.

Remember, our goal in MBE is to streamline the process by removing manual steps. It is best if model-based processes do not rely on craftsmanship and tribal knowledge in the shop to solve the problems of incomplete and inadequate designs. This is especially true today, where it is becoming more difficult to find new hires with applicable experience.

and forth to fill the gaps and to understand what is really required, and sometimes some of that information is actually missing from the source dataset. In the end, their goal is to end up with a satisfactory comfort level that they have the information they need, and they know what the conformance criteria are for the product or process that whatever they are doing must conform to. This is often a circuitous and iterative process, with many informal steps that may yield acceptable results, but often without really understanding how or why those results were obtained.

With MBD and MBE, we need information that:

1. Adheres to a predictable structure and format.
   The information must have the structure and format that the receiving software expects. In MBE, information should be ready-to-use.
2. Has the right data structure.
   Thus, for the information to be ready-to-use, without problems, the information must have the right structure and be in the right format.
3. Is trustworthy.
   A receiving organization should only use information that is trusted, from a trustworthy source, which satisfies authenticity criteria, and preferably has been generated by authenticated processes. It is best practice to follow this procedure if information is used manually or directly.
   When information is used manually, a user can stop the activity to manually authenticate the information and generate trust at the time of use. Manual authentication should be avoided in an MBE as it is inefficient and subjective and requires a person to act as an intermediary between the information and its use case. In MBE workflows, authentication should be performed automatically.
4. Is secure.
   Information and the network and pathways it follows must be secure. In the case of direct-use information, the use case may be automated, and information might be used without authentication at the point of use. Thus, security must be built into systems. This is needed to ensure that information is ready to use.
5. Has the right content.
   While the criteria above ensure information is in the right format, from a trusted source, and has been received securely, it is just as important that it is the right information.
   Consider the case where a company is producing a formed sheet metal part but receives a dataset for a cast part. In a manual process, a recipient would review the information received, and if the user recognizes that it is the wrong information, the user can stop the process to find and obtain the correct information. In a direct-use process, the software will either try to use the incorrect data received to drive the process, which would yield a failure, or it will recognize that it received the wrong information, stop the process, and initiate a mitigation procedure.

6. Has the right logical relationships.

For manual and direct use of digital information, related data elements must be properly related to each other. This facilitates the manual activity of query and response, and it facilitates direct use of sets of related information, such as a logical relationship/link between a specification and the model elements to which it applies. In both cases, the relationships define the context in which information should be used.

7. Is structured optimally for its intended use.

   a. Manual use

      i. The information should be optimized to be used manually.

   b. Direct use

      i. The information should be optimized to be used directly in software.

8. Is of sufficient quality

Information must consistently be of sufficient quality so that the system can use the information and generate an acceptable outcome. The software that manages or the people who manage the model-based process must be able to trust that the information meets the needs of its users.

9. In value stream mapping, *percent complete and accurate (%C/A)* is used as a metric for information quality. %C/A represents the quality of a process output (as measured by users using that output as input). This metric aligns well with the ideas in this text, as it is critical that information is complete enough and accurate enough to be used successfully without requiring clarification or more information, both for manual-use cases and direct use cases. The idea of percent complete and accurate is described well on page 72 of the book *Value Stream Mapping: How to Visualize Work and Align Leadership for Organizational Transformation (see Footnote 3.1)*.

Information must be optimized to satisfy the needs of its intended users. Information quality must be based on the intended use of the information, not on siloed criteria within the source organization. Thus, information quality criteria are defined by the needs of the intended use cases, not only by the authors who create the information. The concept of quality for MBE is one of the most challenging changes that must be worked through during a company's MBE development and implementation. We will address this concept in detail later in the book.

FOOTNOTE

3.1 Martin, K., Osterling, M. (2013). *Value Stream Mapping: How to Visualize Work and Align Leadership for Organizational Transformation*. New York, NY: McGraw-Hill Education Books.

> *Information quality must be based on the intended use of the information, not on siloed criteria within the source organization.*

DUAL USE CASES

DUAL-USE INFORMATION (REPEATED FROM ABOVE)

Information that is intended to be used directly and manually.

It is best to avoid dual-use cases. The best practice is to create and release one version of information, which may be intended and optimized for direct use or manual use – pick one. If information is released that supports direct- and manual-use cases, the user is free to decide whether to use the information manually or directly. A common problem is that different versions of information may be disconnected (they are not linked), and thus they may have different values. I recommend releasing direct-use or manual-use information for a model object, not both (see Figure 3.6).

By contrast, some people think it is a good idea to release information that can be used directly and manually because it lets the user decide which version to use – they believe this flexibility is a good idea. While dual-use information does facilitate flexibility, dual-use information decreases control of the workflow and drives uncertainty about how a result (e.g. final product, process outcome) was obtained – we don't know how we got to that outcome, by directly using

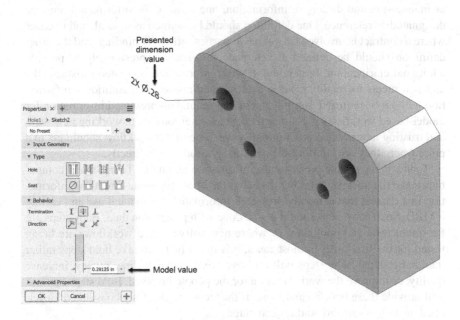

FIGURE 3.6   Product definition data with dual-use information.

information, manually using information, or by some other means. In such cases, it is not clear whether information will be used manually or directly, it is also not clear if the potential savings and quality improvements driven by direct use will be obtained. Such approaches reduce the control over a process and its outcome. A potential consequence is where one type of information is changed to improve processing, but the manufacturer is not using that information to drive the process – for example, a model value is changed but the process is being driven by visual information that is not updated or not equivalent to the model value. The best practice is to pursue and mandate direct use as much as possible.

Not only should information be defined for a single use case (e.g. manual use or direct use), information should only be used as officially designated, and the recipient should agree to use the information as designated. There have been high-profile examples of MBD information not being used as intended. The OEM assumed their information would be used as intended, but the suppliers had another idea. Of course, the use case should be coordinated between source and recipient. There have been some high-profile examples early in the development of MBE where how information was used was not enforced. In fact, it was assumed that the information would be used as intended. Even today, some OEMs think it doesn't matter how the information they release is used by suppliers. This is a mistake. Suppliers, be they internal or external, should use information as intended. In fact, how they use information and how they ensure the information remains linked and satisfies its intended use should be subject to oversight and approval.

If information defining the same thing is provided in multiple forms, such as manual-use and direct-use information, one form of the information must be designated as reference. One definition should be defined as official, and in cases where a contract is involved between two parties, officially binding, and the other definition should be defined as reference information, presumably to provide additional clarification. Preferably, if MBD information is created satisfactorily, and if it meets its quality criteria as described above, then additional information will not be required. Initially, including redundant presented information for model-based workflows is common. People are nervous about working new ways and trusting new processes. Eventually, everyone in the workflow should get to a point where they are comfortable with using information directly.

Today, many people prefer to use information manually. Today, many companies make the mistake of taking the baby step of adding some direct-use information in a dataset that is mostly presented information for manual use and calling it MBD. Such baby steps should not be done with production data. Baby steps are for initial tests and pilot runs, or where new software and workflows are being tested for the first time. In most cases, it is much better to take bold steps rather than baby steps. Baby steps will not save your company time, money, increase quality, or improve the work situation for the people involved. Bold steps can and will provide these benefits and more, if the steps are carefully considered, developed, tested, optimized, and implemented.

As mentioned above, it is common for some information elements in a product definition dataset to be developed and intended for manual use and other information elements in a dataset to be developed and intended for direct use. Note that in this case, best practice is to clearly designate in procedures the official purpose and official use case for each information element in a dataset. For example, a common first step in MBD and MBE is to designate that model geometry in a product definition dataset will be used as the official representation of product geometry. This is a change from the old school approach, where explicit dimensions were the official representation of product geometry. PMI/annotation presented in the product definition dataset will still be used as the official definition representing specifications. The model geometry will be used directly, and the PMI will be used manually. Both types of information, direct-use information and manual-use information, exist side-by-side in the product definition dataset. This is okay and can be managed by training, procedures, and standards. The challenge is when two forms of information representing the same thing are provided, such as the definition of the size of a hole; the size of the hole is defined by model geometry and by an explicit size dimension (PMI). In such cases, it can be confusing for users to know which version of the information to use, and they may use the unofficial version of the information. Human nature is to use the most familiar version of information. Most people in industry are comfortable using explicit dimensions rather than model geometry – they are used to working manually. This is part of the cultural inertia that must be overcome. To close, for each information element in a dataset, pick a use case (direct or manual use) and stick with it.

4 Fundamentals of MBD

Note: While the title of this chapter is Fundamentals of MBD, this material represents new and more useful fundamentals of MBD, based on a comprehensive top-down holistic systematic approach to MBD and MBE. My point? If you have learned about MBD from someone else or even from me, the fundamentals presented here are different from what you learned before. I hope the ideas presented here provide significant improvement in business value and return on investment for your organization.

While the content in this chapter is technical in nature, these concepts must be understood from a business perspective. Our goal must be to increase value for the enterprise and its workforce, not just to make technical changes. As I have stated elsewhere in the text, most implementations of MBD and MBE are inadequate – the implementations changed some of how we work without carefully dissecting how we work, the reasons we work that way, and the cost associated with each step we perform, which merely leads to minor incremental benefits from adopting MBD and MBE.

ACRONYMS AND INITIALISMS

Some definitions for acronyms and initialisms are repeated here, as this chapter may be read separately from preceding chapters, although I recommend reading prior chapters before reading this one. Graphics are included with the definitions below, as appropriate.

3D PDF   Three-Dimensional Portable Document Format
> 3D PDF format is often used as a neutral visualization format. Anyone with Acrobat Reader software can view and use a 3D PDF. A 3D PDF is also a container that may contain other digital files. 3D PDF files are often used as a container and delivery mechanism for TDPs.

BOM   Bill of Materials
> A BOM is a list of the raw materials, sub-assemblies, intermediate assemblies, sub-components, parts, bulk items, and the quantities of each needed to manufacture an end product (adapted from Wikipedia). There are subcategories of BOMs in many companies, such as:

EBOM   Engineering Bill of Materials
> A list and quantities of materials, parts, subassemblies, bulk items, etc. required to define a product. The EBOM is built from design and functional points of view. On drawings, a BOM is called a parts list per ASME Y14 and ISO TC 10 Standards.

DOI: 10.1201/9781003203797-4

MBOM   Manufacturing Bill of Materials

A list and quantities of materials, parts, subassemblies, bulk items, etc. required to manufacture a product. The MBOM is built from a manufacturing point of view. Usually, an MBOM is based on an EBOM that has been restructured to suit the manufacturing process. Additional items are added and more accurately quantified that were inadequately defined in the EBOM, such as bulk items (e.g. paint, adhesive, wire, etc.).

Note that other types of BOMs have evolved in different industries and/or promoted by software companies. Thus, there may be confusion about some BOM-related terms. For example, the term SBOM may be used to mean Software BOM, Service BOM, or Super BOM, some sort of hybrid EBOM/MBOM amalgamation. User beware.

Today BOM information follows a convoluted workflow similar to most other business information – it starts as manual information, it is entered into a digital system, and it may be used digitally or manually. It is easy to think of BOMs as tables or lists of information about the materials that make up a product, system, or assembly. The tables and lists we are familiar with are the presented version of BOM information, which is manual-use information. In CAD, PDM, PLM, and ERP systems, the information exists as digital information, which may be used directly. It is important that we designate one of the versions as the official information and the other version as reference. Note that the official version may be different for different portions of the product lifecycle (e.g. PLM BOM information is the source of the official version in design activities and ERP BOM information is the source of the official version in manufacturing activities), or for different disciplines or product types (e.g. PLM BOM information is the official version for machined parts, sheet metal parts, and physical assemblies, and ERP BOM information is the official version for software and service kits).

BOM information follows a workflow that is similar to MBD information. Note that information management software and EBOM/MBOM are not included in the following sequence, as different companies use CAD, PDM, PLM, and ERP differently (see Figure 4.1).

1. General BOM information (e.g. stock materials, fasteners, components, COTS items, standards components, bulk items) is entered manually into software. General BOM information may be assigned to one or more products or processes.
2. The general BOM information is represented digitally.
3. General BOM information is assigned to and tailored for a product or process.
4. The product-specific BOM information is completed.

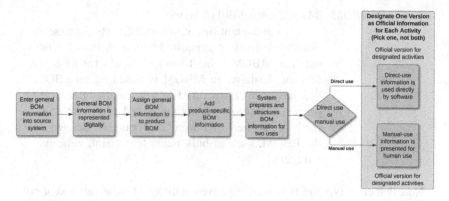

FIGURE 4.1 BOM information lifecycle excerpt.

5. The product-specific BOM information is stored as digital data in software. The BOM information is prepared and structured for two use cases:
   a. The BOM information may be used directly in software.
   b. The BOM information may be presented for manual use.
6. The version of the information to be used is officially designated.
7. The user uses the official version in the manner intended.

CM   Configuration Management

        The methodology of managing both the current and past structure of and changes to a dataset. In the context of MBD and MBE, the dataset is a model-based dataset that may contain ancillary information.

COTS   Commercial Off The Shelf

        Items that are available for purchase without requiring a custom design.

        "A software and/or hardware product that is commercially ready-made and available for sale, lease, or license to the general public."

        From NIST/NSA/CSS Policy 3-14 (online)

DCC   Dataset Classification Code

        Alphanumeric designations from DCC 1 to DCC 6 represent a range of product definition dataset structures, from drawing-only to drawing + model, model-only, and direct-use information only. Dataset Classification Codes DCC 1–DCC 5 are defined in the ASME Y14.41 and Y14.100 and ISO 16792 Standards. Dataset Classification Code DCC 6 is defined in this book. Note that other DCCs could be defined if needed. Dataset structures corresponding to DCCs range from drawing-only (e.g. DCC 1), drawing + model (e.g. DCC 3), model-only (e.g. DCC 5), to data-only (e.g. DCC 6).

        Process definition Dataset Classification Codes (PrDCCs) are defined in Chapter 5.

DDD   Disconnected Derivative Dataset

A dataset that is a copy of or created from (derived from) another dataset but is not digitally linked to the source dataset. DDDs are to be avoided as much as possible.

DMO   Drawing, Model, Other

DOD   Department of Defense (specifically, the U.S. Department of Defense)

DUC   Dataset Use Code

Alphanumeric designations that represent official use cases for product definition datasets and their constituent information assets. The DUC defines the official use for information and information assets, and how information and information assets are supposed to be used. DUCs correspond to and augment DCCs. In this text, DUCs 1.1-6.1 are defined, however, many other DUCs could be generated. DUCs 1.1-6.1 represent common DCC use cases and can be used as a starting point.

DUCs are defined in and unique to this text.

Process definition Dataset Use Codes (PrDUCs) are defined in Chapter 5.

See below for more information.

FEA   Finite Element Analysis

GD&T   Geometric Dimensioning and Tolerancing

A feature-based graphical system for defining nominal product geometry, the relationship between geometric features, and the allowable variation between geometric features. GD&T is defined in the ASME Y14.5 Standard.

GPS   Geometrical Product Specification

GPS is the name of a system of GD&T, surface texture, and measurement defined in ISO standards.

Comment about GD&T and GPS:

ASME GD&T and ISO GPS specifications look similar but are not the same. Specifications based on each system have different rules and defaults and different effects on a product.

GUI   Graphical User Interface

IP   Intellectual Property

MBD   Model-Based Definition

The system and methods used to define digital business information so that it is optimized to be directly used in software. MBD replaces document-based methods, where information is optimized to be presented in documents and used manually.

MBE   Model-Based Enterprise

An organization focused on obtaining maximum value from its information throughout the organization, the supply chain, the product lifecycle, and the information lifecycle by creating and using MBD as much as possible. (See Chapter 5 for a more detailed definition.)

Note: Optimally, MBD data should be used directly in software rather than manually. An MBE prioritizes automated use of information over manual use cases. An optimized MBE obtains the maximum value from MBD and its use of that information.

MBE information management strategy emphasizes efficiency and automation and obtaining the maximum value from business information.

MBSE Model-Based Systems Engineering

Systems engineering that is defined using model-based methods (MBSE is a form of MBD)

MBx Model-Based something (a generic use of MB)

MBx is used generically to encompass all things model-based or MBD and MBE combined.

PMI Product and Manufacturing Information

Annotation, symbols, and attributes in 3D model. PMI may also include metadata.

PMI is also defined in DIN EN/NAS 9300 Long Term Archiving and Retrieval of digital technical product documentation such as 3D, CAD, and PDM data PART 007: Terms and References.

QIF Quality Information Framework

QIF is a data model for manufacturing quality information. QIF is defined in ISO 23952.

ROI Return on Investment

SDO Standards Development Organization

SOP Standard Operating Procedure

SPC Statistical Process Control

STEP STandard for the Exchange of Product model data

STEP is a series of standards defining data models for products and related data. STEP is defined in ISO 10303 Standards.

TDP Technical Data Package

Official definition: "The authoritative technical description of an item."

(Source: MIL-STD-31000 B, a U.S. Government standard)

A more useful definition: Files and information that define a product, product requirements, and supporting information.

It is valuable to differentiate between a product definition dataset and a technical data package. See definition of technical data package and my comments below.

V&V Verification and Validation (of data or a dataset)

For digital data and in MBE, V&V is done by software and is preferably automated.

TERMS AND DEFINITIONS

Some definitions are repeated here, as this chapter may be read separately from preceding chapters, although I recommend reading prior chapters before reading this one. Graphics are included with the definitions below, as appropriate.

ANNOTATED MODEL (2D OR 3D MODEL)

2D or 3D digital product or process definition dataset that contains PMI. Generally, a model includes representation of relevant geometry, associated specifications, and other information represented as PMI.

> Example 1   Annotated product model: A model that defines a machined
> part (see Figure 4.2).

Note that the term and concept *annotated model* is restrictive. We came up with the term in the ASME Y14.41 subcommittee. Previously, the term *model* represented everything that could go into a product definition dataset, such as model geometry, annotation/PMI, metadata (depending on where you look and who you ask, metadata are distinct from PMI or included in PMI), ancillary digital data

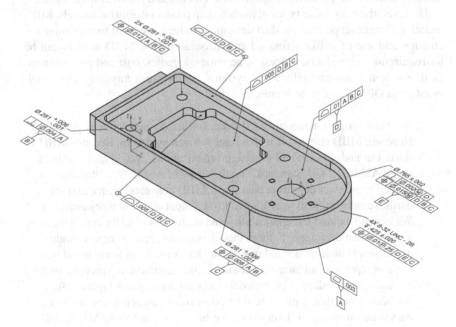

FIGURE 4.2   Annotated product model.

elements, and digital data. While we used the term *model* to mean all the stuff in a model-based product definition dataset, we routinely got tripped up because the conventional definition of model in product definition is 3D CAD or neutral geometry model. We said one thing but were thinking another – our experience using 3D CAD and talking about models as the thing we saw and rotated on the screen biased our thinking and often resulted in confusion. So, we came up with the term *annotated model* to clarify what we meant and to align what we called *model* with what everyone else called it.

Note that the discussion above also applies to ISO 16792, as many concepts were developed in ASME Y14.41 and then adjusted and added to ISO 16792. More recently, some concepts were developed in ISO 16792 before being added to ASME Y14.41, such as support for welded assemblies and surface texture symbols. In 2007, I developed detailed material in the IHS Drawing Requirements Manual for separable and inseparable assemblies. It took another nine years for similar material to show up in standards. Sometimes it is faster to just do it yourself.

An issue, however, is the term *annotated model* implies that there is annotation, and annotation is defined as text and graphical information that is visible. Thus, the term *annotated model* boxes us into a corner. See below for a broader alternative term. *Annotated model* is a fine term to describe a geometric model with annotation, but we need a better term for a model that includes geometry, specifications, and metadata, which may be intended to be used directly in software or to be used manually. For product definition datasets, broadly applicable terms of interest are *geometry + specification model* and *full definition model*.

However, there are more types of models than product definition models. MBD includes a larger scope than product definition. Indeed, we want to maximize the creation and use of MBD across all enterprise activities. MBD models can be for requirements, products, processes, activities, logistics, cost and procurement, facilities, demolition and relocation, systems, pretty much anything that is relevant to an OEM that can be represented digitally.

1. We need terms for model-based datasets that do not define a product. However, MBD started off with a legacy driven by design. Because MBD started in and was limited to design organizations, was initially defined in the ASME Y14.41 Standard, and the scope of ASME Y14 Standards is limited to product definition data, most MBD concepts, terms, and standards we came up with were limited to the output of a design organization. We need to expand the applicability and usefulness of MBx concepts, so it is clear that they apply to more than design and apply enterprise-wide.
   a. Example of an annotated process model. A model of a machined part that represents an intermediate step in the manufacturing process, such as an intermediate hogout model or machining step (see Figure 4.3).
2. We need terms that apply to MBD models and datasets that are not based on visual information. Everything we have done in the ASME Y14.41 Standard is based on 3D CAD and neutral format models that represent product definition data. All of it was derived from and based on drawing

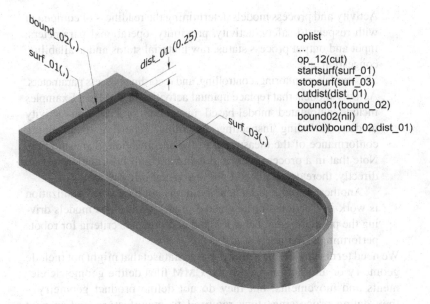

oplist
...
op_12(cut)
startsurf(surf_01)
stopsurf(surf_03)
cutdist(dist_01)
bound01(bound_02)
bound02(nil)
cutvol)bound_02,dist_01)

FIGURE 4.3 Annotated process model (machining step).

standards and practices and CAD and CAD practices. Our scope was limited to redefining what we used to define on drawings and defining it in a 3D product definition dataset. It was all about defining information and creating views of 3D models, showing dimensioning, GD&T, surface texture, title block information, notes, etc. in a 3D format. Our goal, and the goal of most engineers today, is to support visualization.

IMPORTANT FACT: There are many information use cases that do not require visualization.

Note that it is difficult for some people to consider models that are not based on visual information. Many people have a limited idea of what a model is, limited to visually-displayed models that are created using visual tools in visual systems – we are biased that models are visual; that we enter data into software using a GUI and see visual results onscreen increases this bias. Most of our interaction with digital models is by viewing visualization data on a GUI. We create information using graphical systems and we view the results as presented information.

Examples of MBD models that are not based on visual information include

a. Operational models for processes and facilities (e.g. flow, temperature, pressure, salinity, turbidity, dissolved oxygen, resistivity, variation over time, evaluation of variables relative to set points and limits, etc. in purified water systems).

b.   Activity and process models determining the readiness of equipment with respect to safety, activity, proximity, operational parameters, input and output process status, raw material status and availability, congestion, etc.

c.   SPC models monitoring, controlling, and recording process parameters.

d.   Activity models that replace manual activities in an MBE. Examples include automated model-based inspection process and activity models, including inspecting, recording, reporting, determining conformance of the measurand to the specification, dispositioning. Note that in a process such as this in which all information is used directly, there doesn't need to be any visual information.

Another example of an activity that is not based on visualization is work instructions for automated processes, such as models driving the programming, behavior, and conformance criteria for robots performing activities.

3.   We need terms that define a model-based dataset that might not include geometry or annotation. CAM and CMM files define geometric elements and movements, but they do not define product geometry – they define processing steps required to manufacture and inspect products. We can say there's geometry in there, but it isn't the same geometry we describe in the ASME Y14.41 Standard. We can have an MBSE SysML systems requirements model for geometric product definition data. We can have the steps and work instructions for automation that do not include geometry or annotation, but merely contain code.

Note: In addition to the explanations above, there are other types of models that are important for OEMs and related business entities. We need to include them in our discussion and come up with examples and terms related to these other models. Much of the discussion in this book focuses on traditional design models, manufacturing models, and inspection models. While I recognize the importance of addressing the breadth of MBD across the MBE, I assume that I must provide adequate coverage of core MBD topics.

AUTHOR, PRIMARY

An entity that creates a dataset and typically owns the main body of information in the dataset (e.g. design engineer for a product definition dataset.)

Note that, while primary authors create information, their main responsibility is to create value for the organization. Primary authors must ensure that the information they create meets the needs of every lifecycle activity that uses the information and enables them to use the information in the most efficient manner possible.

AUTHOR, SECONDARY

An entity that uses/reuses content created by another group and creates additional content. (e.g. manufacturing engineer for a manufacturing dataset or work instructions based on a product definition dataset.)

Note that while secondary authors create information and add information to existing information, their main responsibility is to create value for the organization. Secondary authors must ensure that the information they create meets the needs of every lifecycle activity that uses the information and enable them to use the information in the most efficient manner possible.

AUTHORING

Creation of content, information, and IP, often for use in a dataset. Note: There are several levels of authoring: primary author and secondary author; and several types of authorship: single-owner, shared ownership, and distributed ownership.

Authoring is generally a manual activity done by people using software. However, authoring is also augmented by software, where software contributes authored content to the authoring process, such as where tolerance analysis or FEA software linked to CAD software defines the optimal thickness for a geometric component. Authoring is also done where the person is partially involved, such as with generative design. With generative design, a person enters parametric values and constraints (e.g. boundary conditions, material, connection points, loads, environmental conditions, etc.) and the software generates alternatives that satisfy the parameters and conditions (see Sidebar/Info Box).

Note that while authors create information and add information to existing information, their main responsibility is to create value for the organization. Authors must ensure that the information they create meets the needs of every lifecycle activity that uses the information and enables them to use the information in the most efficient manner possible.

Generative design (algorithmic design) has been around for a long time, but it has been used more since cloud computing resources have become more readily available. Generative design is typically used as part of the design process. For product design, generative design is mainly used to define perfect product geometry, what ISO GPS calls the nominal model. Note that the nominal model represents a portion of a production-ready product definition dataset. A complete production-ready product definition dataset also requires product metadata, notes, tolerances (that define acceptable levels of variation), and other information. Metadata can be classified as the metadata model. Specification information can be classified as the specification model, which contains specifications, which are generally defined by annotation/PMI in a product definition dataset. Today, generative design is used for geometry, to define the product geometry model. If the geometry is used in a production-ready product definition dataset, specifications are developed and added manually. Someday, specifications will also be created by the generative process. Engineers will no longer have to develop and apply specifications, as the software will do it. With generative systems and AI systems, designers will describe the desired outcome to the generator, and the generator will output optional versions of a complete product definition dataset. This represents a significantly different design process than the one we use today. I see the future design process being a conversation with a generative modeling engine (software), where the designer describes requirements, constraints, environmental conditions, lifecycle requirements, etc. and the software returns options.

Note that while I've made a living writing about and teaching GD&T and GPS, it is clear to me that someday we won't need someone to teach us GD&T or GPS. In fact, someday we won't need GD&T and GPS at all. The sooner we can get there the better. This is a natural progression and is driven by the fact that GD&T and GPS were developed in the mid-twentieth century to be used on drawings in manual workflows. Today, there's a better way. That is a subject for another book…

Axonometric View

A pictorial view, usually with surfaces at an angle to the view, so more than one adjacent surface of a 3D object is presented in the view. Axonometric views provide more context to understand a 3D object, as the view allows the viewer to *see around the corner* of the object. Axonometric views offer a better way to present product geometry and PMI (see Figure 4.4).

With the advent of solid modeling in 3D CAD, it is easy and inexpensive to create and present axonometric views. Pictorial views were time-consuming and expensive to create before 3D CAD. (For comparison, see *Orthographic Views*.)

Binding Information

Official information that is contractually or otherwise designated as the information to be used by a party. In an MBD and MBE context, binding information is information that has been designated as the official information for specified purposes, such as manufacturing, inspection, service, etc. Binding information is official information that has been invoked for one or more use cases.

For example, consider a 3D model that includes the geometry for a group of cylindrical holes and explicit size dimensions for the holes. Thus, two versions or sources of the geometric definition of the holes are provided in the product

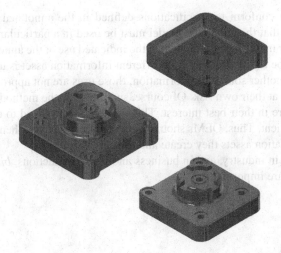

FIGURE 4.4 Axonometric views.

definition. In a formal MBD use case, the model geometry should be designated as the official definition of the nominal size of the holes. A cascading series of events occurs:

1. A contract designates the model geometry representing the holes as the official binding information.
2. A contract and the product definition dataset are issued to a supplier. In the contract, the OEM designates the supplier must conform to the official definition of the hole size in the model geometry, which is the official information for the hole size. Note that the tolerances for the hole size are usually not represented by model geometry but by specification.

In this text, *binding information* is an important concept.

BINDING INFORMATION ASSET

An official information asset that is contractually or otherwise designated as the information asset to be used by a party. In an MBD and MBE context, a binding information asset is an official information asset that has been designated as the official source of information for specified purposes, such as analysis, manufacturing, inspection, service, etc. A binding information asset is an official information asset that has been invoked for one or more use cases.

A Note about Binding Information and Binding Information Asset

If invoked by contract or other means, official information or an official information asset may be legally binding. For example, if an annotated model is defined as an official information asset and it is contractually invoked, then products produced must conform to specifications defined in the annotated model. If a contract states that the annotated model must be used in a particular way, e.g. to manufacture or inspect the product, then the indicated use of the annotated model is intended to be legally binding. If a different information asset is used, such as a drawing or another source of information, those uses are not approved, and the user uses them at their own risk. Of course, suppliers use the methods they deem as those that are in their best interest. This does not always lead to the best outcome for the client. Thus, OEMs should audit supplier behavior, their workflows, and the information assets they create and use.

In this text, in industry, and in business and legal transactions, *binding information assets* are important.

Built for Use

Adjective

Something that has been created for a specific purpose, preferably created in a way that is optimized for how it will be used.

In the context of this text, "built for use" is used to describe data, datasets, information, information assets, etc. that have been designed and developed to serve their intended purpose.

For example, a model-based product definition dataset created for manual use is defined to be used manually and optimized for that use. Thus, it is optimal if data, datasets, information, information assets, etc. are used in the manner that they were developed and intended to be used. If that data, dataset, information, or information asset is used in a different manner, it is likely that the results will be more chaotic and less optimal.

I struggled with the concepts of structured versus unstructured data in the context of MBD and MBE and how I am presenting best practices and how the discrete manufacturing industry tends to operate. We generally create information that is structured in a way that matches how it is used, or in many current MBD implementations, the information matches a more historical way of working (e.g. manual use, visualization, etc.). Nonetheless, when considering structured versus unstructured information, it is clear that much of what is classified as unstructured is actually structured. Case in point, in generative AI, it is written that large language models (LLMs) are built and trained using unstructured data scraped from the Internet. However, most, if not all, of that information is structured – it is HTML information, images, audio files, and video files in various standard formats. Most of the information, the HTML content, the audio, video, and graphical information are all built for how they were being used – that is, they were built or optimized for being used as intended, which is to be included in or provided via a link on a web page. Thus, in the sense of the information's primary or initial use case, being provided as web-based information, the information is built for use. However, if we consider that same information now being scraped, parsed, and used as part of a larger, more random dataset for the purpose of developing LLMs, the information was not built for that purpose. However, the interesting part is that the information has proven to be useful for this new, secondary use case. This can be seen in the usefulness and novelty in the results obtained from queries to LLMs.

Obviously, LLMs are useful and will continue to become more useful to individuals, to industry, to educators, and to governments. The information hoovered up and used in LLMs is structured information with regard to its original use case but is classified as unstructured information with regard to its new use case for developing the LLM. Thus, to me, the terms *structured* and *unstructured* in this case are insufficient. "Built for use" seems to capture a more relevant concept, whether the information was created and optimized for how it is used.

Equally important, and a core concept of the methods defined in this text and my training, is that data, datasets, information, information assets, etc. are used the way they were developed to be used. A common problem in MBE is where data, datasets, information, or information assets are defined to be used one way but are actually used differently. The most common case is where someone tries to use product definition data that was defined to be used manually directly in software. This is common, as the historical purpose of product definition dataset has been to support manual use cases. In fact, most current activities do not sufficiently understand and distinguish between manual and direct use cases or how to optimize data, datasets, information, and information assets for them. Differentiating and optimizing for these use cases is a core topic of this text and my training.

DATA ELEMENT

Individual item contained in a dataset or technical data package. Data element is derived from ASME Y14.100.

Note: Data element refers to an information asset (e.g. model, drawing, file, document, or similar item) in a dataset that contains digital information. Compare with *digital element*.

Data Format/File Format

File structure that is based on a standardized set of rules that define the structure, arrangement, and distribution of content for digital data. Data format is also used here to mean file format, such as native CAD (Creo, NX, Allegro), neutral (JT, PDF, 3D PDF, QIF, STEP, etc.) (see Figure 4.7).

Note: Using standardized formats facilitates use of datasets within an enterprise and throughout its supply chain.

Dataset

A collection of data elements (information assets) that represent a product, process, activity or other important business function or object. Note that dataset is usually used to mean a set of digital data.

Derivative Data

Data derived, translated, or converted from an official information asset or another source. Note: Derivative data is:

1. Converted from an official information asset or original source; it is converted from one version to another (e.g. from Creo 10 to Creo 11 file format) or from one format to another format (e.g. from Creo 10 to STEP AP 242).
2. Created by a secondary author using information defined by someone else (a primary author) (e.g. a manufacturing plan that reuses portions of the product definition dataset).

First Derivative

Data derived, translated, or converted from a primary authored official information asset. e.g.

1. A STEP file converted from a native CAD dataset designated as the official information asset.
2. A reference dimension displayed based on underlying model geometry in a dataset in which the model geometry is the official information for the geometry.
3. The definition of an individual part instance in an assembly.

Second Derivative

Data derived, translated, or converted from a derivative. (e.g. a QIF file or ACIS (.sat) file converted from a STEP file, a STEP file converted from a QIF file, a JT file converted from a STEP file, a 3D PDF created from a STEP file, copying a displayed dimension value as the nominal value in a measurement report).

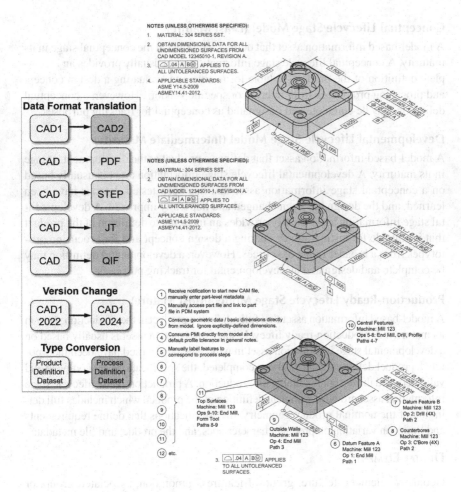

FIGURE 4.5 Derivative data and datasets.

Derivative Dataset

A dataset derived, translated, or converted from an official information asset. Note: Derivative data is usually created by a secondary author and based on other information and often cannot exist without the official information asset (see Figure 4.5).

Design Lifecycle Stage Model

A model-based information asset that defines a product at a particular stage in its maturity. Historically, design lifecycle stages are defined as conceptual, developmental, or production. (Note that these are my definitions of concepts developed by the DOD.) I use the term production-ready to describe product definition datasets that are ready for production.

Conceptual Lifecycle Stage Model (Least Mature)

A model-based information asset that defines a product at the conceptual stage in its maturity. A conceptual lifecycle stage information asset usually provides an incomplete definition of the product suitable for testing and evaluating a design concept and procuring prototypes from internal or specific vendors. However, a conceptual definition may be complete and designated as conceptual for tracking purposes.

Developmental Lifecycle Stage Model (Intermediate Maturity)

A model-based information asset that defines a product at the developmental stage in its maturity. A developmental lifecycle stage information asset is usually based on a conceptual stage information asset that has been tested, lessons have been learned, and the design has been changed and hopefully improved. A developmental stage information asset usually provides an incomplete definition of the product that is suitable for testing and evaluating a design concept and for procuring prototypes from a broader range of sources. However, a developmental definition may be complete and designated as developmental for tracking purposes.

Production-Ready Lifecycle Stage Model (Most Mature)

A model-based information asset that fully defines a product at the production stage in its maturity. A production-ready lifecycle stage information asset is usually based on a developmental stage information asset in which lessons learned have been incorporated, product definition data has been completed, the specifications are complete and validated, and the design is suitable for production. A production-ready lifecycle stage information asset provides complete definition of the product, which includes full definition of the nominal product geometry, the specifications that define requirements and limits on variation on product characteristics, and the product and file metadata.

Digital Element

Geometric element, feature, group of features, annotation, associated group, or attribute that exists in a dataset. (Definition copied from ASME Y14.41-2019). Note: A digital element is a piece of digital information or set of digital information within a dataset or information asset.

Direct-Use Information (Information for Direct Use) (Information for Autonomous Use and Autonomous Operations)

Digitally-defined information that can be read/parsed by software, understood by software, and used directly in software without the need for human intervention. The information is intended to be used autonomously. Direct-use information should be optimized to be used directly in software. In an MBE, direct-use information should be designated as the official definition of the information. Preferably, direct-use information should be the primary goal of MBD activities in an organization.

A mature MBE will use workflows that define all presented information (e.g. PMI and visible metadata) as derived instances of direct-use information, and the presented information will be designated as reference information (e.g. not the official definition of the information).

Direct-use information should be designated as official information as much as possible.

Manual-use information should be designated as reference information as much as possible.

Note that I include "as much as possible" above begrudgingly. I believe that the previous two statements should be absolute, that direct-use information must be official and manual-use information must be reference only. But, I don't think industry is quite ready to go there yet.

DOCUMENT

Discrete, usually static information asset developed to present information for manual use by people.

Note that IP is trapped in documents. Information in documents is optimized for manual use and thus is not easily accessible to systems designed to use information directly in software.

DRAWING

A document that defines something graphically, often with a combination of graphics, text, and symbols.

ENGINEERING DRAWING

A document that defines a product, process, system, activity, or other subject, using engineering drawing format and conforming to some form of engineering drawing practices.

FULL DEFINITION MODEL

A model-based information asset that contains all the data elements needed to fully define a product, process, activity, or other lifecycle object.

Full Definition Model – Product Definition

A model-based information asset that contains the data elements needed to fully define product, which includes the nominal (perfect) product geometry, specifications (e.g. that define allowable variation), conformance criteria, product metadata, and file metadata.

Full Definition Model – Process Definition

A model-based information asset that contains the data elements needed to fully define a process, which includes full definition of the process in context of the environment in which the process occurs. Full definition may include the inputs, goals, starting state, end state, processes, process steps used, operational limits, process interfaces, intermediate steps, actor information, tools and tool information, sensor values and structure, metric reporting and evaluation, etc.

Note that in this context, it is a slippery slope to say that a product definition dataset or other information asset contains full definition. Knowing that the information asset is supposed to include full definition, it is easy to understand that a deficiency is indeed a deficiency.

For example, if I send you a recipe to bake a cake, and I forget to include a critical ingredient (such as flour) or leave out a critical step (such as baking the cake), and you know that a recipe is supposed to define all the ingredients and steps from start to finish, it is clear that there is a deficiency.

The idea of a full-definition model may seem similar to the idea of design lifecycle stage models, but they are distinct concepts that may be used together. The design lifecycle stage indicates where in the design process the design is, and full definition indicates that the information asset is intended to convey complete definition. Many conceptual lifecycle stage information assets do not provide full definition, but a developmental lifecycle stage information asset may provide full product definition, and a production-ready lifecycle stage information asset must fully define the product.

Note that in some cases conceptual and developmental lifecycle stage models may provide complete definition and production-ready lifecycle stage models may be intentionally incomplete.

Being ready for production means that a product definition dataset completely defines a product and its requirements. Many production product definition datasets do not completely define the product. Some datasets are intentionally incomplete (e.g. from deceptive practices), but most are unintentionally incomplete (e.g. mistakes, lack of experience, informal processes in manufacturing that make up for deficiencies). Because so many datasets are intentionally or unintentionally incomplete, it is important to emphasize that the datasets are supposed to be complete, and thus that being incomplete represents a deficiency, a failure mode, a problem, and a potential breach of contract.

GEOMETRY + SPECIFICATION MODEL

A model-based information asset that contains all the data elements needed to fully define the geometry and specifications for a product, process, activity, or other lifecycle object.

INFORMATION (BUSINESS INFORMATION, IN THE CONTEXT OF MBE)

Data that is adequately structured and has context such that it is useful for some purpose.

Information that is relevant to a business, it's products, processes, activities, etc. Business information has value to a business and can be classified as intellectual property.

See Table 4.1 for the relationship between different types of information and information assets.

INFORMATION ASSET

An intangible asset that contains information. In the context of business (e.g. OEMs and similar organizations), information assets contain information that is relevant to their products, processes, activities, inventions, copyrights, trademarks, how they do business, etc.

Examples of information assets include documents, drawings (a type of document), information models, the digital information represented in information model files, such as native CAD, STEP, JT, or QIF files, model-based systems engineering models (SysML models), analytical models, simulation models, digital twins (a type of simulation model), manufacturing, assembly, or inspection work instructions, CNC, CMM, and robot activity programs, additive manufacturing datasets, machine programs and instructions, etc. See Table 4.1 for the relationship between different types of information and information assets.

INTELLECTUAL PROPERTY

"A work or invention that is the result of creativity, such as a manuscript or a design, to which one has rights and for which one may apply for a patent, copyright, trademark, etc."

Google, Oxford Languages (online) (from Google search)

"Intellectual property is a broad categorical description for the set of intangible assets owned and legally protected by a company or individual from outside use or implementation without consent. An intangible asset is a non-physical asset that a company or person owns."

Investopedia (online) ("What Is Intellectual Property, and What Are Some Types?")

INTERMEDIATE

Adjective

"occurring in the middle of a process or series"
"an intermediate stage of growth."
Encyclopædia Britannica, Inc. (online) (from The Britannica Dictionary)

Verb

"act as intermediary; mediate.
'the theory said that by intermediating between buyers and sellers, middlemen lower the costs of transactions.'"
Google, Oxford Languages (online) (from Google search)

The verb definition aligns with how I am describing people's role as recipients of manual-use information and using that information manually. We act as intermediaries, intercepting and trying to make sense of the information, convert it, and transcribe it (RIDICT) so we can use it.

Because we create manual-use information, we *require* people to act as intermediaries between the information and the actual purpose of the information.

E.g. the information may be a material specification for a product. If the material is defined with manual-use information, an intermediary is required to understand the material specification in order to perform a task, such as using software, hardware, and digital systems to procure or inspect the material. By contrast, if the material is defined with direct-use information, the material specification can be used directly by software, thus allowing the use case to be automated. There is no need for a person to act as an intermediary (e.g. read information and manually enter it into another system) in a direct-use workflow.

To be clear, I am not advocating eliminating people from the workplace. I am advocating that we automate mundane low-value tasks and activities and thereby allow people to focus on more interesting and valuable/value-adding tasks and activities. We already do this extensively today. I add this clarification here and in other places in the book because many people are uncomfortable with the idea of automation.

Because we create manual-use information, we require people to act as intermediaries between the information and the actual purpose of the information.

LIFECYCLE ACTIVITY

An activity that occurs during the lifecycle of a product. In many places in this book, I use the term "lifecycle activity" to refer to activities that occur before or after the initial engineering design activity.

This is a subjective definition and useful when we are discussing activities that affect or are affected by the design process. See Figure 4.6 for a concise snapshot of lifecycle activities.

Note: The activities shown, the interaction between the activities, and the flow direction are simplified and idealized. In most companies, there is a lot of iterative interaction between activities.

Earlier in the text, I introduced the concept of the information lifecycle, which is the lifecycle of information or an information asset. The lifecycle of information and information assets may extend beyond or be shorter than an associated product lifecycle – e.g. the information asset has a lifecycle longer than the product, such as where information is retained and maintained after a product line or product instance is retired.

MANUAL-USE INFORMATION (INFORMATION FOR MANUAL USE)

Information that is defined and intended to be used manually. (e.g. information presented in documents, on drawings, in lists, visualization data (e.g. FEA deformation diagram, flowchart, schematic, logic diagram, wiring diagram, P&ID, weather map), audio messages and alarms, audible instructions, haptic feedback, etc.

Manual-use information requires intermediation by one or more people, meaning someone must read, interpret, etc. (RIDICT) the information before the information can be used in a process, task, or activity.

This text focuses on visual manual-use information.

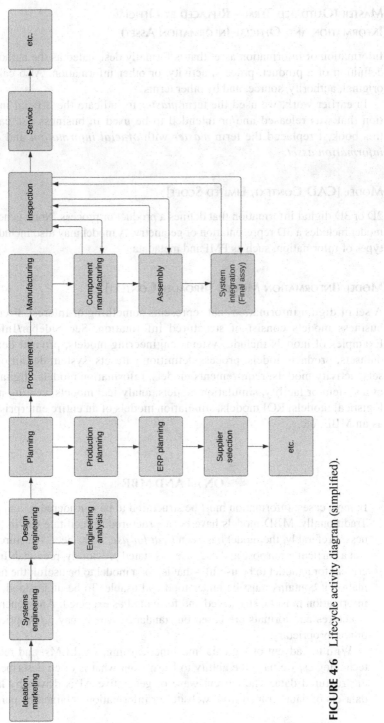

FIGURE 4.6 Lifecycle activity diagram (simplified).

MASTER (OUTDATED TERM – REPLACED BY OFFICIAL INFORMATION AND OFFICIAL INFORMATION ASSET)

Information or information asset that is formally designated as the authoritative definition of a product, process, activity, or other information. Also called the original, authority, source, and by other terms.

In earlier work, we used the term *master* to indicate the official information that was released and/or intended to be used in business use cases. In this book, I replaced the term *master* with *official information* and *official information asset*.

MODEL [CAD CONTEXT, LIMITED SCOPE]

2D or 3D digital information that defines a product or process. Note: generally, a model includes a 3D representation of geometry. A model may also include other types of information, such as PMI and metadata.

MODEL (INFORMATION MODEL) [BROADER CONTEXT]

A set of digital information that represents something of interest. Preferably, business models consist of structured information. See Sidebar/Info Box. Examples of models include systems engineering models, product definition datasets, product models, process definition datasets, system definition datasets, activity models, requirements models, information models, digital twins of a system or facility, simulation models, analytical models, costing models, logistical models, ROI models, simulation models of an entire enterprise, such as an MBE, etc.

ON AI AND MBE

In most cases, information must be structured to be considered as a model. Traditionally, MBD models have been structured to facilitate use in business. Preferably, the model has been *built for use*, it has been structured to suit a particular purpose and use case. As stated previously, predictability is critical for a model to be useful – that is, for a model to be useful, the information it contains must be understood and usable. To be understood, that information must be structured and formatted as expected. And generally, structures and formats are based on standards, which may be proprietary or non-proprietary.

With the advent of big data, machine learning, AI, LLMs, and related technologies, we have the ability to learn from what is often described as unstructured data. The current wave of generative AI is driven by large datasets of data scraped from websites or information obtained by parsing eBooks, articles, and similar information. I don't consider this as unstructured data. The information has context. Much of the data on websites can be classified as information, as it too includes context. However, as a whole, the data and information in a large dataset used to train an LLM follow many formats and structures, and taken as a whole, the information was not created and optimized to be used to develop an LLM – the information was not built for that purpose. The significant issue is whether the information is built for how it is being used, what I call *built for use*. The information making up an LLM was not created/built to be used in an LLM – it was created/built for some other purpose, e.g. for the purpose of a particular webpage from which the information was obtained.

An important premise of this book is that information should be *built for use*. It should be built to satisfy its purpose and its intended use cases. For example, a product definition dataset should be built to define a product and its requirements and to satisfy all the needs of the processes and activities that use that information. It is possible to produce products and perform lifecycle activities without a built-for-use product definition dataset. However, doing so drives informal processes that are not scalable, which are less efficient, more costly, archaic, chaotic, and lead to lower morale, and frustration, less consistent products, take longer, lead to lower customer satisfaction.

MODEL-BASED DEFINITION (MBD)

The system and methods used to define digital business information so that it is optimized to be directly used in software. MBD replaces document-based methods that use information optimized to be presented in documents and used manually.

> **Note:** Historically, MBD has had a limited definition, in which MBD was used in a mechanical design context to describe cases where 3D models are used in place of 2D drawings. This definition leads us to situations where little value is realized by the organization that implements MBD and MBE.

MODEL-BASED ENTERPRISE (MBE)

Simple Definition

An organization focused on obtaining maximum value from its information throughout the organization, the supply chain, the product lifecycle, and the information lifecycle by creating and using MBD as much as possible.

Detailed Definition 1

An organization focused on creating and directly using MBD information as much as possible throughout the organization, the supply chain, the product lifecycle, and the information lifecycle, with the goals of maximizing the lifecycle value of information (LVOI) and minimizing the need to create and use manual-use information.

Detailed Definition 2

An organization that focuses on and prioritizes obtaining the greatest lifecycle value of information (LVOI) for the least cost and burden throughout the organization, the product lifecycle, and the information lifecycle by maximizing the creation and direct use of MBD information.

Detailed Definition 3

An organization that focuses on and prioritizes obtaining the greatest lifecycle value of information (LVOI) for the least cost and burden throughout the organization, the product lifecycle, and the information lifecycle by:

1. Using digital information directly in software and computation (i.e. automation) throughout the organization, the supply chain, the product lifecycle, and the information lifecycle.
2. Optimizing processes and methods to create and use direct-use MBD information throughout the organization, the supply chain, the product lifecycle, and the information lifecycle.
3. Reducing or eliminating the need for, use of, and preference for visually-presented manual-use information throughout the organization, the supply chain, the product lifecycle, and the information lifecycle.
4. Designating direct-use information assets and information as official instead of manual-use information assets and information throughout the organization, the supply chain, the product lifecycle, and the information lifecycle.

Alternate Definition

An organization focused on creating MBD and digital information and using that information directly as much as possible throughout the product lifecycle and information lifecycle, with the goal of maximizing the lifecycle value of information (LVOI) while incurring the least cost and burden from the information.

1. An MBE strives to create the least information needed in order to minimize lifecycle information costs and cognitive loads the information places on staff.
2. An MBE focuses on minimizing manual activities and document-based workflows as much as possible.
3. An MBE is formally structured to achieve these goals.
4. An MBE formally supports and encourages the behavior needed to achieve these goals.

MODEL-BASED PROCESS DEFINITION

Systems and techniques used to define a process as completely as possible using models (digital information models) instead of documents.

Preferred Method: Process-definition models are used directly in software; the process may be performed without using information manually (manual use of information is not required).

MODEL-BASED PRODUCT DEFINITION

Systems and techniques used to define a product as completely as possible using models (digital information models) instead of documents. In the traditional MBD paradigm, a 3D model and related digital data are used to define a product.

Preferred Method: Product-definition models are used directly in software; the product information may be used directly without manual intervention (manual use of information is not required).

MODEL INFORMATION STRUCTURE OR MODEL ORGANIZATIONAL STRUCTURE (SCHEMA)

The naming conventions and organization of information within a model or digital information asset.

I prefer to use the terms "information structure" or "organizational structure" rather than "schema", as information structure and organizational structure more clearly define what is being discussed, and thus are more easily understood. However, the term "schema" was popularized and presented in literature, and in the context of MBD, it is used extensively in an ASME Y14 Standard.

The concept of *model schema (schema)* for product definition datasets is introduced in the ASME Y14.47 Standard. The concepts in this standard provide rudimentary examples and guidance for product definition data but are insufficient for other types of models and information assets.

Product definition data schema were developed to facilitate understanding and increase usability of information in product definition datasets for users. The issue is that the product definition dataset is created by a different group or organization than the receiving organization that must use the product definition dataset to do something, such as manufacture, inspect, assemble, disassemble, install, or service the product, and sometimes the receiving organization cannot easily find or use the required information.

By contrast, many people are comfortable using engineering drawings. People and industry have been using engineering drawings for many years. While engineering drawings are abstract and inefficient, mature systems and training are in place to help people prepare and understand engineering drawings.

Model-based product definition methods and information models are relatively new. Where engineering drawings are static, MBD information assets are dynamic – visibility of information within an MBD dataset can be turned on or off, possibly inadvertently. While traversing orthographic views on an engineering drawing may be tedious and abstract, it can be relatively easy for experienced workers trained in how to do it. Presenting views in an MBD product definition dataset is also easy and provides information in an easy-to-use and easy-to-understand manner. However, the way the information is presented and traversed is different from how people are used to doing it with drawings. Thus, model schemas are also needed because model-based methods are different from document-based methods. It isn't that MBD information is more complex; it is that people aren't used to model-based methods yet.

Using model schemas also standardizes the way datasets are structured. Using model schemas formalizes and amplifies the requirement for consistently structured predictable datasets. Using model schemas enables direct use of information.

In many cases, contracting entities (e.g. government contractors) are large companies with many divisions, some of which were acquired from other companies that do things differently or use different software; thus, the information they provide may be structured differently than the recipients expect.

For example, a government agency paid a contracting company to design something and provide information assets, such as model-based product definition data, but upon receipt of the information, the information was difficult to use because the recipient didn't have a legend and/or map of the information received. Another reason the information was difficult to use was that the contracting company didn't use a standardized approach to constructing the information within the information asset.

A model schema has several purposes and is important for several reasons. See *model structure (model schema)* below.

MODEL STRUCTURE (MODEL SCHEMA)

The structure and organization of information within an MBD information asset.

Schema is used by some to describe information structure, information naming conventions, and how information is organized within a product definition dataset.

Native Format

The default data file format used by a software application. Note that native data formats are usually proprietary and may hinder interoperability between MBD and MBE lifecycle applications, especially applications created by different companies.

Native file formats include extensions such as .docx, .xlsx, .dwg, .prt, .asm, .catpart, .catproduct, .sldprt, .slddrw, .sldasm, .iam, .ipt, .idw, etc. (see Figure 4.7).

Neutral Format

A non-proprietary data file format developed to aid interoperability between software applications. Note that neutral data formats are often developed by consortia and/or standards development organizations.

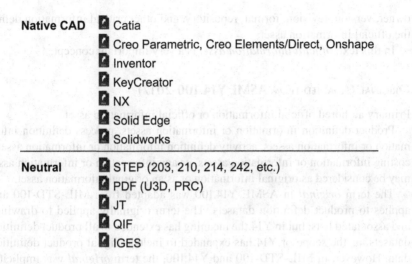

FIGURE 4.7   Native CAD and neutral information formats.

Neutral file formats include extensions such as. pdf, .dxf, .stp, .jt, .qif, .iges, .qif, .jpg, etc. (see Figure 4.7).

OFFICIAL INFORMATION

Information that is designated as the legal, authoritative definition of a subject, such as a product, process, activity, results, etc. Official information is usually defined in an official information asset. Other terms for official information include master, original, or authoritative version.

In this text, *official information* is an important concept.

OFFICIAL INFORMATION ASSET

The information asset that is designated as the legal, authoritative asset for designated use cases. It is common for many versions of a document, model, dataset, or other information asset to exist in a business. The official information asset is the version that is designated as the official version that is supposed to be used in business activities.

Example 1 – Corporate document, such as the standard operating procedure for inspection. There may be several drafts and revisions on company servers, on computers, and on printed copies. The official information asset should be designated as a specific version and/or revision that resides in an officially designated repository, such as in the company's PLM system.

Example 2 – Engineering 3D model and 2D drawing. The official information asset will be designated as a model, a drawing, or both. Assume the model is designated as the official information asset. The name, originating activity, owner, version, revision, format, repository, and other critical information define the official information asset.

In this text, *official information asset* is a very important concept.

ORIGINAL (ADAPTED FROM ASME Y14.100-2017)

Primary authored official information or official information asset.

Product definition information or information assets, process definition information or information assets, activity definition information or information assets, costing information or information assets, or any information or information asset may be considered as original information or as an original information asset.

The term *original* in ASME Y14.100 was adapted from MIL-STD-100 and applies to product definition datasets. The term originally applied to drawings and associated lists, but in Y14 the meaning has extended to all product definition datasets, as the scope of Y14 has expanded to include digital product definition data. However, in MIL-STD-100 and Y14.100, the term *original* was implicitly limited to product definition datasets. There has never been a prohibition against using the term *original* for other types of information assets, such as process definition datasets or activity definition datasets. It's just that the scope of Y14 is limited to product definition data practices, and thus process, activity, costing, and other information assets are outside of Y14's scope. Note that I have tried to get Y14.41 to accept that some of our work can apply to more than product definition data, but my suggestions were disregarded.

In this text, *original* is defined more broadly, such that the term and concept apply to any type of information that meets the criteria in the definition.

Thinking about original and in the context of product definition datasets:

- Original information assets are usually saved in the native data format of the authoring software used to define the product (e.g. native CAD models, native CAD drawings, lists, documents, and ancillary information in the appropriate native format, other digital data in its native format, etc.). Native formats are the formats in which the datasets were saved internally, usually the default data format of the authoring software. Original information assets (e.g. native files) are generally used as the basis for revisions (e.g. design organizations prefer to start a revision from an existing dataset in native format, preferably one they developed and own).
- The released official information asset is a TDP, which is derived from and may also contain some or all of the product definition dataset. TDPs often include DDDs that have been converted into neutral or other data formats for interoperability and enhanced usefulness.

  From this standpoint, a TDP is a derivative dataset that may contain original information, such as a product definition dataset, but the TDP is often derived from the original product definition dataset and ancillary information.

Note 1: In the early 1990s, I was part of the group that converted MIL-STD-100 (a U.S military standard) into ASME Y14.100. Today, most of the content in ASME Y14.100 applies to any sector that uses discrete manufacturing, although some of the content is still skewed toward supporting DOD acquisition and DOD's suppliers. The definitions of *original, design activity, current design activity*, and *original design activity* come from the DOD business model. Defense contractors and their supply chains design and build products for some portion of the DOD. In this case, per the definitions in this section, the original versions of product definition data reside with the defense contractor. The DOD receives derivative copies as TDPs to use as official information assets, but these copies are not the original versions in most cases. This business model is also common in consumer electronics, which uses large contract manufacturers, such as Foxconn and Flextronics. Hence the need in these cases to differentiate between the original and another version.

Note 2: The definition for original in ASME Y14.100-2017 is still not optimized for MBD and MBE, thus I improved the definition.

ORIGINATING ACTIVITY

The organization or group that is responsible for the creation, revision, and submission of a dataset for release by releasing organization (e.g. configuration management). Note that an external organization may be the originating activity for an external dataset, such as a manufacturing plan or an inspection plan developed by a supplier for work done by the supplier.

ORTHOGRAPHIC VIEWS

Views that present an object at 90° angles to one another. Surfaces are usually presented aligned to the view, with one or more selected surfaces of the part perpendicular to or parallel to the viewing direction. Orthographic projection and orthographic views are the main way views of products have been and continue to be presented on mechanical engineering drawings. It is common to name orthographic views, such as top view, front view, side view, etc. (see Figure 4.8).

There are pros and cons of orthographic views. On engineering drawings and rendered solid models, pictorial views provide more context to understand a 3D object, as the view allows the viewer to *see around the corner* of the object.

With the advent of solid modeling in 3D CAD, it is now easy and inexpensive to present pictorial views. Pictorial views were time-consuming and thus expensive to create before 3D CAD. Today, pictorial views, such as axonometric views, offer a better way to present product geometry and PMI. (For comparison, see *Axonometric Views*.)

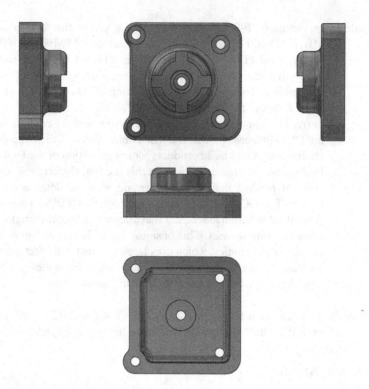

FIGURE 4.8   Orthographic views.

OWNER

Entity that is responsible for the content within a dataset, its release from the originating activity, and its maintenance throughout the lifecycle of the dataset.

PRESENTED INFORMATION

Information that is presented in a manner for a human to use as sensory input. In the context of MBE today, most presented information is visual information. Visual information may be presented on hard copy media, it may be presented in 2D or 3D, it may be presented on a 2D drawing or a document, it may be a view of a rendered 3D model onscreen, it may be a color-coded FEA deflection map, it may be static or dynamic, it may allow user interaction, etc.

If information is intended to be understood by a person by interacting with the information, it is presented information. Note that some information may be visible or audible but not intended to be viewed or understood by a person, such as smudged ink on a printed drawing, malfunctioning pixels on a display, static in an audio recording, and unintentional visual artifacts in a video recording.

Note that presented information can take many forms, such as audio information, visual information, tactile information, and other types of information. This text primarily discusses visual presented information.

PROCESS DEFINITION DATA (PROCESS DEFINITION INFORMATION)

Information that defines a process.

Process definition data includes definition of the process, process components (e.g. materials, tools, steps, sequence, dependencies, permissions, restrictions, actors, activities, etc.), process constraints, process targets, limits, SPC requirements, process performance parameters and requirements, metadata, process item and subcomponent identification and quantities, safety, security, and other information as needed.

PROCESS DEFINITION DATASET

An information asset that defines a process using process definition data.

A process definition dataset defines processes required in an enterprise. Process definition datasets are often derivative datasets created from information in a product definition dataset, e.g. a manufacturing plan, inspection plan, assembly work instructions, quality plan, etc. Note that process definition datasets may be created for other purposes and may be primary authored and not derivative datasets. Process definition datasets may also contain primary authored and secondary authored information.

PRODUCT AND MANUFACTURING INFORMATION (PMI)

Annotation, symbols, and attributes in a 3D model. See note below. PMI conveys information necessary for understanding product requirements in a product definition dataset and process requirements in a process definition dataset. PMI may define geometry, it may define restrictions or limits on geometry (tolerances, surface texture requirements), it may define product requirements such as performance requirements, it may define product characteristics such as material, hardness, finish, and allowable variation thereof, it may define tests and validation procedures, it may define items, alternates, and substitutes in assemblies, it may define steps in processes such as work instructions, it may refer to other models and documents, and other uses. PMI is defined in the context of information that is conveyed in a product definition dataset, such as GD&T, notes, thread specifications, etc. However, PMI applies to any annotation or attribute data that is used in a product definition or process definition dataset (see Figure 4.9).

PMI is also defined in DIN EN/NAS 9300 Long Term Archiving and Retrieval of digital technical product documentation such as 3D, CAD, and PDM data PART 007: Terms and References.

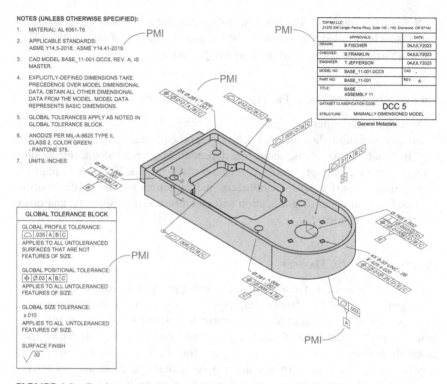

FIGURE 4.9 Product and manufacturing information (PMI) (visually-presented).

Note: PMI is defined in the DIN EN/NAS 9300 Standard and the term originated in the context of 3D CAD. In ISO TC 10, ISO TC 213, and ASME Y14 Standards, we define *annotation* and generally don't use the term *PMI*. Thus, overlap and contradiction between the definitions are possible. I use the term annotation in the context of 2D drawings and PMI in the context of 3D models, although annotation is one of the types of information included in PMI. PMI can be considered as annotation, symbols, and attributes in a 3D model that may be used manually, PMI may be used directly in software, or PMI may be direct-use information that is also used as the basis for the presentation of manual-use information. Thus, information does not necessarily have to be visible or intended to be visible to be PMI.

Product Definition Data (Product Definition Information)

Information that defines a product.

Product definition data includes the nominal product definition (e.g. geometric definition, material composition), constraints and allowable variation from nominal, product performance parameters and requirements, product configuration information, metadata, item and subcomponent identification and quantities, finish and appearance information, safety, security, proprietary and rights in use information, and other information as needed.

PRODUCT DEFINITION DATASET (PRODUCT DEFINITION INFORMATION ASSET)

An information asset that defines a product using product definition data. Note that in some cases, processing information is included in a product definition dataset.

QUALITY (IN AN MBD CONTEXT)

See the following definitions of *data quality*, *dataset quality*, and *model quality*.

Data Quality (Information Quality)

The overall integrity and accuracy of data relative to its intended purpose and use cases. Data quality includes relevance of data to its intended purpose, structure of the data, content of the data, format of the data, etc. For derivative data, data quality includes a measure of how well the derivative data corresponds to the source data from which it was derived (e.g. data content, structure, format, associativity, etc.). As an example, data quality could be a measure of how well a specification and model geometry represent a .250-20 UNC threaded hole in a model.

Dataset Quality (Information Asset Quality)

The overall integrity and accuracy of a dataset relative to its intended purpose, use cases, and file format. Dataset quality includes relevance of a dataset to its intended purpose, structure of the dataset, content of the dataset, format of the dataset, adherence to applicable standards, etc. For a derivative dataset, dataset quality includes a measure of how well the derivative dataset corresponds to the source dataset from which it was derived (e.g. dataset content, structure, format, associativity, etc.). As an example, dataset quality could be a measure of how well and how completely a dataset represents a product, such as a machined part, including representation of the part, its specifications, file metadata, and referenced data elements.

Model Quality

The overall integrity and accuracy of a model relative to its intended purpose and use cases. Model quality includes relevance of a model to its intended purpose, structure of the model, content, format, etc. For a derivative model, model quality includes a measure of how well the derivative model corresponds to the source model from which it was derived (e.g. content, structure, format, associativity, geometry, topology, etc.). As an example, model quality could be a measure of how well and how completely a model represents product geometry, such as a machined part.

REPRESENTED INFORMATION

Digitally encoded information that can be understood and used by computational systems.

Represented information may be direct-use information (e.g. to be used directly in software or computation), it may be manual-use information represented by digital information (e.g. underlying data driving the display of onscreen graphics intended as official information), or it may be information that is represented digitally but not useful in a particular context.

SPECIFIC USE DATASET (SPECIFIC USE INFORMATION ASSET)

A dataset created for a single purpose within the product lifecycle. This may be an original dataset or a derivative dataset or TDP that is used and/or optimized for a particular use case, for example:

1. A model-based work instruction for a particular task
2. A model-based machining model for a part at a particular supplier (supplier A). Another supplier that manufactures the same part (supplier B) may create a different machining model because supplier B has different processing machinery available.
3. A model-based request for proposal for a specific supplier.

TECHNICAL DATA PACKAGE (TDP)

An information asset consisting of or more files to be distributed outside of the authoring organization. Traditionally, a TDP includes product definition data derived from a product definition dataset and is intended to be used outside of the design engineering organization. A TDP contains the information needed to define a product and its configuration for acquisition, production, engineering, and logistics. TDP data is usually converted into neutral formats, but may also contain proprietary formats, such as native CAD. A TDP may be equivalent to the source product definition dataset, it may be simplified from the product definition dataset, or it may contain additional information.

Note that TDP is officially defined in MIL-STD-31000. My definition differs from MIL-STD-31000 in several ways, specifically clarifying that a TDP is not necessarily a product definition dataset, and TDP is not necessarily equivalent to a product definition dataset. Note that the term and concept of *product definition dataset* was developed by members of ASME Y14 activities and released as part of Y14 Standards, which focus on product definition data (the stuff created by design activities). The term *technical data package* was developed by people working for the DOD and its contractors and released in MIL-STD-31000 and its predecessors, which focus on the information needed for the US government to acquire materiel.

These use cases present different requirements; thus, the terms were developed for different uses and have had different meanings for many years. I find it useful for these terms to have different meanings, as product definition datasets have intended uses and are used, maintained, and have lifecycles that are distinct from the intended uses, use, maintenance, and lifecycle of TDPs. While a product definition dataset may be part of a TDP, a product definition dataset may be retained inside an OEM for internal use, such as being used as the basis for the next revision, where a TDP contains a version of a product definition dataset and other information that has been optimized for distribution.

INFORMATION AND INFORMATION ASSET HIERARCHY IN MBE

Figure 4.10 shows the relationships between and hierarchy of MBE information and information assets.

TABLE 4.1
Information and Information Asset Terms Hierarchy and Relationships

Term	Describes	Context	Use
Asset	Something perceived to be useful, valuable, and relevant to an organization.	Applies in a business context, specifically things that are valuable to an organization.	Describes things that are valuable or important to a business or organization.
Information	Information that is relevant and useful for a business activity.	Applies in a business context and to what is important to an organization. Information is often defined in information assets.	Describes information that is relevant to a business or similar organization.
Information Asset	An asset (e.g. a set of one or more documents, models, datasets, and/or objects) that contains business information and/or IP.	An information asset contains information that is relevant to a business or organization.	Describes an asset that contains information relevant to a business or organization.
Official Information	Official version of information.	Information designated by the source or owner of the information as the official version of the information.	Indicates information that is intended to be used as the official version of information.
Official Information Asset	Official version of an information asset.	Information asset designated by the source or owner of the information asset as the official version of an information asset.	Indicates an information asset that is intended to be used as the official version of an information asset.
Official Specification	Official specification that describes something and its requirements.	A specification in an official information asset.	Indicates a specification that is intended to be used as the official version of a specification.
Binding Information[1]	Contractually binding version of information.	Information that is contractually or otherwise invoked.	Binding information may contain contractually-binding requirements.
Reference Information	Non-binding version of information.	Usually redundant information (the official version of the information is defined elsewhere).	Should never be used as official binding information (should not be used for official definition or conformance).
Unofficial Information	Unofficial version of information.	Information that is not designated as official information.	Indicates information that is not the official version of information – Unofficial information is for reference only.

(Continued)

TABLE 4.1   (Continued)

Information and Information Asset Terms Hierarchy and Relationships

Term	Describes	Context	Use
Unofficial Information Asset	Unofficial version of an information asset.	Information asset that is not designated as the official version of an information asset.	Indicates an information asset that is not the official version of an information asset – Unofficial information assets are for reference only.
Use Designation: Designated for use	Version of official information that is intended and approved to be used.	Official information that is designated for use in lifecycle processes, activities, datasets, etc.	Designates official information that is intended and approved to be used in or to drive lifecycle processes, activities, datasets, etc.
Use Designation: Designated as reference	Version of official information that can only be used as reference information.	Official information that is designated as reference in lifecycle processes, activities, datasets, etc. – It is not and cannot be binding. information.	Designates official information that can only be used as reference in lifecycle processes, activities, datasets, etc.
Use Designation: Designated not for use	Information that is not intended or approved to be used.	Information that is designated not to be used in lifecycle processes, activities, datasets, etc.	Designates information that is not intended or approved to be used in or to drive lifecycle processes, activities, datasets, etc. or as reference.
Binding Information Asset	A contractually binding information asset.	Information asset that is contractually or otherwise invoked.	An information asset that contains contractually-binding requirements.
Binding Specification[1]	A contractually binding specification.	Specification that defines contractually binding requirements for recipient.	A specification that is contractually binding, which may contain contractually-binding requirements.
Mandatory Information[1]	Information that defines conformance criteria.	Information defines a mandatory requirement.	Information that affects conformance of something to some criteria.
Non-mandatory Information[1]	Information that does not define conformance criteria.	Information does not define a mandatory requirement.	Information that does not affect conformance of something to some criteria.

[1] Binding information may be mandatory information or non-mandatory information. For example, a binding information asset (e.g. a drawing or annotated model) contains binding information, some of which is mandatory information (it defines mandatory requirements) and some of which is non-mandatory information (it does not define mandatory requirements). Note that binding information and binding information assets are usually official information.

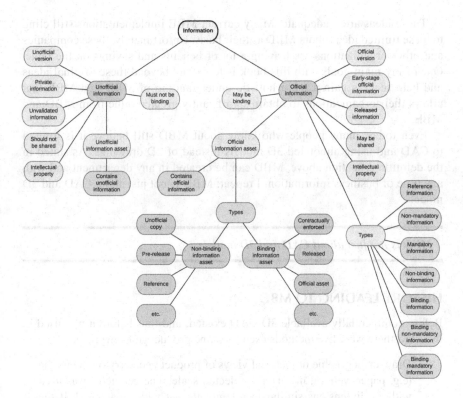

FIGURE 4.10 Information and information asset hierarchy in MBE.

TRADITIONAL CONCEPT OF MBD

The original idea of MBD was using 3D models instead of drawings, using 3D CAD or neutral model geometry instead of dimensioned drawing views – MBD was a method that design engineers used. Design engineers would share 3D CAD model geometry with other departments to use instead of the traditional information that defined product geometry on 2D drawings. There were three parts to the original idea:

1. MBD was limited to using 3D CAD or neutral model geometry to represent product geometry. MBD information was limited to product definition. (Note that this limited idea was and continues to be underutilized, as most organizations continue to partially or fully define model-based product geometry with explicit dimensions.)
2. MBD was limited to the design organization. MBD information was only created by the design organization.
3. The primary use case for MBD information was visualization – it was a manual use case.

These ideas are inadequate. Many current MBE implementations still cling to these limited ideas about MBD as their basis. Unfortunately, these companies and other organizations are leaving a lot of benefits and savings on the table. One of my main goals with this book is to extend beyond these original ideas and help readers gain a more holistic understanding of MBD and MBE that allows their organizations to obtain significantly greater value from MBD and MBE.

Even today, many people who know about MBD still believe it is limited to CAD and using annotated 3D models instead of 2D drawings. As stated in the definitions section above, MBD can be created in any department and with any type of business information. I repeat: MBD is not just about CAD and 3D models.

MBD is not just about CAD and 3D models

HISTORY LEADING TO MBD

Before commercially available 3D CAD existed, and thus before it was used in industry, there were five methods used to define product geometry:

1. Draw orthographic or pictorial views of product geometry on 2D media (e.g. paper, vellum, mylar) at a selected scale. The geometry was used without dimensions similar to a template, but with greater risk if the scale was not 1:1.
2. Draw orthographic views of product geometry on 2D media and define the geometry using dimensions. The dimensions represent product geometry in this case. In fact, the dimensions are the official information defining the product geometry. More on this later.
3. Draw orthographic sections, with contour lines and elevation and station lines, etc. to represent portions of product geometry on stable media (e.g. mylar, aluminum sheet) full scale to be used as a template.
4. Create a physical 3D model to represent the product geometry, often made from clay, wood, metal, or other material.
5. Use an existing part as physical representation, such as a part being reverse engineered, or an ideal (e.g. a golden part).

For most parts, method 2 became the preferred method, especially parts with simpler geometry. For parts with complex geometry, such as complex curved shapes in auto body, aerospace structures, and nautical structures, method 3 was common. Note that for these parts with complex curved shapes, a combination of methods 2 and 3 was often used. Early ideas for MBD came from the challenges posed by defining complex shapes using traditional drafting methods.

Originally, MBD was based on the idea that some or all of the geometry defined by traditional means on 2D mechanical engineering drawings could be replaced by

3D CAD model geometry. Using the terminology defined in this book, the idea was to use CAD model geometry as the official information representing product geometry instead of using views and dimensions on 2D drawings to officially represent product geometry. This implied that the CAD model would be used in lifecycle activities to represent the product, alongside the drawing and BOM. However, it took a while for companies to realize that this meant the 3D model dataset should be classified as an official information asset and formally controlled and released along with or in place of a drawing, BOM, and other information. Note that many companies still don't do a particularly good job of identifying and managing official information when models are involved. See Chapter 6 for recommendations and best practices for this process.

Originally, given the scope of MBD was limited to design activities, people thought that MBD information was only created by a design organization. The design organization used MBD techniques to define product geometry using 3D CAD data. It was a logical idea, as design engineers were using 3D CAD to create solid models of the product, and inside the design organization, those solid models were the official information assets representing product geometry. In fact, as 3D CAD became more powerful and commonplace, it was quickly apparent that rendered 3D CAD models were orders of magnitude better than 2D drawings for visualizing product geometry. Soon, everyone who could use rendered 3D model views to understand product geometry did so. Drawings were still created, and in most environments, the drawings were still the official information assets for product definition, but if someone really wanted to understand the product, they looked at onscreen views of the model.

The process was (and still is in most companies) convoluted. Design engineers created models during the design process to understand the model and if/how well it solved the problems they were trying to solve. When the model was complete enough and the design trusted enough, the design engineer used the model as the basis to create a 2D drawing. (Note that the process of generating views on a 2D drawing of a 3D model in modeling software is amazingly easy compared to the methods available before 3D solid modeling software.) Drawing views were laid out/arranged and annotated. Most orthographic views and sections on mechanical engineering drawings are needed for dimensioning. We need a lot of views and sections, sometimes redundant views of the same geometry, to be able to clearly show all the dimensions needed to fully define geometry. It takes up a lot of drawing real estate to do this. Note that while having many views and sections on a drawing may make certain information easier to understand, having more views and sections increases the amount of information on a drawing and makes the overall drawing harder to learn and navigate. It also increases the cost of the drawing, both in terms of its initial cost (creation cost) and in terms of its lifecycle cost. I will discuss the cost of information and information assets in greater detail later in the book.

In contrast, MBD allows us to bypass using dimensions completely. We can use model geometry to represent product geometry, and this can be good for all involved if they are trained and provided efficient tools and methods to use the MBD information. Later in this chapter, I include examples showing how minimal dimensioning methods used in MBD require fewer views than methods where parts are fully-dimensioned.

Note that most mechanical engineers entering the workforce today were not trained in school how to make engineering drawings. The part of the design process where drawings are created is often painful, not very productive, and is often hurried through at the end of the design process – it is part of the design process most design engineers do not like. Even for many experienced mechanical engineers, creating drawings is not their favorite part of the design process. For some, it is the dreaded thing right before the design is released, as there are lots of Is to dot and Ts to cross, there are lots of arcane drawing standard rules, creating good drawings can take a long time, and thus creating drawings is much less enjoyable than modeling (see Sidebar/Info Box). Contrast that against 1990 when we still had huge cohorts of designers and drafters in the workforce. As I mentioned, my first degree was a design drafting degree. Drafters were taught how to make drawings and how to communicate technical information. Many of us liked to do it and were therefore good at it, or at least we tried harder to do a good job than what comes from the ambivalence toward drawing quality that prevails today.

In GD&T, tolerance analysis, and MBE classes I have taught over the years I have asked more than a thousand engineers whether they prefer modeling or making drawings. I don't recall anyone ever saying they like making drawings more than modeling. This illuminates the lack of incentive and interest people have in making drawings. It is simply something most people dislike. Many people feel it is a waste of time and they don't like doing it – they were zipping along, creatively solving problems and coming up with something to be proud of during the modeling process, only to slow down into a lower gear and spend a lot of time doing something that they perceived as not fun and not as important as the model. This situation leads to lower drawing quality, lower job satisfaction, and according to some of my clients, staff churn as engineers leave for other jobs, so they don't have to work on drawings so much. It also leads to confusion about what is important, as people see the thing they don't like (the drawing) must be released and shared with the supply chain, and the thing they and others like (the model), is not released or informally shared behind the scenes. One thing carefully executed MBD and MBE do for us is to eliminate or mitigate these problems.

Today, in companies that use a traditional mechanical design process, a 3D model is created, the model is used as the basis to make a 2D drawing, and the drawing is released as the official information asset. See Figure 4.11 for a

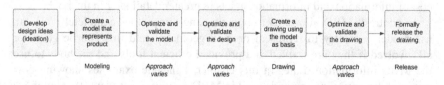

FIGURE 4.11 Traditional model-to-drawing process for manual-use workflows 1.

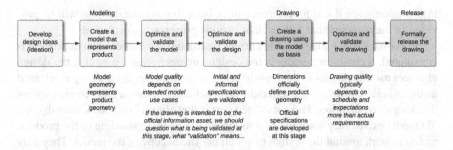

FIGURE 4.12   Traditional model-to-drawing process for manual-use workflows 2.

simplified traditional approach. Note that the drawing is often the information asset that is formally released. People are supposed to use the drawing to understand the product and its requirements. Before 3D solid modeling CAD software was commercially available, adding an isometric or pictorial view to a drawing was uncommon, and it was even frowned upon. "We're not artists here" was a common refrain from the drafting manager. We used orthographic views and sections to depict product geometry. Today, the first information asset we create in mechanical design is usually a 3D CAD solid model. We can view a 3D CAD solid model from any angle we want, zoom in and out, etc. Today, it is extremely easy and requires little time to create pictorial views. Before 3D CAD, it took a long time to manually create pictorial views.

Figure 4.12 adds more information to the simplified approach shown in Figure 4.11. Using drawings as official information assets does not support model-based use cases. Likewise, annotated models intended to be used in manual workflows are not well suited for direct-use.

By comparison and looking ahead to later discussions, Figure 4.13 shows a simplified process for creating model-based product definition datasets for direct use. The goal of this workflow is to fully support model-based use cases. Thus, it is critical that relevant information is optimized for direct-use. Notice that I did not include steps describing using information manually in Figure 4.13. This is intentional, as I am trying to get people thinking about whether presented information is required. Of course, in many cases, we will still want to have presented

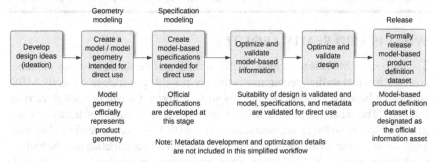

FIGURE 4.13   Model-based process for direct-use workflows.

information, even if it is for reference only. Note that many examples will be provided in the material that follows that include presented information.

To get back to the idea that the current traditional mechanical design process is convoluted, a common process is the design engineer creates a model, the design engineer uses the model as the basis to create a 2D drawing, the drawing is released as the official information asset, and the model is not released and remains within the design organization. The convoluted part is that almost everyone knows that the 3D model exists; they want to use it to improve their understanding of the product, and thus work around the system to get an unofficial copy of the model. They may want to use it manually, for visualization purposes, or they might want to use it directly in software, say, to generate tool paths for machining or probe paths for inspection. And, since the drawing is the official information asset and the model is not released, using the model in this context is informal and risky, as the model and drawing may have inconsistent and conflicting information.

MBE and proper use of MBD correct these issues. With MBE, we designate the model and/or a drawing as the official information asset, we release the model and/or drawing, we formally designate what they are for and how they are supposed to be used, and which one is the boss for each aspect of the product definition.

Note that this discussion has been very design-centric. This is traditionally where MBD discussions originate. In fact, many people believe that MBD is only used by design organizations, and that model-based information is only defined by design engineers. This is very much a misunderstanding. MBD information can be and should be created in many areas of an enterprise and its supply chain, in many lifecycle activities throughout the lifecycle. Any group or activity within an organization may create model-based information. It is a fundamental tenet of MBE that MBD information is created and used throughout the enterprise and across the lifecycle. In fact, we want to create and use MBD as much and as efficiently as possible and in as many activities and organizations across the enterprise as possible.

Model-based information can be and should be created in many areas of an enterprise and its supply chain, in many lifecycle activities throughout the life-cycle. Any group or activity within an organization may create model-based information. MBD is not limited to design engineering or to product definition.

WHAT IS MBD REALLY ABOUT?

The purpose of MBD is to facilitate MBE methods and the dramatic savings in time, cost, burden, and staffing requirements that are possible *if it is done correctly*. MBD is defining business information digitally so the information can be used directly in software. The initial purpose of MBD is to facilitate automation of legacy manual activities that are no longer necessary, too slow, and no longer cost-effective. More generally, in the longer term, the purpose of MBD is to enable automation and the highest productivity possible. While MBD is about automation through direct use of information, MBD also facilitates better visualization where manual workflows are still used. Remember, manual use of presented MBD information does not yield savings as great as direct-use workflows. However, companies often choose to move toward direct use in stages. Thus, in many cases, some MBD information will be defined to facilitate direct-use workflows and other MBD information will be defined to facilitate manual-use workflows. That said, I am strongly advocating direct-use workflows in this book. Many people are too comfortable with baby steps. They need a nudge…

The purpose of MBD is to facilitate MBE methods and the dramatic savings in time, cost, burden, and staffing requirements that are possible if it is done correctly.

Take a look at my definition for *model (CAD context)*:

MODEL (CAD CONTEXT, LIMITED SCOPE)

2D or 3D digital information that defines a product or process.
This definition is close to the original idea of MBD, except I expanded it to also include process definition information.

A definition that more closely captures the original limited idea of MBD is:

MODEL (CAD CONTEXT, LIMITED SCOPE)

2D or 3D digital data that defines a product or process.
Note that I include 2D information/data because that is another use of CAD data representing product data. In the earliest MBD attempt I witnessed, 2D IGES data was used in a plasma cutter's software to cut stainless steel plate. 2D data is also commonly used to represent product geometry in printed circuit board and routing operations.

Take a look at my definition for *model (broader context)*:

MODEL (INFORMATION MODEL) [BROADER CONTEXT]

A set of digital information that represents something of interest. Preferably, business models consist of structured information. Examples of models include systems engineering models, product definition datasets, product models, process definition datasets, system definition datasets, activity models, requirements models, information models, digital twins of a system or facility, simulation models, analytical models, costing models, logistical models, ROI models, simulation models of an entire enterprise, such as an MBE, etc.
This definition is intentionally broad, so broad, in fact, that it encompasses any digital dataset or information model that represents something of interest. This is intentional and necessary to support the full concept of MBE. See the following definitions.

Now consider more rigorous definitions for MBD and MBE. These definitions state the preference for creating and using information directly in software and for decreasing reliance on document-based workflows and manual methods.

Model-Based Definition (MBD)

The system and methods used to define digital business information optimized to be directly used in software. MBD replaces document-based methods that use information optimized to be presented in documents and used manually.

MBD applies to business information, such as product definition information, process definition information, activity definition information, engineering analytical information, simulation and digital twin information, and other information related to products and the processes used to design, produce, service, and maintain them. Other types of business information are usually not considered MBD, such as financial information, human resources information, and legal and contract information. However, the line between what should and should not be considered MBD is fuzzy, especially if we accept that future enterprises and all their data will be modeled. (We should accept that this will occur.)

Model-Based Enterprise (MBE)

An organization focused on creating and directly using MBD information as much as possible throughout the organization, its supply chain, the product life-cycle, and the information lifecycle. An MBE has the following goals:

Maximize

- The value of information
- The value the organization obtains from information
- Focus and emphasis on direct use information
- Creation and use of direct-use information
- Information-enabled automation

Minimize

- The cost of information incurred by the organization
- Focus and emphasis on manual-use information
- Creation and use of manual-use information
- Manual activities and the need for manual-use information
- Historical inefficiencies affecting how we work

THE PRIMARY AND SECONDARY PURPOSES OF MBD

The primary purpose of MBD is to:

- Create business-critical information to be used directly in software, that:
  - Is optimized for direct use

- Is as lean as possible, providing only the necessary information for intended use cases
- Facilitates dramatic improvement in the value of information and reduces the cost of creation and cost of ownership of information.

The secondary purpose of MBD is to:

- Ensure that the direct-use information created can be used as the basis for presented information, to support manual workflows (visualization) in cases where direct-use workflows are not developed yet or offline.

Using presented information should be limited, such as for backup work methods, unavoidable meetings, clarification, troubleshooting, emergencies, and escapes, as a short-term solution to use during MBE development and deployment stages in which segments of the workflow are not ready to use information directly in software. Wherever possible, presented information should be derived from direct-use information.

The reason MBD has primary and secondary purposes is that during development and in certain parts of the workflow the tools and methods might not be in place to support using information directly in software. A common first-stage MBE implementation is to use model geometry directly to represent product geometry and use PMI manually to represent specifications. The stages of MBE implementation will be addressed in Chapter 6.

I recommend trying to create as little presented information as possible. (Note that this is counter to how most work is currently done.) When trying to adopt MBE, trying to get people and organizations to change from manually using information presented in documents to directly using information in software is difficult. If presented information is available, people will use it, as it is what they are used to. Thus, I recommend not creating redundant versions of the same information, such as direct-use and manual-use information for the same thing.

MBD TOPICS

Dataset Classification and Dataset Classification Codes (DCCs)

Product definition DCCs were developed by members of the ASME Y14.41 sub-committee. Before the advent of CAD, product definition datasets were limited to 2D drawings and ancillary information such as schematics, parts lists, wiring lists, instruction documents, and other documents. Thus, before CAD, product definition datasets were structured using documents.

The proliferation of 2D and 3D CAD in industry led to new ways to structure product definition datasets. In the 1990s, companies began to experiment with using 3D data in product definition datasets. Over time, this practice evolved into MBD. In 2003, ASME released the first edition of the ASME Y14.41 Standard, which addressed aspects of using 3D data in product definition datasets. Dataset classification and DCCs were developed and released in the ASME Y14.100-2013 Standard. The classifications were published in ASME Y14.100 because dataset class 1 (DCC 1) only contains drawings and documents and did not necessarily include digital data. Dataset classifications DCC 2–DCC 5 are based on or include digital data. As stated in earlier sections, the meaning and understanding of MBD have evolved over time. Initially, MBD was treated as a design engineering concept and primarily in terms of replacing 2D drawings with 3D CAD data. Over time, it has become clear that MBD is much broader than design engineering and is much more than drawings versus CAD. As I clarify many places in this text, how the information intended to be used and how it is actually used is most important.

As previously stated, dataset classification was developed by people primarily thinking about MBD in terms of *drawings versus CAD models.* Thus, only DCC 1–DCC 5 were developed. I added DCC 6 in this text to address datasets that are not based on engineering drawings or annotated models, such as other information models intended to be used directly in software or computation, including compiled object code, G-code for machining, STL for 3D printing, and many others.

Table 4.2 presents a graphical view of DCC 1–DCC 6. The concepts listed here and the need to indicate how datasets are structured should be extended to other datasets created and used by other disciplines, such as analysis, simulation, procurement, quality, manufacturing, assembly, inspection, installation, twins, service, training, and others.

Definition

DCC   Dataset Classification Code (repeated here from beginning of chapter)

Alphanumeric designations from DCC 1 to DCC 6 represent a range of product definition dataset structures from drawing-only to drawing + model, model-only, and direct-use information only. Dataset Classification Codes DCC 1–DCC 5 are defined in the ASME Y14.41 and Y14.100 and ISO 16792 Standards. Dataset Classification Code DCC 6 is defined in this book. Note that other DCCs could be defined if needed. Dataset structures corresponding to DCCs range from drawing-only (e.g. DCC 1), drawing + model (e.g. DCC 3), model-only (e.g. DCC 5), to data-only (e.g. DCC 6).

Dataset classifications facilitate different ways of doing business.

TABLE 4.2
Dataset Classification and Dataset Classification Codes (DCCs)

Dataset Class	Official Information Assets		
	Drawing	Model	Ancillary Docs
DCC 1 2D Drawing Complete definition	Drawing	No Model	Ancillary Docs
DCC 2 2D Drawing based on 3D Model, model is reference	Drawing	Model not released	Ancillary Docs
DCC 3 2D Drawing + 3D Model Partial definition in both	Drawing + Model		Ancillary Docs / Info 01000100 01101001 01110010 01100101 01100011 01110100 00101101 01110101 01110011 01100101 00100000 01101001 01011110 01100110 01101111
DCC 4 2D Drawing + 3D Model Complete definition in both *DCC 2 + DCC 5 = DCC 4*	Drawing or Model		Ancillary Docs / Info 01000100 01101001 01110010 01100101 01100011 01110100 00101101 01110101 01110011 01100101 00100000 01101001 01011110 01100110 01101111
DCC 5 3D Annotated Model Complete definition	Model	No Drawing	Ancillary Info 01000100 01101001 01110010 01100101 01100011 01110100 00101101 01110101 01110011 01100101 00100000 01101001 01011110 01100110 01101111
DCC 6 Direct-use Information Complete definition	Direct-use Model Data	No Drawing	Data only (display is optional) 01000100 01101001 01110010 01100101 01100011 01110100 00101101 01110101 01110011 01100101 00100000 01101001 01011110 01100110 01101111 Ancillary Info 01000100 01101001 01110010 01100101 01100011 01110100 00101101 01110101 01110011 01100101 00100000 01101001 01011110 01100110 01101111

Dataset Classification Code 1 (DCC 1)

Official information: 2D drawing (e.g. drawing or sketch not derived from a model, raster scan of a drawing) and ancillary documents [drawing and ancillary documents are released; there is not a model].

Product definition dataset contains a drawing that is not derived from a digital model. The dataset may include ancillary documents with information that is not presented on the drawing, such as a separate parts list. The drawing and ancillary documents should completely define the product (see Figure 4.14).

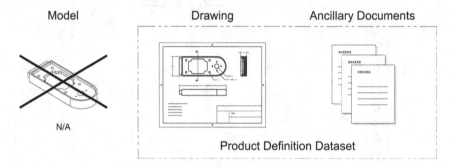

FIGURE 4.14   Information assets in DCC 1.

Dataset Classification Code 2 (DCC 2)

Official information: 2D drawing (derived from a model) and ancillary documents [drawing and ancillary documents are released]

Product definition dataset contains a drawing (derived from a model) and ancillary documents. The dataset may include ancillary documents with information that is not presented on the drawing. The drawing and ancillary documents are supposed to completely define the product. The model should not be transmitted outside the design organization. The premise is the model is not required for lifecycle activities, as the drawing and ancillary documents adequately support the activities (see Figure 4.15).

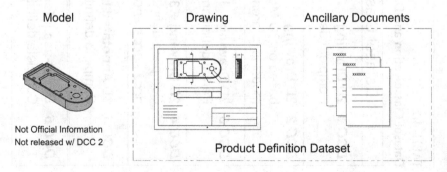

FIGURE 4.15   Information assets in DCC 2.

Dataset Classification Code 3 (DCC 3)

Official information: 2D drawing, model, and ancillary information

[drawing, model, and ancillary information are released]

Product definition dataset contains a drawing, a model, and ancillary documents or information. Neither the drawing nor the model alone completely defines the product. The dataset may include ancillary documents or information that are not included on the drawing or in the model. Full product definition requires the drawing, the model, and ancillary documents or information (see Figure 4.16).

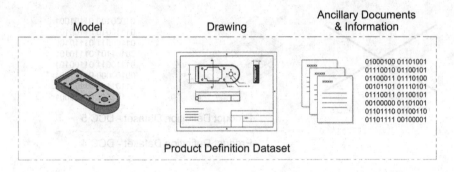

Model | Drawing | Ancillary Documents & Information

```
01000100 01101001
01110010 01100101
01100011 01110100
00101101 01110101
01110011 01100101
00100000 01101001
01101110 01100110
01101111 00100001
```

Product Definition Dataset

FIGURE 4.16 Information assets in DCC 3.

Dataset Classification Code 4 (DCC 4)

Official information: 2D drawing and ancillary documents

and

Model and ancillary information

(dual official information)

[duplicate versions of product definition data, equivalent to DCC 2 and DCC 5, are released]

Product definition dataset contains redundant versions of a drawing, a model, and ancillary information. Separately, the DCC 2 version of the product definition data and the DCC 5 version of the product definition data are supposed to be equivalent and completely define the product. The drawing and ancillary documents, and the model and ancillary information, completely define the product. DCC 4 is not recommended, as this method does not yield optimal business value (the cost and burden are greater than the benefits).

Note that a DCC 1 dataset could be used in place of a DCC 2 dataset in DCC 4. I assume this is unlikely, as a company that is advanced enough to have a model-based version of a product (e.g. a 3D CAD model), would create a 2D drawing based on that model rather than rely on a likely older DCC 1 version of a drawing that is not based on a model (see Figure 4.17).

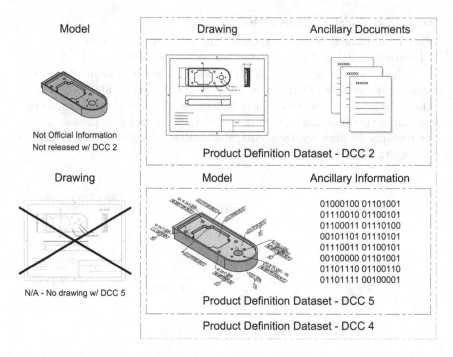

FIGURE 4.17   Information assets in DCC 4.

Dataset Classification Code 5 (DCC 5)

Official information: model and ancillary information (no drawing, model-only)
   [model and ancillary information are released]

Product definition dataset contains a model and ancillary information. The model and ancillary information are supposed to completely define the product. Note that the 3D model in this dataset structure is typically an annotated model with metadata and other data elements. I recommend DCC 5 instead of DCC 1– DCC 4 (see Figure 4.18).

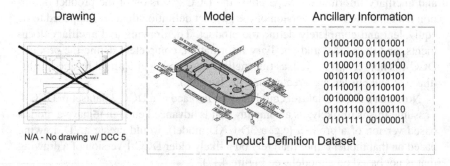

FIGURE 4.18   Information assets in DCC 5.

Dataset Classification Code 6 (DCC 6)

[DCC 6 is currently only defined in this book]

Official information: direct-use model information and direct-use ancillary information (presentation is optional)

[direct-use model information and direct-use ancillary information are released]

Product definition dataset contains direct-use model information and direct-use ancillary information that are not intended for manual use. The model and ancillary information are supposed to completely define the product. This dataset class is intended to be used directly in software and to facilitate automation. If information is displayed (i.e. for manual use), that displayed information is reference information and is not used as binding information.

DCC 6 is best where processes are automated and human review of dataset contents is not critical or is performed at an earlier lifecycle stage, such as during development of source code in cases where object code is released in a product definition dataset or TDP. If released information is not intended to be used manually, as with code or other types of digitally-formatted information, being able to read, view, or otherwise manually use the information is unnecessary. Remember, manual-use information drives cost, both during its creation and during the information lifecycle (see Figure 4.19).

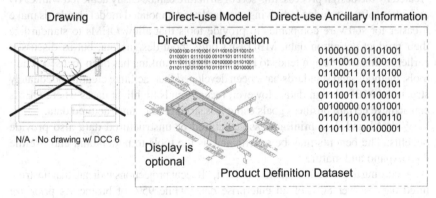

FIGURE 4.19 Information assets in DCC 6.

If we want to ensure the intended official information is used in lifecycle activities, or that information is used the right way, merely indicating a product definition dataset's dataset classification is insufficient. DCCs 3 and 5 may be structured and intended for manual use, direct use, mixed use, or dual use. Knowing a dataset's classification only indicates which information assets are officially part of the dataset. The DCC does not indicate how the information should be used. DCC 2 indicates that the drawing is official information and must be used in lifecycle activities. However, with DCC 2, models tend to informally leak out of the design organization and are used in lifecycle activities regardless of their designation. Such informal use can lead to problems with legal consequences (e.g. non-conforming parts are produced using unofficial information, such as a leaked model). To solve this problem, the official use cases for the product definition dataset must be indicated by including the appropriate DUC, which are defined in this book. This ensures that recipients of information assets know how the information is intended to be used and for which use cases the information has been optimized, and conversely, which use cases would incur additional risk because the information has not been optimized for those uses.

STRUCTURED VERSUS UNSTRUCTURED DATA

Given that the primary goal for MBD information should be to use it directly in software (e.g. automated versus manual use of information), it is critical that the information is ready to be used directly. Thus, MBD information should be optimally structured for direct use. The information is structured to represent what is being modeled (the subject), it is structured to represent how it is intended to be used (generally its primary use), it is structured to facilitate incorporation of other information that enhances its primary use, or it is structured by the software used to create the model.

In MBD, information is structured. Given that MBD information may be used manually and/or directly, information should be optimized for its intended use. The format of presented information should conform to presentation standards and norms, and the format and structure of direct-use information should conform to data structure format standards and norms. In general, for product definition data created by the design process, this sort of structure can be easily achieved with CAD and other software. Indeed, drawing standards and annotated model standards make it easier for software companies to provide tools that allow OEMs to standardize their product definition data. MBD isn't only about design data, but as discussed earlier in this text, design tends to be where MBD thinking has been focused. Thus, tools, methods, and standards have been developed that facilitate creating adequately structured MBD design data. However, recent trends highlight cases where OEMs can achieve certain business goals using less structured or unstructured data.

With AI and data mining, less structured or unstructured data also provide benefits. The benefits may be significant, as the tools to understand and use the data expand and mature.

According to projections quoted by IBM, "Recent projections indicate that unstructured data is over 80% of all enterprise data, while 95% of businesses prioritize unstructured data management."[4.1] Thus, while traditional information used in OEMs for defining requirements, products, processes, activities, and supporting information use structured data, the amount, role, and importance of unstructured data is increasing quickly. Recently released chatbots using unstructured data demonstrate the power and importance of unstructured data and provide a glimpse into how such information may be used in a corporate setting to improve productivity. These technologies are still nascent, but their promise is easy to see. Structured information will continue to be used as the primary model for OEMs and the discrete manufacturing industry, but other uses for unstructured or multi-structured data will increase rapidly in this space.

FOOTNOTE

4.1 IBM Cloud Education (2021, June 29). Structured vs Unstructured Data: What's the Difference? [Web log post]. Retrieved from https://www.ibm.com/blog/structured-vs-unstructured-data/

As stated in several places in this chapter, I am using the term *built for use* to better describe the preferred information structure for MBD. I cannot ignore the terms structured information and unstructured information, but I don't think those terms adequately address what is most important in MBD and MBE. To repeat, it is important for information to be structured and optimized for how it will be used. It is equally important for the information to be used in the manner in which it was intended to be used. The idea of *built for use* means that information is created, structured, and optimized for a particular use and the information is indeed used that way. A common blunder is when information is created to be used one way but is actually used another way. The most common example today is where a company uses models or other information that was developed to support manual use and visualization directly in software (e.g. they take models that support drawing-based workflows and use the models as the basis for direct use cases, such as CAM, CAE, or CMM activities). In most cases, models developed for manual use are insufficient to be used directly, as they have insufficient quality (e.g. include errors, discrepancies, etc.). This is a very common problem in MBE implementations. We help our clients understand, recognize, and mitigate the risks of migrating legacy information to MBD.

INFORMATION STRUCTURE (SCHEMA)

The structure of information in a dataset should be optimized for its intended purpose and use cases. This means that information should be structured to benefit its users and not just its authors. This is often overlooked. Careful planning and execution ensure that we suit users' needs. In Chapter 6, we will consider the needs of users, and thus how information should be structured to suit their needs. We will also address what their needs actually are, rather than their preferences and how they do business today. Most MBD implementations have been severely limited by focusing on supporting current work methods and information use cases – that is, trying to provide information in a manner that allows people to work the same way they have been working for decades (e.g. using drawings, reading information, interpreting visual information, using information manually, etc.).

Information should be structured to benefit its users and not just its authors. This is often overlooked.

In the sections under Types and Examples of MBD later in this chapter, examples of structured MBD datasets will be presented and discussed.

THE MEANING OF THE TERM MODEL-BASED DEFINITION (MBD)

As MBD is a relatively new discipline, there are different ideas about what it is and what it should be. Some people believe that MBD is a thing, an information asset, which can be sent and received in a transaction. This is an incorrect and inadequate definition of MBD, as it unnecessarily limits the applicability and usefulness of the term and of the concept. Other people, including me, believe that MBD is a set of techniques or methods that are used to describe information that is model-based, information that is intended to be used in model-based workflows. As with many terms, MBD is used in more than one way, in this case as a noun and as an adjective.

In this book, and in the way I am describing the discipline, MBD is used as both a noun and an adjective.

MBD

noun
The system and methods used to define digital business information so that it is optimized to be directly used in software. MBD replaces document-based methods, where information is optimized to be presented in documents and used manually.

adjective
Indicating information or an information asset is model-based or has been created using model-based methods.

In Chapter 1, I explained other uses of the initialism MBD. As I mentioned, in this book *MBD* means *Model Based Definition*. Nonetheless, outside of this book, *Model-Based Definition* means different things to different people. Part of the purpose of this book is to pick a sensible broadly-applicable meaning or develop one. See the following list of several ideas people have of what *Model-Based Definition* is, its domain, and what its limits are.

1. Some people believe that MBD exists solely in the domain of mechanical 3D CAD and product definition data.
2. Some people believe that MBD information is only created by design engineering activities.
3. Some people believe that MBD is about defining information as native CAD or neutral information models, such as STEP, JT, and QIF data formats.
4. Some people believe that MBD is limited to cases of primary authored data in native format, information assets that have not been translated into a derivative format.
5. Some people believe that MBD is a model-based information asset that defines a product and is created by a design activity, consisting of an annotated model, related information, and information assets.
6. Some people believe that MBD is "an annotated model and its associated data elements that define a product in a manner that can be used effectively without a drawing graphic sheet" as defined in the ASME Y14.47-2023 Standard. (Quote reprinted from ASME Y14.47-2023 by permission of The American Society of Mechanical Engineers. All rights reserved.)

7. 01010011 01101111 01101101 01100101 00100000 01110000 01100101
   01101111 01110000 01101100 01100101 00100000 01110100 01101000
   01101001 01101110 01101011 00100000 01110100 01101000 01100001
   01110100 00100000 01110100 01101000 01100101 00100000 01110000
   01110101 01110010 01110000 01101111 01110011 01100101 00100000
   01101111 01100110 00100000 01001101 01000010 01000100 00100000
   01101001 01110011 00100000 01110100 01101111 00100000 01110110
   01101001 01110011 01110101 01100001 01101100 01101100 01111001
   00100000 01110000 01110010 01100101 01110011 01100101 01101110
   01110100 00100000 01101001 01101110 01100110 01101111 01110010
   01101101 01100001 01110100 01101001 01101111 01101110 00100000
   01100110 01101111 01110010 00100000 01101101 01100001 01101110
   01110101 01100001 01101100 00100000 01110101 01110011 01100101
   00101110 00100000 00111010 00101000

Note that the items in the preceding list are much narrower than the definition I provided above the list. The definition I provided is broadly applicable, it is based on structured digital information, it is not limited to a particular discipline (e.g. mechanical design), it is not limited to a particular information asset (e.g. product definition dataset), it is not limited to particular format (e.g. native CAD format), it is not limited to whether it is primary authored information or derivative information, and it is not biased toward or restricted to manual use of digital information.

THOUGHTS ABOUT MBD STANDARDS

With the many different ideas about what MBD is listed above, it is no surprise that MBD is addressed differently in different standards. For example, product definition standards address MBD from a product definition point of view, while information modeling standards address MBD from an information modeling point of view. It is okay that the topic is addressed from multiple points of view until one of the standards tries to limit the topic and use of the term. Following that line of thinking, the definition of MBD in ASME Y14.47-2019 is extremely limiting and shortsighted. It is unfortunate that this restrictive and shortsighted definition was included in the released standard.

THOUGHTS ABOUT STANDARDS

I have been involved in standards throughout college and my entire professional career, since 1982. Standards development is an imprecise practice, often marred by turf battles, preference, ladder climbing, ignorance, and sometimes just pure obstinance. Standards are presented as records of consensus, but they are rarely based on consensus. Standards are often based on the least objectionable alternative. We usually don't select the best option; we end up selecting the option that is the least painful, in terms of conflict, argument, and obstruction within the group developing the standard. This is an unfortunate side effect of human nature and freedom of choice.

When I decided to change the focus of my career from mechanical design to GD&T in the 1990s, a good friend (who was also a mechanical designer and had recently changed from designing systems of machinery to designing injection molded parts for automobile interiors), pointedly asked me, "How could you switch from mechanical design to GD&T? GD&T is boring!" I thought about it and said, "Good point. But I disagree about GD&T, I don't find it boring at all – geometry is a very deep subject. And how could you do what you're doing? After designing complex machinery systems, focusing on injection molded parts seems very boring." Yes, our agreed conclusion was *different strokes for different folks*. Find something you like, hopefully something you are good at, hopefully something you can make a living doing, and go for it.

DEFINITION AND DIMENSIONING STRATEGY FOR MBD

Considering MBD for product definition, the information that design engineers create and release, we need to decide whether we want to use traditional methods to define products or model-based methods. In a traditional sense, the distinction between these approaches becomes clear when considering dimensioning strategy.

Traditionally, in the drawing-based workflows we've used for many decades, we rely on dimensions to formally represent product geometry. To say that a different way, dimensions shown on drawings are used as the official definition of product geometry – the geometry of the product has been defined by views of geometry and dimensions on drawings. The object lines and hidden lines in 2D drawing views depict product geometry and dimensions attached to that geometry represent the official definition of the geometry. For example, two parallel lines are shown in a view, and we know they are 15 mm apart because the distance between them is defined by an explicit dimension (see Figure 4.20).

Using explicit dimensions seems to be the biggest barrier to achieving value from an MBD implementation. That is, whether we decide that we need to show dimensions on models, annotated models, or class 3 datasets, directly affects the potential value that can be achieved by MBD and MBE. If we cling to the manual drawing-based methods we've used for centuries and continue to believe that we need to see dimensions to understand a part or assembly and to do our jobs, such as analyze, manufacture, inspect, and assemble the product, then we will miss a significant opportunity to increase productivity and decrease the amount of documentation, information, and work required to achieve our goal.

Our goal should be to create a functional product that satisfies the customer and other stakeholders for the minimum cost, cost in terms of time required to

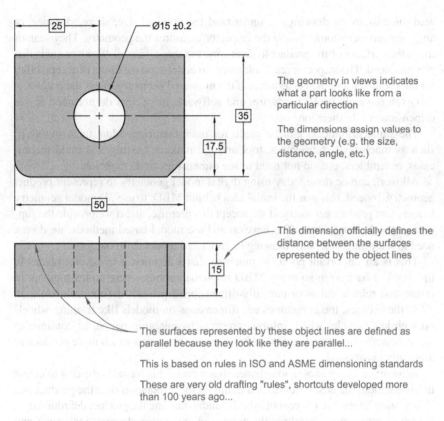

The geometry in views indicates what a part looks like from a particular direction

The dimensions assign values to the geometry (e.g. the size, distance, angle, etc.)

This dimension officially defines the distance between the surfaces represented by the object lines

The surfaces represented by these object lines are defined as parallel because they look like they are parallel...

This is based on rules in ISO and ASME dimensioning standards

These are very old drafting "rules", shortcuts developed more than 100 years ago...

FIGURE 4.20   The role of dimensions in drawings and annotated models.

define the product, to develop the product definition data, to produce the product, and evaluate the product, to maintain the product definition data, to deploy the product, to service the product, to use the product, to retire the product, etc.

Many people think that they must see dimensions on drawings or annotated models so they can do their jobs. Many design professionals believe that if they don't explicitly define (show) dimensions on drawings or annotated models that other people will not be able to do their jobs. Many people think that their job requires dealing with dimensions. Many inspectors have been taught that their job is to measure dimensions.

People do not need to see dimensions to do their jobs.

People do not need to see dimensions to do their jobs.

Inspectors do not measure dimensions. Among other activities, inspectors measure geometry on parts and assemblies. Because of historical methods, inspectors read dimensions on drawings to understand the nominal size, shape, distance, or angle for product geometry, and the inspector measures the geometry. They scan or touch the surfaces of the product to determine the shape, size, distance, or angle that was produced. The inspector reads tolerances to understand the limits of acceptability for the geometry and then determines if the measured geometry meets the criteria.

Given the right tools, hardware, and software, inspectors do not need to see dimensions to do their job.

The same arguments can be made for manufacturing. Manufacturing staff, such as machinists, fabricators, tool and die makers, casting and mold technicians, assemblers, etc. do not need to see dimensions to do their job.

All work can be done today using digital model geometry to represent product geometry. Indeed, this was the initial idea behind MBD: to use 3D model geometry to represent product geometry. If we accept this premise, and if we provide the support and technology for staff to develop and use model-based methods, we do not need to see dimensions on a drawing or annotated model to do our jobs.

This is an important point, so much so that I mention it in several places in this book. I have seen so many MBD implementations where companies opt to create and release fully- or partially-dimensioned product definition models. In all of these cases, the companies use dimensions on models like training wheels on a child's bicycle, which inhibits progress by enabling people to continue to work as inefficiently as they always have, even when a new much more productive alternative is at hand.

Remember, every annotation element has a cost. This includes the cost to create the annotation, the cost to revise and maintain the annotation over the product and information lifecycles, the cost of initially understanding the product definition data (which increases non-linearly as the number of annotation elements increases), and the cost the annotation drives for every person that must use the product information.

Remember, every annotation element has a cost

We should separate how we approach product geometry from product specifications in our approach to MBD and MBE. We should immediately strive to use digital model geometry as the official definition of product geometry. We should ensure that we create MBD datasets that facilitate using model geometry as the official definition of product geometry. The systems, methods, tools, and software to do this are mature and readily commercially available. For specifications, it may be a good idea to take a more cautious and measured approach, as the systems, methods, tools, and software to do this are less mature and less widespread. While initially people tend to be nervous about using model geometry instead of dimensions to officially represent product geometry, they tend to be much less comfortable with the idea of directly using machine-readable specifications such as GD&T, surface texture, material specifications, coating and plating, etc. in software. People tend to feel that they need to read the specifications and (mis)interpret them like they always have – it makes them more comfortable...

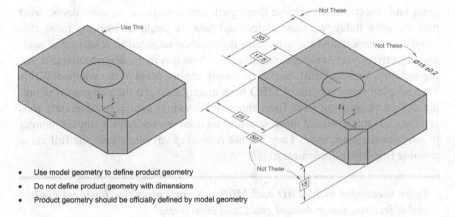

- Use model geometry to define product geometry
- Do not define product geometry with dimensions
- Product geometry should be officially defined by model geometry

FIGURE 4.21   Model geometry represents product geometry.

My best advice for defining product geometry, using model geometry, and avoiding dimensions is shown in Figure 4.21.

MBD FOR DIRECT USE

The goal of MBD for direct use is to create information that is optimized to be used directly in or by software. The information is defined such that it can be used without the need for a person to process, preprocess, read, interpret, organize, or otherwise intervene before the information can be used by software in an automated or semiautomated process, such as CAE, CAI, CAM, CMM, CAx…. MBD for direct use does not require a person to act as an intermediary between the information and its ultimate use – the information can be used directly in software.

Direct-use information is defined and optimized to be used directly by software. Our goal when creating information for direct use is to create digital information that is in the correct format, structured properly, includes the right information, that the digital information elements are properly and adequately related to one another (associated), is suitable for the intended direct use cases, and of sufficient quality, trustable, secure, and fit for its intended use.

As stated throughout the text, the best approach for MBD and MBE is to create model-based information that can be used directly by software, without the need for a person to read, interpret, convert, or otherwise intervene and manually use the information (the need for a person to use the RIDICT process on the information in target activities has been disintermediated). Our goal should be to create the least information necessary to achieve our business objectives. To be successful with MBD and MBE, we must prioritize business goals and objectives, not technical goals and objectives. To achieve these goals and objectives, we must deconstruct how we work today, we must understand how we could be working more efficiently, and understand the minimum information required to achieve our goals and objectives. Very few organizations have done this in the discrete manufacturing industry, such as OEMs and their supply chains. Most of the companies that have adopted some version of MBD have merely changed their engineering output and manufacturing input from annotated drawings to annotated models, thus enabling the model-based information to be used manually in a similar manner to how drawings are used. They are not achieving anywhere near the full value possible from their implementations.

To be successful with MBD and MBE, we must prioritize business goals and objectives, not technical goals and objectives.

We should be working hard to eliminate all low-value manual activities, with particular focus on eliminating high-skill low-value manual activities, such as creating and reading engineering drawings and complex specifications, such as GD&T and GPS.

We have technology available to do many things automatically that we do manually today. We do a lot of things manually that could be easily automated.

Some things have already been automated, but many of these we take for granted. For example, we no longer require the drafting skills to project orthographic views, as our powerful 3D CAD software automatically creates views for us. All we have to do is drag and drop the views we want. Do you remember that I mentioned that my first degree was a design drafting degree? Many of the skills I learned and developed are no longer needed. The current workforce benefits from this simple automation. There are many more benefits and savings possible by pursuing MBD for direct use.

There is an imbalance between the skills of our incoming workforce and the skills needed in industry. Many of the skills that we think we need are not really necessary anymore. Many of our current "needs" are holdovers and remnants from past practices that we haven't updated. We can train new hires and impart the skills needed to perform the outdated low-value activities, or we can modernize our workflows so that we no longer require these archaic skills. This scenario is playing out in many areas of our lives, much of it occurring via smartphones interacting with the Internet and the cloud. The scenario is also playing out in many areas of industry.

The purpose of MBD for direct use is simple: to increase efficiency, productivity, quality, throughput, return on capital, competitiveness, and market share, while reducing cost, development time, time to market, overhead, labor requirements, training requirements, amount of documentation, maintenance costs and burden, lifecycle costs and burden, mistakes, RFIs, scrap, and rework.

As stated earlier in the text, and as shown below in the MBD dataset examples, it is critical that direct-use information is defined as legally binding information, carrying equivalent legal weight as an engineering drawing or other contractually-binding documentation commonly used in business today. It is important that this concept is also defined on purchasing orders and in engineering standards. I have created a section that states this in a new standard I am developing.

Note that there are few standards that govern direct-use information. The closest may be information modeling standards such as ISO 10303 series (STEP standards), ISO 23952 (QIF standard), ISO 14306 (JT), ISO 6983-1 (G-code), ISO/ASTM 52915 (AMF format for additive manufacturing), native CAD, CAM, CMM, and other proprietary data formats. Note that standards that define presentation methods and rules for engineering product definition and specifications, such as ITO TC 10, ISO TC 213, and ASME Y14 Standards, provide little benefit for direct-use information and direct-use workflows.

HERE ARE SOME RECOMMENDATIONS FOR DIRECT-USE INFORMATION:

Best practice 1: Direct-use information should be optimized for direct use and should not be burdened by considerations of how the information looks when presented.

Best practice 2: Direct-use information should be defined as the official representation of products, processes, activities, requirements, conformance criteria, etc. in company best practices, SOPs, contracts, and in datasets. Official direct-use information should be treated as binding in cases where it is contractually invoked.

MBD FOR MANUAL USE

The goal of MBD for manual use is to create information that is optimized to be used manually. *Manually* means used by a person in a traditional manner, such as reading or viewing. The information is defined so that a person can use it. The information is read or viewed, interpreted, organized, translated, converted, and transcribed (RIDICT) so it can be used in a manual activity or process, such as estimating, machining, fabricating, inspecting, assembling, servicing, etc. Sometimes manual-use information is converted into direct-use information or is used as the basis for an automated or semiautomated process. But, in most cases, manual information in = manual information out = manual activities to create manual-use information to be used manually in manual activities. Documents in = documents out = a self-reinforcing cycle of inefficiency, mismatch between new hires and work requirements, low quality, mistakes, scrap and rework, and excessive cost.

While I strongly believe and advocate that we should focus on creating direct-use information and using that information directly in software, I can safely assume that no company is ready to completely stop using visualization and manual activities today. It is much more likely that people prefer taking small steps toward a model-based future. It is much more likely that they will continue to rely on manual activities and visualization, which means, in many cases, clinging to outdated, inefficient methods. So be it. Thus, there is a struggle between what is possible and how work is currently performed.

In light of the above, and that an organization might not have all software and hardware needed to fully create direct-use information and support direct-use workflows, I often recommend a stepwise or tiered approach to replacing manual workflows and manual-use information. That said, people continue to think they need manual-use information long after they no longer need it. They also retain manual activities and use cases in their new workflows that require manual-use information. Thus, for various unproductive reasons, they still want manual-use information, regardless of how strongly I try to influence them to modern up.

While it is acceptable to create manual-use information, information supporting visualization and other manual use cases should be based on information for direct use. Preferably, manual-use information is derived from and equivalent to the direct-use information from which it was derived. This best practice is loosely implied in the ASME Y14.41 and ISO 16792 Standards.

Manual-use information should be created and optimized to be used manually by people, to be read or visualized. Information should be formatted, arranged, and otherwise presented in a manner that is easy to use manually. Visual format and arrangement are important for manual-use information. Thus, the format and arrangement of similar documents are critical. Drawings, annotated models, lists, and other documents should be consistently formatted. Many standards exist that support consistent methods for presenting technical information, such as the many drafting standards of ISO TC 10, TC 213, ASME Y14, AWS A2.4, and others. Following standard formatting rules and methods is important in manual-use information.

HERE IS A RECOMMENDATION FOR MANUAL-USE INFORMATION:

Best practice 1:   Manual-use information should be based on and derived from direct-use information, and where possible, the manual-use information should be equivalent to the direct-use information on which it is based.

See Figure 4.22 for a comparison of the importance of how direct-use and manual-use information looks (e.g. its presentation). To recap, how direct-use information looks, how it is presented, is irrelevant; how manual-use information looks, how it is presented, is critically important.

MBD ORGANIZATIONAL STRUCTURE (MBD SCHEMA)

One of the challenges of digitally-defined information assets, specifically information assets that are not necessarily stored and used in tangible physical form, such as printed documents, is that it may not be clear to recipients what

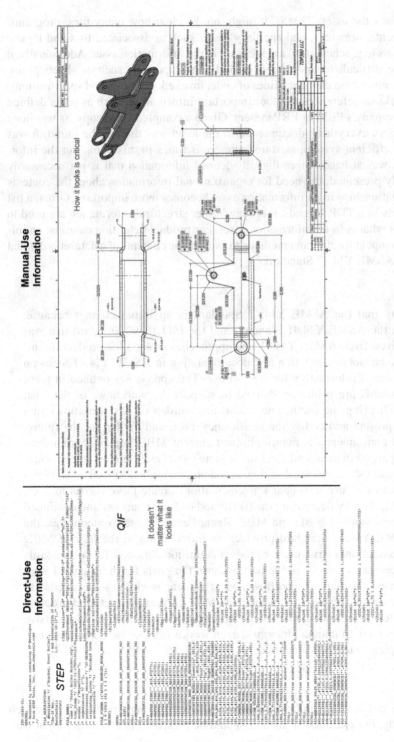

FIGURE 4.22 Importance of appearance of direct-use versus manual-use information.

comprises the asset. That is, it might not be clear how many files, programs, documents, drawings, and ancillary assets such as associated lists and logical diagrams (e.g. schematics) are included in the information asset. Additionally, it may be difficult to understand or find subcomponents such as sheets/pages, views, annotation elements, lines of code, invoked standards and specifications, and links or references to other important information, such as parts defined in a company's PLM or ERP system. Given a complete hard copy, as was done years ago, everything documented in the asset was there to be found. It was not an efficient system, as stated numerous times in this book, but the information was at hand. Given digitally-defined information that is not necessarily visually-presented, the need for organizational information about the contents and relationships in an information asset becomes more important. Often, a list of assets in a TDP, legends, maps, and other structural information are used to explain what is in an information asset. For product definition datasets, an initial attempt at formalizing methods to explain the contents of a dataset is defined in the ASME Y14.47 Standard.

I say that the ASME Y14.47 Standard is an initial attempt because, like the ASME Y14.41 Standard and the ISO 16792 Standard that was derived from ASME Y14.41, the approaches defined for product definition dataset organization and documentation in ASME Y14.47 focus on presented information for manual use. The approaches defined in these standards are primarily defined to support the workflows of the past, which bring the inefficiency, cost, and burden of past workflows into the present and codify the inefficiency, cost, and burden for the future. The consequence is manifold: most current MBE implementations are subpar, inefficient, and yield inadequate value; most current implementations are slow, awkward, difficult, and place a high-skill load on staff and new hires; current subpar implementations create more inertia to overcome, thereby decreasing the likelihood of future success and continued commitment to MBD and MBE. Regardless of these shortcomings, the ASME Y14.47 Standard provides an introduction to the idea of MBD organizational structure for product definition datasets. These standards represent workflows of the past because the goals for the standard were focused on manual workflows. Most current workflows are similar to how companies worked in the past. Most current product- and process-related workflows at OEMs can be traced through a series of recursive steps back to manual drawings, even workflows that have been recently adapted to use MBD methods.

It is critical for similar information assets to be structured and formatted similarly. For example, all memos should be presented in a similar format with similarly structured information elements, such as TO, FROM, DATE, SUBJECT, contact information, etc. Likewise, engineering drawings in an organization, sector, industry, or country should also be structured and formatted similarly. This is an underlying reason for drawing standards, which evolved long ago. (Note that the oldest US mechanical engineering drawing standard I have is ASA Z14.1-1935 Drawings and Drafting Room Practice, from 1935.) Models, annotated models, and other digital product definition datasets within an organization, sector, industry, or country should also be structured and formatted similarly. This facilitates ease of use, as the information is available in a consistent and predictable manner. Consistency and predictability are essential for all business information, as they are prerequisites for being able to understand and use the information, both manually and directly in software.

Figures 4.23a–4.23c show an overview of common information elements in information assets, such as product definition datasets. Note that Figure 4.23 is split into three parts: 4.23a, 4.23b, and 4.23c. Figure 4.23a shows the header of the diagram, which leads into Figure 4.23b, which shows common information elements in information assets intended for manual use, and Figure 4.23c, which shows common information elements in information assets intended for direct use in software. The diagram is split into three parts because the complete diagram is much too large to be visible within the format of this book. Other diagrams in this section have also been split into multiple parts for the same reason of space constraint.

ORGANIZATIONAL STRUCTURE FOR MANUAL-USE INFORMATION ASSETS

Information in an information asset should be structured and optimized for its intended use. Thus, information intended to be used manually, e.g. visually, by reading, viewing, and otherwise manually interacting with the information, should be structured and optimized to be used manually. The content of the information

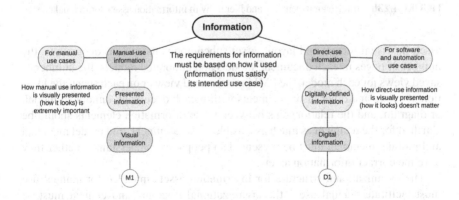

FIGURE 4.23a Information elements and hierarchy in information assets: header.

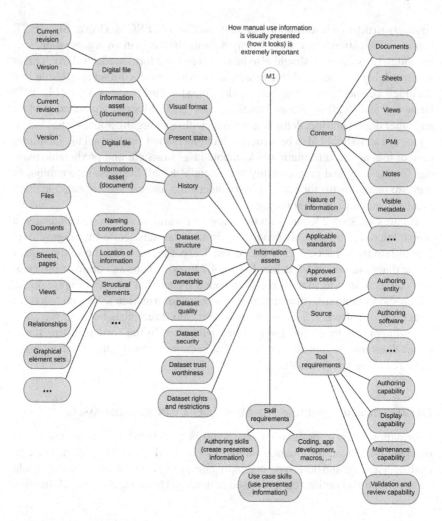

FIGURE 4.23b Information elements and hierarchy in information assets for manual use.

asset, its visual format and layout, asset subcomponents such as documents, the number of pages in each document, models and annotated models, the number of saved views for each model, the PMI in each saved view or on each sheet, the location of views and other visual elements within each document, annotated model, or diagram, and the relationships between these information elements should be clarified for the recipient as much as feasible. File, document, and model metadata and product metadata must be presented so people can understand whether they have the correct information assets.

The organizational structure for information assets intended for manual use must facilitate manual use – the organizational structure and content must be optimized to make the information as easy to use manually as possible. The

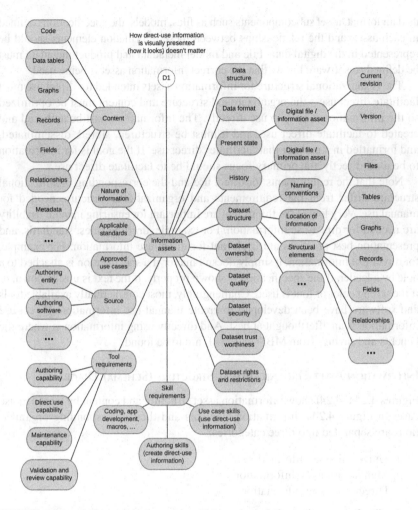

FIGURE 4.23c   Information elements and hierarchy in information assets for direct use in software.

information must be authored and created to facilitate manual use, and it must be presented, stored, disseminated, and displayed in a manner that facilitates manual use. If the goal is for information to be used manually, the primary focus should be to facilitate manual use.

ORGANIZATIONAL STRUCTURE FOR DIRECT-USE INFORMATION ASSETS

Information in an information asset should be structured and optimized for its intended use. Thus, information intended to be used directly by software should be structured and optimized to be used directly. The content of the information asset, its data format, asset subcomponents such as files, models, the specifications defined in each asset, and the relationships between these information elements should be represented in the digital data. File and model metadata and product metadata must be defined so software knows that the correct information asset is being used.

The organizational structure for information assets intended for direct use must facilitate direct use – the organizational structure and content must be optimized so the information is easy to use directly. The information must be authored and created to facilitate direct use, and it must be structured, stored, disseminated, and formatted in a manner that facilitates direct use. If the goal is for information to be used directly, the primary focus should be to facilitate direct use.

Note that the requirements for direct use and the corresponding organizational structure differ from the requirements and organizational structure needed for manual use. Very few of the things that are important for ensuring manual usability are needed for direct-use information. For example, drafting rules, standards, and presentation best practices are irrelevant for direct-use information. For example, for direct-use information, it simply doesn't matter if a specification is attached to a hole by a leader, if the specification is shown properly, if the text is upside down, or if the wrong color or font is used. Challengingly, most commercially available tools and software have been developed to create manual-use information. Direct-use information is an afterthought at best. And directly using information is where the benefits and savings from MBD and MBE are to be found.

INFORMATION ASSET ORGANIZATIONAL STRUCTURE (SCHEMA) DIAGRAMS

Figures 4.24a–4.24l show information asset structure and content by intended use case. In Figure 4.24a, information elements and their corresponding organizations are separated into three categories:

1. Universal asset information
2. Manual-use asset information
3. Direct-use asset information

Upper diagram i. in Figure 4.24a shows a high-level view of information asset organizational structure for the three use case categories above. Lower diagram ii. in Figure 4.24a shows a high-level view of information asset content conforming to the intended use case for the information for information assets for manual use, information assets for direct use, information assets for mixed use, and information assets for dual use. Note that in each case, the required information consists of universal asset information and information specific to the particular use case (e.g. for direct-use information assets, both universal information and direct-use information are required). The schemas required for each use case are also listed. I segregated the universal information because it is common for all the intended use cases. Note that given a larger format, I would present this diagram in fewer pieces. Again, the small footprint of the book requires that diagrams be compact to ensure legibility.

While the organizational structures for information assets in Figure 4.24a depict multiple schema elements for each information asset, there is a single

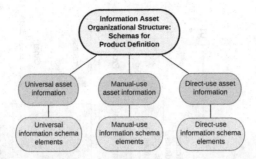

i. Information Asset Organizational Structure - By Information Use Case

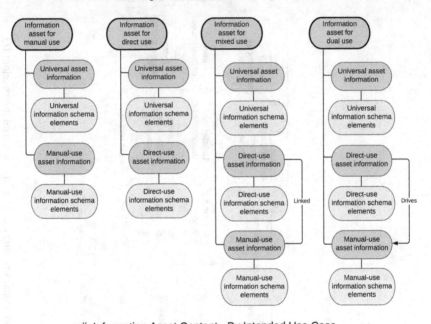

ii. Information Asset Content - By Intended Use Case

FIGURE 4.24a Information asset organizational structure and content – header.

schema for each information asset. The information has been separated into three categories to aid understanding:

Universal asset information applies to all product definition information assets regardless of their intended use case.

Manual-use asset information applies to product definition information assets that include information intended for manual use.

Direct-use asset information applies to product definition information assets that include information intended for direct use.

Figures 4.24b–4.24d show universal asset information structure and content (schema elements).

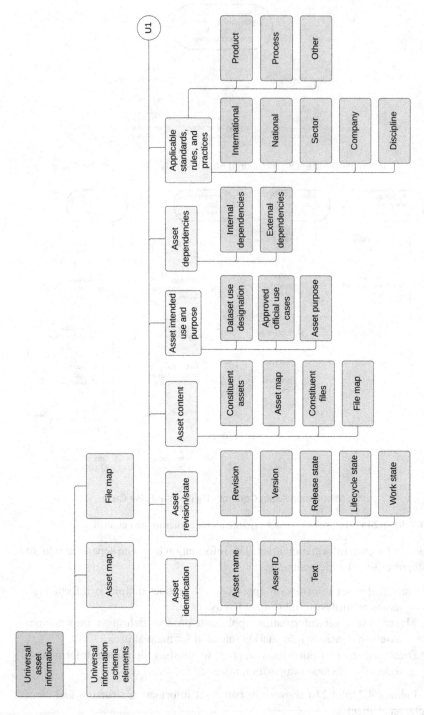

FIGURE 4.24b Information asset organizational structure and content – universal information: 1/3.

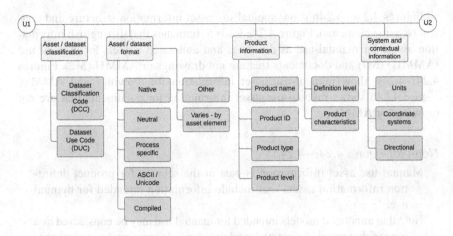

FIGURE 4.24c   Information asset organizational structure and content – universal information: 2/3.

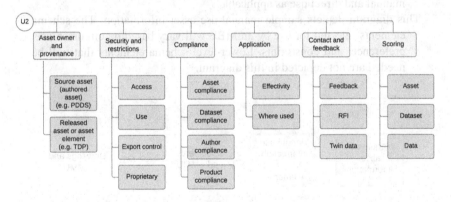

FIGURE 4.24d   Information asset organizational structure and content – universal information: 3/3.

Notes for Figures 4.24b–4.24d

These figures depict sample universal asset information.

Universal asset information is part of the schema for an information asset.

Universal asset information applies to all product definition information assets regardless of their intended use case.

The schema elements selected for use by an MBE will vary based on circumstance, preference, and constraints. Many lower-level schema elements are not depicted in this diagram.

Figures 4.24e–4.24h show manual-use asset information structure and content (schema elements). Figure 4.24e leads to branches that distinguish information assets for manual-use as drawings and annotated models for manual use (AMMU) (M1) and documents that are not drawings or AMMU (M2). Figures 4.24f–4.24g show manual-use asset information for drawings and AMMU. Figure 4.24h shows manual-use asset information for documents that are not drawings or AMMU.

Notes for Figures 4.24e–4.24h

Manual-use asset information is part of the schema for product definition information assets that include information intended for manual use.

Note that annotated models intended for manual use may be considered as a type of document. Annotated models intended for manual use must conform to format, structure, and quality requirements that are optimized for manual use. Requirements for annotated models intended for direct use are different from requirements for manual use. Cases where annotated models are intended for hybrid or dual use must be optimized for manual and direct use as applicable.

This diagram depicts sample manual-use asset information. The schema elements selected for use by an MBE will vary based on circumstance, preference, and constraints. Lower-level schema elements that will be needed are not depicted in this diagram.

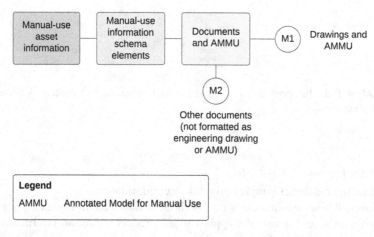

FIGURE 4.24e Information asset organizational structure and content – manual-use information: 1/4.

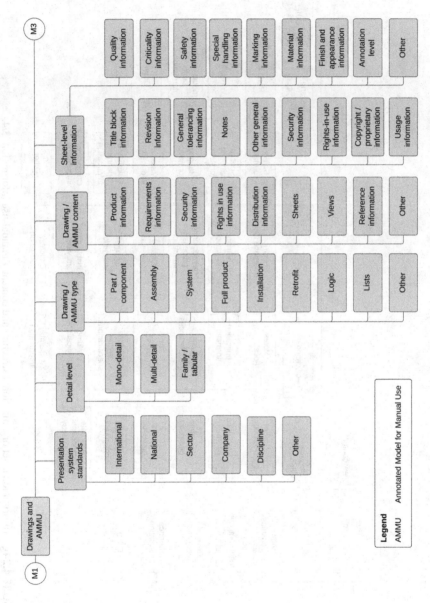

FIGURE 4.24f Information asset organizational structure and content – manual-use information: 2/4.

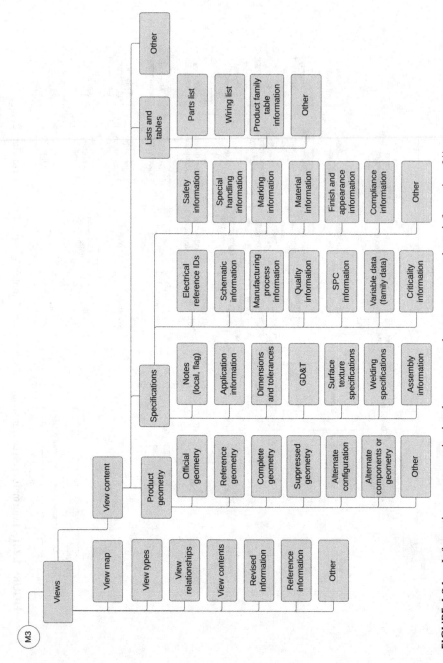

FIGURE 4.24g Information asset organizational structure and content – manual-use information: 3/4.

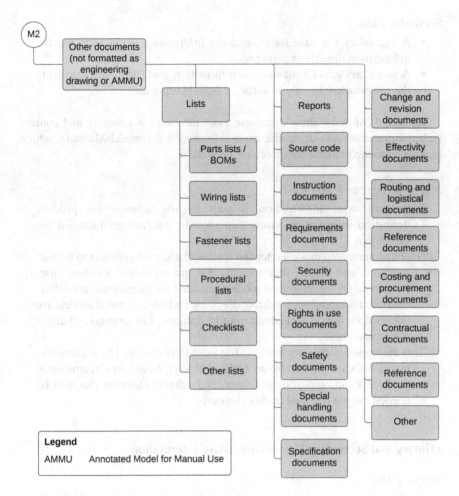

FIGURE 4.24h Information asset organizational structure and content – manual-use information: 4/4.

Primary and Secondary Use of Manual Information

Primary Use

- The primary use case for manual-use information is to be viewed, read, and manipulated by people. Thus, manual-use information should be structured and classified so people can use the information (RIDICT).
- The primary goal for manual-use information is to allow people to use the information manually (official use).

Secondary Use

- A secondary use case for manual-use information could be to use the information directly in software.
- A secondary goal for manual-use information could be to facilitate using the information directly in software (non-binding use).

Figures 4.24i–4.24l show direct-use asset information structure and content (schema elements) and direct-use asset information for models/datasets (which include annotated models for direct use (AMDU)).

Notes for Figures 4.24i–4.24l

Direct-use asset information is part of the schema for product definition information assets that include information intended for direct use.

Note that annotated models intended for direct use must conform to format, structure, and quality requirements that are optimized for direct use. Requirements for annotated models intended for manual use are different from requirements for direct use. Cases where annotated models are intended for hybrid or dual use must be optimized for manual and direct use as applicable.

This diagram depicts sample direct-use asset information. The schema elements selected for use by an MBE will vary based on circumstance, preference, and constraints. Lower-level schema elements that will be needed are not depicted in this diagram.

Primary and Secondary Use of Direct-Use Information

Primary Use

- The primary use case for direct-use information is to be used by software. Thus, direct-use information should be structured and classified so software can use the information.
- The primary goal for direct-use information is to be readily accessible and structured for use by software (official use).

Secondary Use

- The secondary use case for direct-use information is to be used manually (viewed, read, manipulated) by people.
- The secondary goal for direct-use information is to allow people to use the information manually for reference use cases (non-binding use).

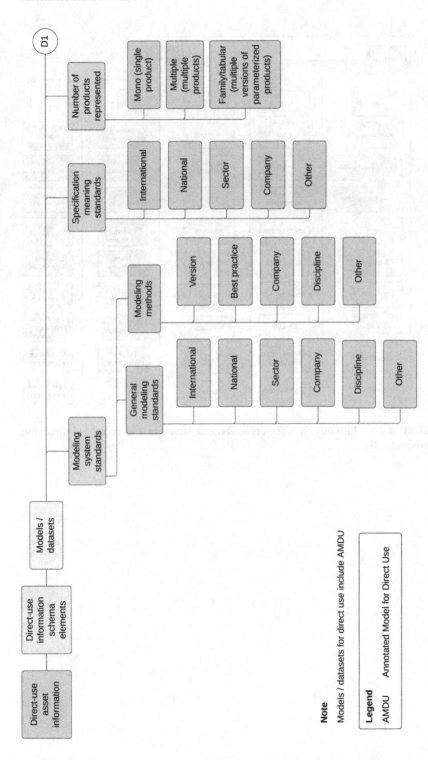

Note

Models / datasets for direct use include AMDU

Legend

AMDU   Annotated Model for Direct Use

FIGURE 4.24i Information asset organizational structure and content – direct-use information: 1/4.

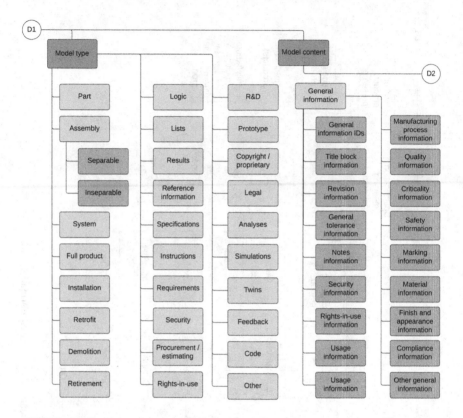

FIGURE 4.24j Information asset organizational structure and content – direct-use information: 2/4.

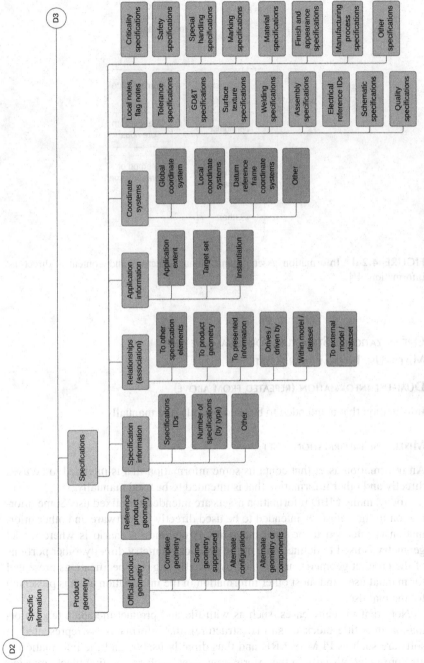

FIGURE 4.24k Information asset organizational structure and content – direct-use information: 3/4.

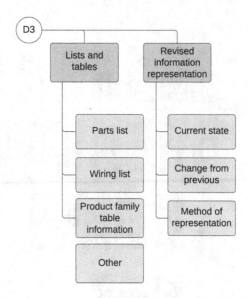

FIGURE 4.24l Information asset organizational structure and content – direct-use information: 4/4.

ORGANIZATIONAL STRUCTURE FOR DUAL-USE AND MIXED-USE INFORMATION ASSETS

DUAL-USE INFORMATION (REPEATED FROM ABOVE)

Information that is intended to be used directly and manually.

MIXED-USE INFORMATION ASSET

An information asset that contains some information that is intended to be used directly and other information that is intended to be used manually.

Today, many MBD information assets are intended for mixed use. Some information in the dataset is intended to be used directly by software and other information is intended to be used manually. A common scenario is where model geometry is used to define some of the product geometry directly, other portions of the product geometry are defined by visually (e.g. by specifications presented for manual use), and most other information in the information asset is presented for manual use.

Note that in many cases, such as with file and product metadata (e.g. information in a title block or similar structure), the information is represented in software such as PLM or ERP, and thus directly usable, and the information is also presented visually in the information asset, such as in a title block, revision block, parts list, wiring list, etc. In such cases, the information may inadvertently be dual-use information. Even in cases where rules are added for approved use cases (see DUCs), information provided in more than one form may be used in an uncontrolled and unofficial manner – whether approved or not, mixed-use information becomes dual-use information.

As stated earlier in this chapter, I do not recommend creating dual-use information. That is, I do not recommend creating and releasing information that can be used manually and directly, allowing the recipient to decide whether to use the information manually or directly. For example, if the size of a hole is defined in model data and it is also defined by an explicit (visibly displayed) dimension, two definitions of the hole size are included in the product definition data. In many cases, these two definitions are not the same. The hole may be modeled as Ø11.2 mm, but the explicit dimension may show a different value, such as Ø11, Ø10.9–11.3, or Ø11 F6. This discrepancy appears more frequently when defining products using inch units, as many drill sizes and design practices are still based on fractions and converted to decimal presentation or representation. See Figure 3.6 in Chapter 3 and the corresponding text. Another problem is more common in ISO, but also occurs in ASME using tolerance class codes, such as Ø11 F6 as shown above. In some cases where tolerance class codes are used, the dimension value is outside the range of allowable sizes – for example, in the example of Ø11 F6, the tolerances from F6 in ISO 286-2 are +.027 +.016, which yields size limits of Ø11.016–11.027. A hole produced at the nominal size of Ø11 is outside the range of acceptable sizes. People who are used to working with tolerance class codes are comfortable with this issue. However, it is important in such cases to remember that the hole will probably be modeled at the nominal size (Ø11) and must be used in conjunction with a tolerance class code to really understand the allowable size.

Most people I talk with are uncomfortable with the idea of directly using information without some way to visualize the information. There are many, many examples in our lives where we use information directly and we don't need or expect to see the underlying information. For example, we don't need or expect to know the RPM of the tires on our car to understand how fast we are going. Fast forward a little, and consider self-driving cars, and we won't be looking at the speedometer anymore. It will be automated. In fact, in cases where someone else is driving, such as another adult, an Uber driver, a taxi driver, or bus driver, we generally are not looking at the speedometer in those cases either. In all those cases, the task of managing speed has been delegated and/or automated – it is managed by another actor, be it a person or software and sensors.

Most people will want to be able to view some direct-use information. We might not want to know how fast our tires are rotating, but we might want to know the speed at which we are traveling. For product definition data, this translates to direct-use workflows where we will expect to be able to see some information and there will be other information that we will not expect to see. I recommend that we do not persistently present direct-use information, in direct-use workflows, that the information could be available to be presented on demand if needed, but the display is suppressed. In cases where direct-use information is presented, be it persistently or on-demand, or where presented information also exists as represented information, it must be legally clear which version of the information is the official information.

We should not persistently present direct-use information in direct-use workflows. The information could be available to be presented on demand if needed, but the display should be suppressed.

The MBD organizational structure for dual-use and mixed-use workflows must encompass which information is intended for manual use, which information is intended for direct use, which information may be used manually or directly (please avoid doing this), which information is presented for manual use, which information is represented for direct use, which version of the information is the official binding information and which version of the information is for reference only. This may sound like a pain, but if you don't do these things, you will encounter problems and conflict with your supply chain and internal customers.

FORMALIZING HOW INFORMATION ASSETS ARE USED – DATASET USE CODES (DUCs)

While many companies have tried to use MBD and implement MBE, they often overlook a crucial requirement – they do not formally define how their information assets are supposed to be used. That is, they do not define which version of their information is official and binding and they do not define the official methods for using their information assets and the information they contain. Information is often provided in multiple forms (e.g. apparently suitable for manual use and direct use), and the way the information is used is left to the recipient. This leads to uneven results and chaos, as the manual-use and direct-use information are not equivalent, and using one version leads to different results than from using the other version. Such practices also decrease control of the process and decrease likelihood of achieving adequate statistical process control. The methods for using manual-use information are different from the methods for using direct-use information, and manual processes should be controlled differently than direct-use processes. To clarify, the steps and overall process to achieve a result differ if using information manually or directly. The skills required, risks, and failure modes, also differ between these methods.

Thus, companies must define how their information and information assets are to be used. Specifically, companies must formally designate how their information and information assets must be used, and that information must be shared with all users. For product definition datasets, this means that the official use case for information and information assets must be included in the product definition dataset, and the official use case must be provided in a form that matches the intended use. The official use case is represented by a DUC. The DUC is defined and presented visually for manual-use information and information assets intended to be used manually, and the DUC is defined digitally/logically for direct-use information and information assets intended to be used directly in software. Note that DUCs are defined in this text, not in a standard.

Table 4.3 includes ten DUCs used to designate how an information asset and how specific information elements in an information asset are officially supposed to be used. The table includes different combinations of information use cases. The table does not include an exhaustive list, as other combinations may be desired. Note that the DUC defines how an organization wants to work, how it wants participants in its workflows to use certain information. DUCs are decidedly an MBE concept, but they are presented here because it is critical that the officially-sanctioned use for information assets and their information elements are formally defined in the asset itself or as asset metadata.

Note that there are multiple DUCs possible for DCCs 3 and 5. (I did not include DCC 4 in this list because DCC 4 is merely a combination of DCC 1 or 2 and DCC 5.) Early on, as we worked with companies trying to adopt and implement MBD and MBE, it was apparent that merely deciding on a dataset class as defined by the DCCs was not enough. A company could decide to define a dataset conforming to DCC 5, which is an MBD approach, but the information in the dataset could be used manually if the information was presented visually. In fact, this is a common occurrence. If the company expected people to use the digital version of the information, but they also provided corresponding visual information, the recipient could use manual-use information or the direct-use information at will. As stated above, this leads to loss of control and conflict, as the digital version and presented version of corresponding information are often not equivalent. Thus, for example, conforming to the presented definition of product geometry and associated specifications may yield different results than conforming to the digital definition of product geometry and associated specifications. Again, this is a common occurrence and often overlooked.

Indeed, many companies believe that allowing people to use whichever version of information they prefer (e.g. manual-use information or direct-use information) is a feature and not a bug. They see that doing this allows flexibility in the workflow and causes less complaints early in the process as people can continue to work as they always have. "You give me presented manual-use information and digital direct-use information, well, I'll use the presented information as I'm used to working that way." While this is a flexible approach, it does not yield the benefits expected from adopting MBD and MBE. This apparent flexibility is, in fact, a failure mode for MBD and MBE.

DUCs solve a longstanding problem and close a loophole of how MBD may be used in an MBE, clarifying how MBD information is intended to be used, which form of the information must be used (which form is binding), managing how it is used, and controlling its use. Without DUCs, companies find problems with their products and processes too late in the lifecycle because the wrong version of information was used in earlier process stages.

TABLE 4.3

Product Definition Dataset Classification Codes (DCCs) and Dataset Use Codes (DUCs)

Information Asset Designation				Official Information Asset – Official Product Definition Dataset Content									
				Geometry		Specifications		Product Metadata		File Metadata		Ancillary Info	
DCC	DUC	Schema Level	Use Type	Official Information	P/R$_1$	Official Information	P/R$_2$	Official Information	P/R$_3$	Official Information	P/R$_4$	Official Information	P/R$_n$
1	1.1	1.1	Manual	Drawing - Dims	P	Drawing - Ano	P	Drawing - Ano	P	PDM/PLM**	P	Manual use	P
2	2.1	2.1	Manual	Drawing - Dims	P	Drawing - Ano	P	Drawing - Ano	P	PDM/PLM**	P	Manual use	P
3	3.1	3.1	Manual	Model - Dims	P	Drawing - PMI	P	Drawing - Ano	P	PDM/PLM**	P	Manual use	P
3	3.2	3.2	Mixed	Model - Geometry	R	Drawing - PMI	P	Drawing - Ano	P	PDM/PLM**	P	Manual use	P
3	3.3	3.3	Mixed	Model - Geometry	R	Model - PMI	P	Model - PMI	P	PDM/PLM**	P	Manual use	P
5	5.1	5.1	Manual	Model - Dims	P	Model - PMI	P	Model - PMI	P	PDM/PLM**	P	Manual use	P
5	5.2	5.2	Mixed	Model - Geometry	R	Model - PMI	P	Model - PMI	P	PDM/PLM**	P	Manual use	P
5	5.3	5.3	Mixed	Model - Geometry	R	Model - PMI	R	Model - PMI	P	PDM/PLM**	P	Manual use	P
5	5.4	5.4	Direct	Model - Geometry	R	Model - PMI	R	Model - PMI	R	PDM/PLM**	R	Direct use	R
6	6.1*	6.1	Direct	Direct-use geometric definition	R	Direct-use specifications	R	Direct-use metadata	R	PDM/PLM**	R	Direct-use information	R

Notes

Official information and P/R columns for geometry, specifications, metadata, and ancillary info designate the source of official information in the product definition dataset that is to be used in lifecycle activities.

P/R Indicates if presented or represented information is the official information

P Presented information is official

R Represented information is official

* DUC 6.1 Information is created solely for direct use. Corresponding presented information may or may not be defined. Information may be in pre-compiled form, such as ASCII or source code (which are human readable), or it may be compiled to facilitate direct use, such as object code (which is generally not human readable).

** While PDM or PLM are likely sources for the information, it is not mandatory that they are the source. Information in linked digital platforms and datasets may drive or be driven by another system. How this is done is a business decision.

DATASET USE CODES

The following explanation is limited to product definition datasets. DUCs can and should also be defined for other types of MBD information assets, such as assets supporting analysis, assembly, estimating, inspection, logistics, manufacturing, purchasing, quality, service, simulation, and other disciplines or activities. The DUCs for these other dataset types are called Process definition Dataset Use Codes (PrDUCs) and are defined in Chapter 5.

DATASET USE CODE 1.1 – MANUAL-USE CASE

All information elements are officially designated for manual use. Manual-use information is designated as the official information. DUC 1.1 applies to DCC 1. The schema level for DCC 1 using DUC 1.1 is 1.1.

- Product geometry is defined by manual-use information (explicit dimensions).
- Specifications are defined by manual-use information (drawing annotation).
- Product metadata are defined by manual-use information.
- File metadata are defined by manual-use information.
- Ancillary information is defined by manual-use information.

DATASET USE CODE 2.1 – MANUAL-USE CASE

All information elements are officially designated for manual use. Manual-use information is designated as the official information. DUC 2.1 applies to DCC 2. The schema level for DCC 2 using DUC 2.1 is 2.1.

Note that a drawing is the official information asset for DCC 2. There should not be a model released with the product definition dataset or TDP, thus, there should not be any digital information available for direct use. However, many companies do not use this best practice, and release a model or other form of a digital data along with the drawing. Thus, digital information is often used as official information even though the digital information was released informally (leaked) and not designated as official information.

- Product geometry is defined by manual-use information (explicit dimensions). Digital information representing product geometry is not binding and may only be used as reference information.
- Specifications are defined by manual-use information (drawing annotation). Digital information representing specifications is not binding and may only be used as reference information.
- Product metadata are defined by manual-use information. Digital information representing product metadata is not binding and may only be used as reference information.
- File metadata are defined by manual-use information. Digital information representing file metadata is not binding and may only be used as reference information.

- Ancillary information is defined by manual-use information. Digital information representing ancillary information is not binding and may only be used as reference information.

DATASET USE CODE 3.1 – MANUAL-USE CASE

All information elements are officially designated for manual use. Manual-use information is designated as the official information. DUC 3.1 applies to DCC 3. The schema level for DCC 3 using DUC 3.1 is 3.1.

- Product geometry is defined by digital information and manual-use information. Manual-use information (explicit dimensions) is the official definition of product geometry. Digital information representing product geometry is not binding and may only be used as reference information.
- Specifications are defined by digital information and manual-use information. Manual-use information (drawing annotation) is the official definition of specifications. Digital information representing specifications is not binding and may only be used as reference information.
- Product metadata are defined by digital information and manual-use information. Manual-use information is the official definition of product metadata. Digital information representing product metadata is not binding and may only be used as reference information.
- File metadata are defined by digital information and manual-use information. Manual-use information is the official definition of file metadata. Digital information representing file metadata is not binding and may only be used as reference information.
- Ancillary information is defined by digital information and manual-use information. Manual-use information is the official definition of ancillary information. Digital information representing ancillary information is not binding and may only be used as reference information.

DATASET USE CODE 3.2 – MIXED-USE CASE

Some information elements are officially designated for direct use, and other information elements are designated for manual use. Direct-use information is designated as the official information for product geometry, and manual-use information is designated as the official information for all other information. DUC 3.2 applies to DCC 3. The schema level for DCC 3 using DUC 3.2 is 3.2.

- Product geometry is defined by direct-use information (model geometry). Direct-use information is the official definition of product geometry. Manual-use information representing product geometry is not binding and may only be used as reference information.
- Specifications are defined by manual-use information. Manual-use information (drawing annotation) is the official definition of specifications.

Digital information representing specifications that could be used directly is not binding and may only be used as reference information.

- Product metadata are defined by digital information and manual-use information. Manual-use information is the official definition of product metadata. Digital information representing product metadata that could be used directly is not binding and may only be used as reference information.
- File metadata are defined by digital information and manual-use information. Manual-use information is the official definition of file metadata. Digital information representing file metadata that could be used directly is not binding and may only be used as reference information.
- Ancillary information is defined by digital information and manual-use information. Manual-use information is the official definition of ancillary information. Digital information representing ancillary information that could be used directly is not binding and may only be used as reference information.

Dataset Use Code 3.3 – Mixed-Use Case

Some information elements are officially designated for direct use, and other information elements are designated for manual use. Direct-use information is designated as the official information for product geometry, and manual-use information is designated as the official information for all other information. DUC 3.3 applies to DCC 3. The schema level for DCC 3 using DUC 3.3 is 3.3.

- Product geometry is defined by direct-use information (model geometry). Direct-use information is the official definition of product geometry. Manual-use information representing product geometry is not binding and may only be used as reference information.
- Specifications are defined by manual-use information. Manual-use information (PMI presented on model) is the official definition of specifications. Digital information representing specifications that could be used directly is not binding and may only be used as reference information.
- Product metadata are defined by digital information and manual-use information. Manual-use information is the official definition of product metadata. Digital information representing product metadata that could be used directly is not binding and may only be used as reference information.
- File metadata are defined by digital information and manual-use information. Manual-use information is the official definition of file metadata. Digital information representing file metadata that could be used directly is not binding and may only be used as reference information.
- Ancillary information is defined by digital information and manual-use information. Manual-use information is the official definition of ancillary information. Digital information representing ancillary information that could be used directly is not binding and may only be used as reference information.

Dataset Use Code 5.1 – Manual-Use Case

All information elements are officially designated for manual use. Manual-use information is designated as the official information. DUC 5.1 applies to DCC 5. The schema level for DCC 5 using DUC 5.1 is 5.1.

- Product geometry is defined by digital information and manual-use information. Manual-use information (explicit dimensions) is the official definition of product geometry. Digital information representing product geometry is not binding and may only be used as reference information.
- Specifications are defined by digital information and manual-use information. Manual-use information (PMI presented on model) is the official definition of specifications. Digital information representing specifications is not binding and may only be used as reference information.
- Product metadata are defined by digital information and manual-use information. Manual-use information is the official definition of product metadata. Digital information representing product metadata is not binding and may only be used as reference information.
- File metadata are defined by digital information and manual-use information. Manual-use information is the official definition of file metadata. Digital information representing file metadata is not binding and may only be used as reference information.
- Ancillary information is defined by digital information and manual-use information. Manual-use information is the official definition of ancillary information. Digital information representing ancillary information is not binding and may only be used as reference information.

Dataset Use Code 5.2 – Mixed-Use Case

Some information elements are officially designated for direct use, and other information elements are designated for manual use. Direct-use information is designated as the official information for product geometry, and manual-use information is designated as the official information for all other information. DUC 5.2 applies to DCC 5. The schema level for DCC 5 using DUC 5.2 is 5.2.

- Product geometry is defined by direct-use information (model geometry). Direct-use information is the official definition of product geometry. Manual-use information representing product geometry is not binding and may only be used as reference information.
- Specifications are defined by digital and manual-use information. Manual-use information (PMI presented on model) is the official definition of specifications. Digital information representing specifications that could be used directly is not binding and may only be used as reference information.

- Product metadata are defined by digital information and manual-use information. Manual-use information is the official definition of product metadata. Digital information representing product metadata that could be used directly is not binding and may only be used as reference information.
- File metadata are defined by digital information and manual-use information. Manual-use information is the official definition of file metadata. Digital information representing file metadata that could be used directly is not binding and may only be used as reference information.
- Ancillary information is defined by digital information and manual-use information. Manual-use information is the official definition of ancillary information. Digital information representing ancillary information that could be used directly is not binding and may only be used as reference information.

DATASET USE CODE 5.3 – MIXED-USE CASE

Some information elements are officially designated for direct use, and other information elements are designated for manual use. Direct-use information is designated as the official information for product geometry and specifications, and manual-use information is designated as the official information for all other information. DUC 5.3 applies to DCC 5. The schema level for DCC 5 using DUC 5.3 is 5.3.

- Product geometry is defined by direct-use information (model geometry). Direct-use information is the official definition of product geometry. Manual-use information representing product geometry is not binding and may only be used as reference information.
- Specifications are defined by direct-use information. Direct-use information (digital representation of PMI on model) is the official definition of specifications. Digital information representing specifications that could be used manually is not binding and may only be used as reference information.
- Product metadata are defined by digital information and manual-use information. Manual-use information is the official definition of product metadata. Digital information representing product metadata that could be used directly is not binding and may only be used as reference information.
- File metadata are defined by digital information and manual-use information. Manual-use information is the official definition of file metadata. Digital information representing file metadata that could be used directly is not binding and may only be used as reference information.
- Ancillary information is defined by digital information and manual-use information. Manual-use information is the official definition of ancillary information. Digital information representing ancillary information that could be used directly is not binding and may only be used as reference information.

Dataset Use Code 5.4 – Direct-Use Case

All information elements are officially designated for direct use. Direct-use information is designated as the official information. DUC 5.4 applies to DCC 5. The schema level for DCC 5 using DUC 5.4 is 5.4.

- Product geometry is defined by direct-use information (model geometry). Direct-use information is the official definition of product geometry. Manual-use information representing product geometry is not binding and may only be used as reference information.
- Specifications are defined by direct-use information. Direct-use information is the official definition of specifications. Manual-use information representing specifications that could be used manually is not binding and may only be used as reference information.
- Product metadata are defined by direct-use information. Direct-use information is the official definition of product metadata. Manual-use information representing product metadata that could be used directly is not binding and may only be used as reference information.
- File metadata are defined by direct-use information. Direct-use information is the official definition of file metadata. Manual-use information representing file metadata that could be used directly is not binding and may only be used as reference information.
- Ancillary information is defined by direct-use information. Direct-use information is the official definition of ancillary information. Manual-use information representing ancillary information that could be used directly is not binding and may only be used as reference information.

Dataset Use Code 6.1 – Direct-Use Case

All information elements are officially designated for direct use. Direct-use information is designated as the official information. DUC 6.1 applies to DCC 6. The schema level for DCC 6 using DUC 6.1 is 6.1.

Presented information may be omitted from a DCC 6 product definition dataset, as the dataset is intended for direct use and direct-use information is designated as the official information. Alternate forms of the information that may be used manually may exist, such as source code as the basis for object code. However, the human-readable source code would not be released as a product definition dataset or TDP – only the object code is released, which is not human readable.

- Product geometry is defined by digital direct-use information. Direct-use information is the official definition of product geometry. If any manual-use information representing product geometry exists, it is not binding and may only be used as reference information.

- Specifications are defined by direct-use information. Direct-use information is the official definition of specifications. If any manual-use information representing product specifications exists, it is not binding and may only be used as reference information.
- Product metadata are defined by direct-use information. Direct-use information is the official definition of product metadata. If any manual-use information representing product metadata exists, it is not binding and may only be used as reference information.
- File metadata are defined by direct-use information. Direct-use information is the official definition of file metadata. If any manual-use information representing file metadata exists, it is not binding and may only be used as reference information.
- Ancillary information is defined by digital information and manual-use information. Direct-use information is the official definition of ancillary information. If any manual-use information representing ancillary information exists, it is not binding and may only be used as reference information.

EXAMPLES: MBD INFORMATION ASSETS – PRODUCT DEFINITION DATASETS

The following examples depict product definition datasets constructed to serve various purposes. This illustrates a critical point, which should be repeated here: information assets and their content must be created such that it is optimized for their intended use.

- Information intended to be used manually should be optimized for manual use.
- Information intended to be used directly in software should be optimized for direct use.
- Information intended to be used manually and directly in software (dual use) should be optimized for both use cases. To repeat an earlier admonition, I recommend not designating dual-use information as official binding information – don't define two versions of information as the official version.

Information should be optimized to suit how it will be used, not necessarily what the information will be used to do. The primary purpose of a product definition dataset is to define a product and its requirements. This is its most important quality criterion. While a product definition dataset will be used by many people in many environments and activities during the product lifecycle, it is most important that the product definition dataset defines the product and its requirements. The product definition dataset should not focus on the requirements of other departments or activities if those requirements detract from the primary purpose of the dataset. In a multidisciplinary environment where MBD information is created, shared, and used as the basis for as many activities as possible, other disciplines can create derivative datasets to suit their needs. There are many benefits to this approach. I am not advocating ignoring the needs of other activities – I am stating that the primary purpose of a product definition dataset must be prioritized.

The examples below are shown on sample template formats for MBD product definition datasets. Some information elements in these examples are optimized to facilitate direct use of information and other information elements are optimized for manual use of information. While the figures are static images, and the format may look like an engineering drawing, the product definition datasets and corresponding TDPs would be digital, presented on a display, and dynamically interactive. Various dialogs and interactive toggles are shown in the figures. Note that in model-based information assets, the display represents not only the graphical item being depicted (in this case product definition datasets) but also includes a graphical user interface (GUI) by which the user interacts with the information asset. Note that none of the following examples are for DCC 5.4 or 6.1, where all information in the dataset is intended to be used directly. DCC 5.4 and 6.1 datasets do not emphasize or include presented information respectively, thus, they are not well suited to be shown as graphical figures. While I believe DCC 5.4 and 6.1 provide the greatest potential for obtaining business value, they don't make for very interesting figures!

Figures 4.25–4.37 depict discrete parts. Figures 4.38 and 4.39 depict separable assemblies.

The first sheet of Figures 4.25–4.37 includes the following information elements.

- Title block – contains file and product metadata
- Revision History block – contains revision history
- Official Information Assets table – lists official information assets for this product definition dataset/TDP
- Global Tolerances block – contains globally applicable tolerance and related specifications
- Specifications table – lists specifications officially-invoked by the product definition dataset/TDP
- Notes – contains general notes and specifications
- Information Services block
- View(s) – geometry-only – includes In View Services dialog for each view
- Bounding box – sizes of a square-prism that bounds the product relative to the global coordinate system
- Global coordinate system

Figure 4.25a shows sheet 1 of 3 of machined part 1.

The product definition dataset is structured as DCC 5, a minimally dimensioned model officially defined as DUC 5.2, which is a mixed data MBE use case.

Item	Machined Part 1
Dataset Class	DCC 5
Dimension State	Minimally-dimensioned (MD)
Dataset Use Code	DUC 5.2
Use Case	Mixed

View A is geometry only.

As this is a minimally dimensioned product definition dataset, most dimensions are not persistently displayed in the product definition dataset.

Note: If a product definition dataset is incompletely dimensioned or undimensioned, it is critical to indicate that in the dataset. This is particularly important for product definition datasets that have traditionally been fully dimensioned, such as mechanical parts traditionally defined on engineering drawings. It is important that the dataset indicates that it is intentionally minimally dimensioned or undimensioned. Without this information, people receiving a manual- or mixed-use information asset may think that some of the information they need is missing or that the information is improperly presented on their display. Direct-use software should also be set to expect an attribute or metadata tag indicating the dimension state of the information asset.

Figure 4.25b shows sheet 2 of 3 of machined part 1.

Annotated View B includes datum feature specifications and other annotation.

Figure 4.25c shows sheet 3 of 3 of machined part 1.

Annotated View C includes datum feature specifications, GD&T, and other annotation.

Figure 4.26a shows sheet 1 of 3 of machined part 2.

The product definition dataset is structured as DCC 5, a minimally-dimensioned model officially defined as DUC 5.2, which is a mixed data MBE use case.

Item	Machined Part 2
Dataset Class	DCC 5
Dimension State	Minimally-dimensioned (MD)
Dataset Use Code	DUC 5.2
Use Case	Mixed

View A is geometry only.

As this is a minimally-dimensioned product definition dataset, most dimensions are not persistently displayed in the product definition dataset.

Figure 4.26b shows sheet 2 of 3 of machined part 2.

Annotated View B includes datum feature specifications and other annotation.

Figure 4.26c shows sheet 3 of 3 of machined part 2.

Annotated View C includes datum feature specifications, GD&T, and other annotation.

Figure 4.27a shows sheet 1 of 3 of machined part 3.

The product definition dataset is structured as DCC 5, a minimally dimensioned model officially defined as DUC 5.2, which is a mixed data MBE use case.

Item	Machined Part 3
Dataset Class	DCC 5
Dimension State	Minimally-dimensioned (MD)
Dataset Use Code	DUC 5.2
Use Case	Mixed

View A is geometry only.

As this is a minimally dimensioned product definition dataset, most dimensions are not persistently displayed in the product definition dataset.

Figure 4.27b shows sheet 2 of 3 of machined part 3.

Annotated View B includes datum feature specifications and other annotation.

Figure 4.27c shows sheet 3 of 3 of machined part 3.

Annotated View C includes GD&T and other annotation.

Figure 4.28a shows sheet 1 of 3 of sheet metal part 1.

The product definition dataset is structured as DCC 5, a minimally-dimensioned model officially defined as DUC 5.2, which is a mixed data MBE use case.

Item	Sheet Metal Part 1
Dataset Class	DCC 5
Dimension State	Minimally-dimensioned (MD)
Dataset Use Code	DUC 5.2
Use Case	Mixed

View A is geometry only.

As this is a minimally-dimensioned product definition dataset, most dimensions are not persistently displayed in the product definition dataset.

Best Business Practice: This product definition dataset intentionally does not include a CAD-generated flat pattern. Flat patterns should be derived from the product definition data by manufacturing engineers and stored in an MBD process definition dataset. Manufacturing engineering or a similar organization should own and manage their process definition datasets.

Figure 4.28b shows sheet 2 of 3 of sheet metal part 1.

Annotated View B includes datum feature specifications.

Figure 4.28c shows sheet 3 of 3 of sheet metal part 1.

Annotated View C includes datum feature specifications, GD&T, and other annotation.

Figure 4.29a shows sheet 1 of 2 of sheet metal part 2.

The product definition dataset is structured as DCC 5, a minimally-dimensioned model officially defined as DUC 5.2, which is a mixed data MBE use case.

Item	Sheet Metal Part 2
Dataset Class	DCC 5
Dimension State	Minimally-dimensioned (MD)
Dataset Use Code	DUC 5.2
Use Case	Mixed

View A is geometry only.

As this is a minimally dimensioned product definition dataset, most dimensions are not persistently displayed in the product definition dataset.

This product definition dataset intentionally does not include a CAD-generated flat pattern. See Best Business Practice that follows Figure 4.28a.

Figure 4.29b shows sheet 2 of 3 of sheet metal part 1.

Annotated Views B and C include datum feature specifications and other annotation.

Figure 4.30a shows sheet 1 of 3 of simple part 1, machined.

The product definition dataset is structured as DCC 5, a fully-dimensioned model officially defined as DUC 5.1, which is a manual data MBE use case.

Item	Simple Part 1, Machined
Dataset Class	DCC 5
Dimension State	Fully dimensioned (FD)
Dataset Use Code	DUC 5.1
Use Case	Manual

View A is geometry only.

As this is a fully dimensioned product definition dataset, all dimensions must be persistently displayed in the product definition dataset.

Figure 4.30b shows sheet 2 of 3 of simple part 1, machined.

Annotated Views B and C include datum feature specifications, GD&T, and other annotation.

Figure 4.30c shows sheet 3 of 3 of simple part 1, machined.

Annotated Views D and E include dimensions. In a DCC 5 DUC 5.1 product definition dataset, the product geometry should be fully defined by explicit dimensions.

Figure 4.31a shows sheet 1 of 3 of simple part 2, sheet metal.

The product definition dataset is structured as DCC 5, a fully-dimensioned model officially defined as DUC 5.1, which is a manual data MBE use case.

Item	Simple Part 2, Sheet Metal
Dataset Class	DCC 5
Dimension State	Fully-dimensioned (FD)
Dataset Use Code	DUC 5.1
Use Case	Manual

View A is geometry only.

As this is a fully-dimensioned product definition dataset, all dimensions must be persistently displayed in the product definition dataset. Sheet metal thickness is officially designated in the material specification in note 1.

Figure 4.31b shows sheet 2 of 2 of simple part 2, sheet metal.

Annotated Views B and C include datum feature specifications, GD&T, and other annotation. As this product definition dataset is optimized for manual use, the A-side of material is indicated graphically by the <AS> symbol in View B.

Annotated View D includes dimensions.

In a DCC 5 DUC 5.1 product definition dataset, the product geometry should be fully defined by explicit dimensions.

Figure 4.32a shows sheet 1 of 2 of simple part 3, purchased.

The product definition dataset is structured as DCC 5, a fully-dimensioned model officially defined as DUC 5.1, which is a manual data MBE use case.

Item	Simple Part 3, Purchased
Dataset Class	DCC 5
Dimension State	Fully-dimensioned (FD)
Dataset Use Code	DUC 5.1
Use Case	Manual

View A is geometry only.

This is a purchased part. Some or most dimensions are defined and owned by the manufacturer. Even though the dataset is designated as fully-dimensioned, only dimensions that are modified or customized for this item should be persistently displayed in the product definition dataset. If desired, reference dimensions may be included to provide additional non-binding information. In this example, all external part surfaces are defined by explicit dimensions except for the electrical connector geometry.

Figure 4.32b shows sheet 2 of 2 of simple part 3, purchased.

Annotated View B includes datum feature specifications and GD&T.

Annotated View C includes dimensions and other annotation.

In a DCC 5 DUC 5.1 product definition dataset of a purchased part (e.g. a modified commercial off the shelf (COTS) part), any modified or customized product geometry should be fully defined by explicit dimensions.

Note about Figures 4.33 and 4.34: Most of the examples in this section show product definition datasets created using a method that supports one use case. Some examples were created to support manual use and structured as DUC 5.1, and other examples were created to support mixed use and structured as DUC 5.2.

To facilitate deeper understanding of these methods, Figures 4.33 and 4.34 use different methods to define the same part, thereby allowing the reader to compare the methods.

Figure 4.33 shows a fully-dimensioned DCC 5 product definition dataset. The dimensions are the official definition of product geometry, and the dataset is optimized for manual use (DUC 5.1).

Figure 4.34 shows a minimally-dimensioned DCC 5 product definition data-set. Model geometry is the official definition of product geometry, and the dataset is optimized for mixed use (DUC 5.2).

Figure 4.33a shows sheet 1 of 3 of the fully-dimensioned pivot roller bracket.

The product definition dataset is structured as DCC 5, a fully-dimensioned model officially defined as DUC 5.1, which is a manual data MBE use case.

Item	Pivot Roller Bracket
Dataset Class	DCC 5
Dimension State	Fully-dimensioned (FD)
Dataset Use Code	DUC 5.1
Use Case	Manual

View A is geometry only.

As this is a fully dimensioned product definition dataset, all dimensions must be persistently displayed in the product definition dataset. Sheet metal thickness is officially designated in the material specification in note 1.

This product definition dataset intentionally does not include a CAD-generated flat pattern. See Best Business Practice that follows Figure 4.28a.

Figure 4.33b shows sheet 2 of 3 of pivot roller bracket.

Annotated Views B and C include datum feature specifications, GD&T, and other annotation.

Figure 4.33c shows sheet 3 of 3 of pivot roller bracket.

Annotated Views D and E include dimensions and other annotation. As this product definition dataset is optimized for manual use, the A-side of material is indicated graphically by the <AS> symbol in Views D and E.

In a DCC 5 DUC 5.1 product definition dataset, the product geometry should be fully defined by explicit dimensions.

Figure 4.34a shows sheet 1 of 2 of the minimally-dimensioned pivot roller bracket.

The product definition dataset is structured as DCC 5, a minimally-dimensioned model officially defined as DUC 5.2, which is a mixed data MBE use case.

Item	Pivot Roller Bracket
Dataset Class	DCC 5
Dimension State	Minimally-dimensioned (MD)
Dataset Use Code	DUC 5.2
Use Case	Mixed

View A is geometry only.

As this is a minimally-dimensioned product definition dataset, most dimensions are not persistently displayed in the product definition dataset. Sheet metal thickness is officially designated in the material specification in note 1.

This product definition dataset intentionally does not include a CAD-generated flat pattern. See Best Business Practice that follows Figure 4.34a.

Figure 4.34b shows sheet 2 of 2 of pivot roller bracket.

Annotated Views B and C include datum feature specifications, GD&T, and other annotation.

Figure 4.35a shows sheet 1 of 2 of pivot roller bushing.

The product definition dataset is structured as DCC 5, a minimally-dimensioned model officially defined as DUC 5.2, which is a mixed data MBE use case.

Item	Pivot Roller Bushing
Dataset Class	DCC 5
Dimension State	Minimally-dimensioned (MD)
Dataset Use Code	DUC 5.2
Use Case	Mixed

View A is geometry only.

As this is a minimally-dimensioned product definition dataset, most dimensions are not persistently displayed in the product definition dataset.

Figure 4.35b shows sheet 2 of 2 of pivot roller bushing.

Annotated View B includes datum feature specifications, GD&T, and other annotation.

Figure 4.36a shows sheet 1 of 2 of pivot roller keyed bushing.

The product definition dataset is structured as DCC 5, a minimally dimensioned model officially defined as DUC 5.2, which is a mixed data MBE use case.

Item	Pivot Roller Keyed Bushing
Dataset Class	DCC 5
Dimension State	Minimally dimensioned (MD)
Dataset Use Code	DUC 5.2
Use Case	Mixed

View A is geometry only.

As this is a minimally dimensioned product definition dataset, most dimensions are not persistently displayed in the product definition dataset.

Figure 4.36b shows sheet 2 of 2 of pivot roller keyed bushing.

Annotated View B includes datum feature specifications, GD&T, and other annotation.

Figure 4.37a shows sheet 1 of 2 of pivot roller wheel.

The product definition dataset is structured as DCC 5, a minimally-dimensioned model officially defined as DUC 5.2, which is a mixed data MBE use case.

Item	Pivot Roller Wheel
Dataset Class	DCC 5
Dimension State	Minimally-dimensioned (MD)
Dataset Use Code	DUC 5.2
Use Case	Mixed

View A is geometry only.

As this is a minimally dimensioned product definition dataset, most dimensions are not persistently displayed in the product definition dataset.

Molding data such as draft angle and parting line(s) are defined in the model geometry and attributes in the CAD data.

Figure 4.37b shows sheet 2 of 2 of pivot roller wheel.

Annotated View B includes datum feature specifications, GD&T, and other annotation.

The first sheet of Figures 4.38 and 4.39 includes the following information elements.

- Parts list – contains list of items with supporting information and displays toggles for each item
- Title block – contains file and product metadata
- Revision History block – contains revision history
- Official Information Assets table – lists official information assets for the product definition dataset/TDP
- Global Tolerances block – contains globally applicable tolerance and related specifications
- Specifications table – lists specifications officially-invoked by the product definition dataset/TDP
- Notes – contains general notes and specifications
- Information Services block
- View(s) – annotated – includes In View Services dialog for each view
- Bounding box – sizes of a square-prism that bounds the product relative to the global coordinate system
- Global coordinate system

Figure 4.38a shows sheet 1 of 2 of simple assembly.

The product definition dataset is structured as DCC 5, a fully-dimensioned model officially defined as DUC 5.2, which is a mixed data MBE use case.

Item	Simple Assembly
Dataset Class	DCC 5
Dimension State	Minimally-dimensioned (MD)
Dataset Use Code	DUC 5.2
Use Case	Mixed

Annotated View A shows the assembly in the assembled state and includes assembly annotation.

This is a separable assembly. In a separable assembly, each item's product geometry, specifications, and other requirements are defined in the item's product definition dataset. In some assemblies, items are modified or applied at the assembly level, changing item geometry (e.g. removing item material by machining) or obscuring item geometry (e.g. covering by over-molding, plating, painting, etc.). In this assembly, there is no assembly-level machining, over-molding, coating, etc.

As this is a minimally dimensioned product definition dataset, and because it is an assembly, most dimensions are not persistently displayed in the product definition dataset.

Figure 4.38b shows sheet 2 of 2 of simple assembly.

Annotated View B shows the assembly in the exploded state and includes assembly annotation. Toggles are included in the In View Services dialog box to display animations of assembly and disassembly processes.

Figure 4.39a shows sheet 1 of 2 of pivot roller assembly.

The product definition dataset is structured as DCC 5, a fully-dimensioned model officially defined as DUC 5.2, which is a mixed data MBE use case.

Item	Pivot Roller Assembly
Dataset Class	DCC 5
Dimension State	Minimally-dimensioned (MD)
Dataset Use Code	DUC 5.2
Use Case	Mixed

Annotated View A shows the assembly in the assembled state and includes assembly annotation.

This is a separable assembly. In a separable assembly, each item's product geometry, specifications, and other requirements are defined in the item's product definition dataset. In some assemblies, items are modified or applied at the assembly level, changing item geometry (e.g. removing item material by machining) or obscuring item geometry (e.g. covering by over-molding, plating, painting, etc.). In this assembly, there is no assembly-level machining, over-molding, coating, etc.

As this is a minimally-dimensioned product definition dataset, and because it is an assembly, most dimensions are not persistently displayed in the product definition dataset.

Figure 4.39b shows sheet 2 of 2 of pivot roller assembly.

Annotated View B shows the assembly in the exploded state and includes assembly annotation. Toggles are included in the In View Services dialog box to display animations of assembly and disassembly processes.

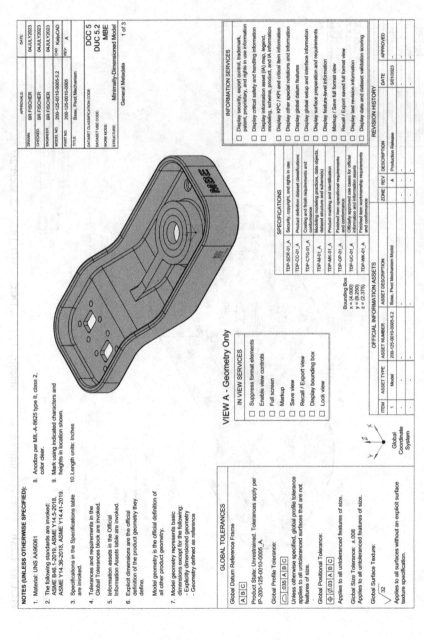

FIGURE 4.25a Product definition dataset: machined part 1: DCC 5, MD, DUC 5.2, sheet 1/3.

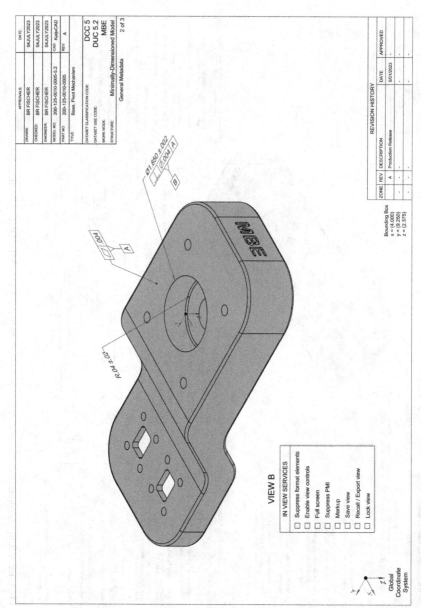

FIGURE 4.25b   Product definition dataset: machined part 1: DCC 5, MD, DUC 5.2, sheet 2/3.

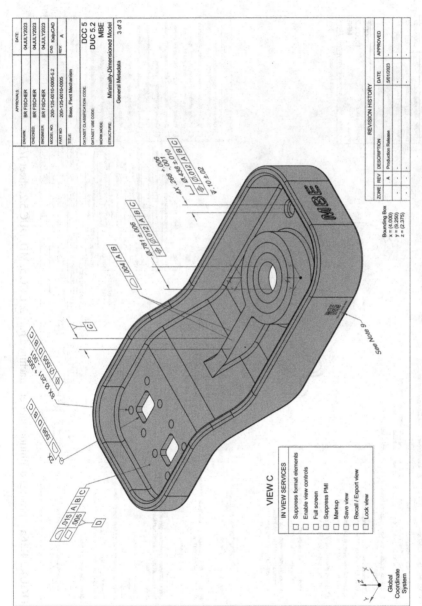

FIGURE 4.25c   Product definition dataset: machined part 1: DCC 5, MD, DUC 5.2, sheet 3/3.

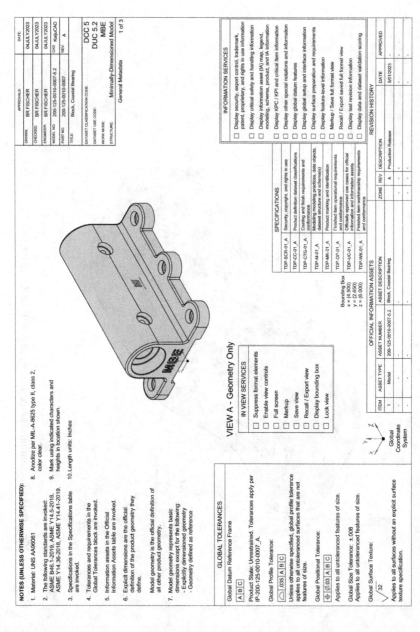

FIGURE 4.26a Product definition dataset: machined part 2: DCC 5, MD, DUC 5.2, sheet 1/3.

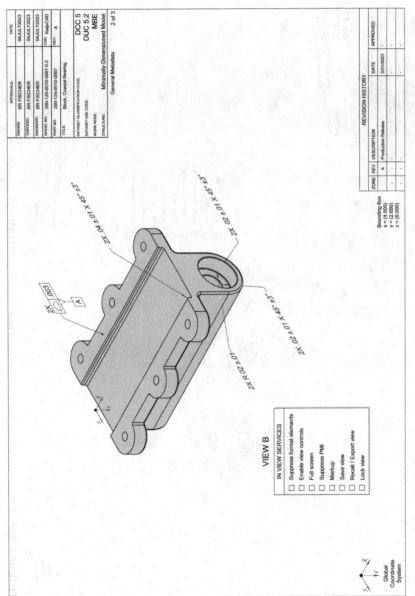

FIGURE 4.26b Product definition dataset: machined part 2: DCC 5, MD, DUC 5.2, sheet 2/3.

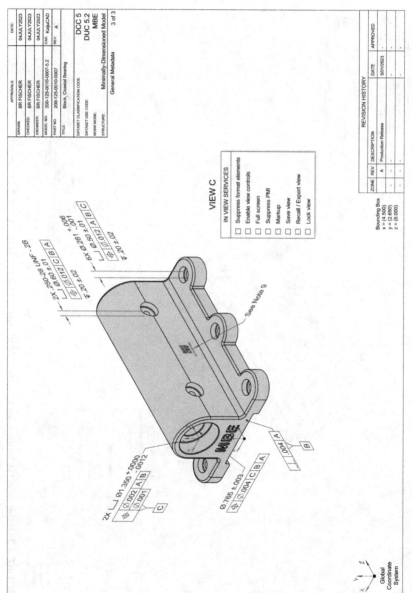

FIGURE 4.26c Product definition dataset: machined part 2: DCC 5, MD, DUC 5.2, sheet 3/3.

NOTES (UNLESS OTHERWISE SPECIFIED):

1. Material: UNS AA96061
2. The following standards are invoked: ASME B46.1-2019, ASME Y14.5-2018, ASME Y14.36-2018, ASME Y14.41-2019.
3. Specifications in the Specifications table are invoked.
4. Tolerances and requirements in the Global Tolerances block are invoked.
5. Information assets in the Official Information Assets table are invoked.
6. Explicit dimensions are the official definition of the product geometry they define.

   Model geometry is the official definition of all other product geometry.
7. Model geometry represents basic dimensions except for the following:
   - Explicitly dimensioned geometry
   - Geometry defined as reference
8. Anodize per MIL-A-8625 type II, class 2, color clear.
9. Mark using indicated characters and heights in location shown.
10. Length units: Inches

GLOBAL TOLERANCES

Global Datum Reference Frame
A B C

Product State: Unrestrained. Tolerances apply per IP-200-125-0010-0012_A.

Global Profile Tolerance:
⌓ .035 A B C

Unless otherwise specified, global profile tolerance applies to all untoleranced surfaces that are not features of size.

Global Positional Tolerance:
⊕ ⌀.03 A B C

Applies to all untoleranced features of size.

Global Size Tolerance: ±.008
Applies to all untoleranced features of size.

Global Surface Texture:
√32

Applies to all surfaces without an explicit surface texture specification.

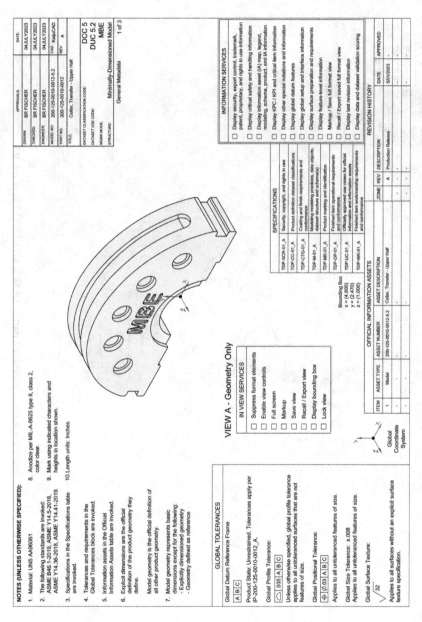

VIEW A - Geometry Only

IN VIEW SERVICES
☐ Suppress format elements
☐ Enable view controls
☐ Full screen
☐ Markup
☐ Save view
☐ Recall / Export view
☐ Display bounding box
☐ Lock view

Global Coordinate System

Bounding Box
x = (4.800)
y = (2.470)
z = (1.000)

SPECIFICATIONS

TDP-SCR-01_A	Security, copyright, and rights in use
TDP-CC-01_A	Product definition dataset classifications
TDP-CTG-01_A	Coating and finish requirements and conformance
TDP-M-01_A	Modeling modeling practices, data objects, dataset structure and schema(s)
TDP-MK-01_A	Product marking and identification
TDP-QP-01_A	Finished item operational requirements and conformance
TDP-UC-01_A	Officially approved use cases for official information and information assets
TDP-WK-01_A	Finished item workmanship requirements and conformance

OFFICIAL INFORMATION ASSETS

ITEM	ASSET TYPE	ASSET NUMBER	ASSET DESCRIPTION
1	Model	200-125-0010-0012-5.2	Collar, Transfer - Upper Half

	APPROVALS:	DATE:
DRAWN	BR FISCHER	04JULY2023
CHECKED	BR FISCHER	04JULY2023
ENGINEER	BR FISCHER	04JULY2023
MODEL NO	200-125-0010-0012-5.2	CAD KaijuCAD
PART NO	200-125-0010-0012	REV A

TITLE: Collar, Transfer - Upper Half
DATASET CLASSIFICATION CODE: DCC 5
DATASET USE CODE: DUC 5.2
WORK MODE: MBE
STRUCTURE: Minimally-Dimensioned Model

General Metadata   1 of 3

INFORMATION SERVICES
☐ Display security, export control, trademark, patent, proprietary, and rights in use information
☐ Display critical safety and handling information
☐ Display information asset (IA) map, legend, modeling, schema, product, and IA information
☐ Display KPC / KPI and critical item information
☐ Display other special notations and information
☐ Display global datum features
☐ Display global setup and interface information
☐ Display surface preparation and requirements
☐ Display feature-level information
☐ Markup / Save full format view
☐ Recall / Export saved full format view
☐ Display last revision information
☐ Display data and dataset validation scoring

REVISION HISTORY

ZONE	REV	DESCRIPTION	DATE	APPROVED
-	A	Production Release	5/01/2023	-

FIGURE 4.27a Product definition dataset: machined part 3: DCC 5, MD, DUC 5.2, sheet 1/3.

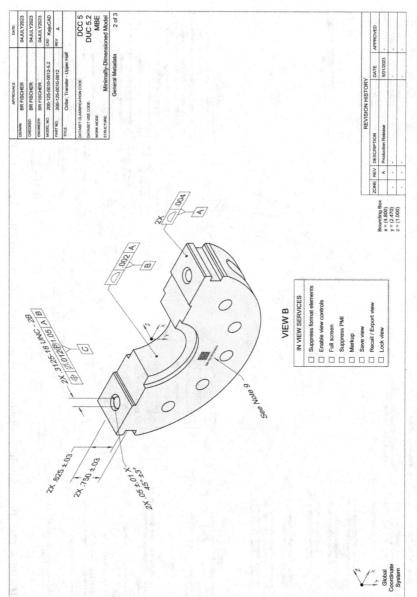

FIGURE 4.27b Product definition dataset: machined part 3: DCC 5, MD, DUC 5.2, sheet 2/3.

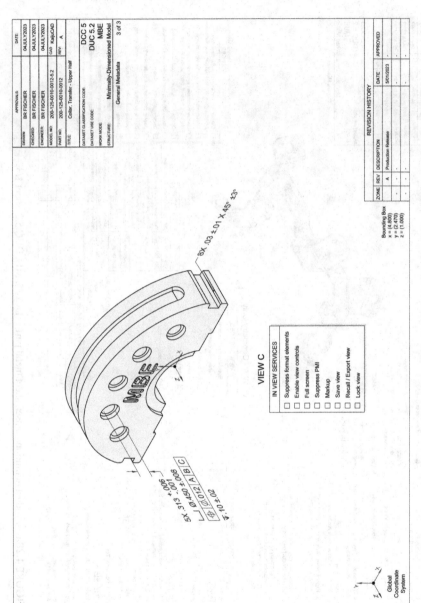

FIGURE 4.27c Product definition dataset: machined part 3: DCC 5, MD, DUC 5.2, sheet 3/3.

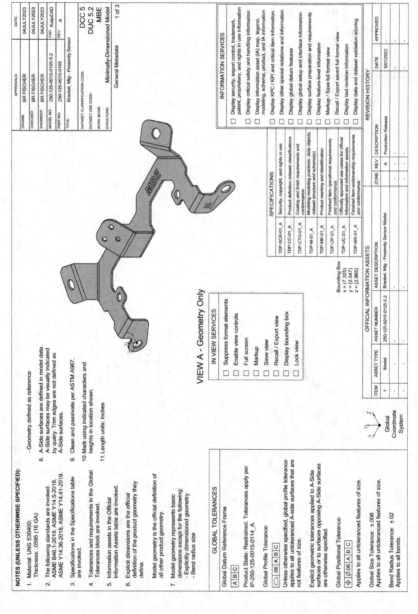

FIGURE 4.28a Product definition dataset: sheet metal part 1: DCC 5, MD, DUC 5.2, sheet 1/3.

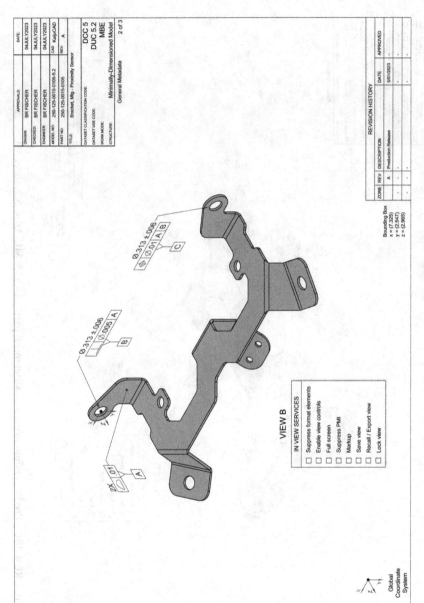

FIGURE 4.28b Product definition dataset: sheet metal part 1: DCC 5, MD, DUC 5.2, sheet 2/3.

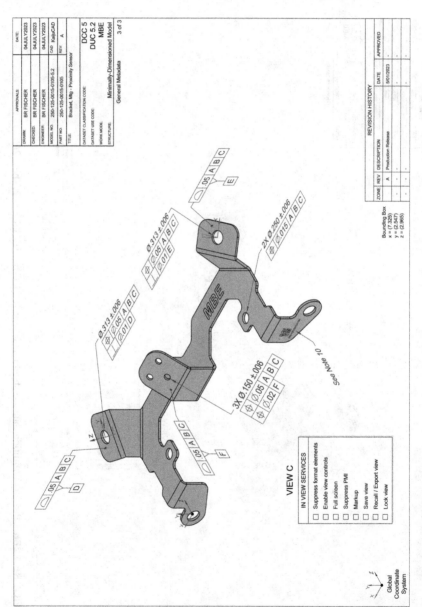

FIGURE 4.28c   Product definition dataset: sheet metal part 1: DCC 5, MD, DUC 5.2, sheet 3/3.

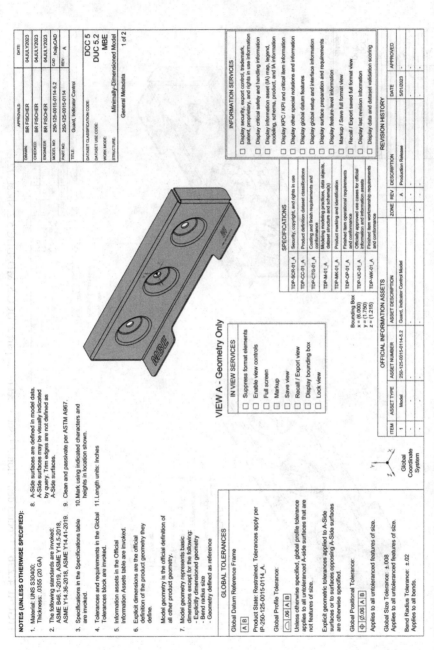

FIGURE 4.29a Product definition dataset: sheet metal part 2: DCC 5, MD, DUC 5.2, sheet 1/2.

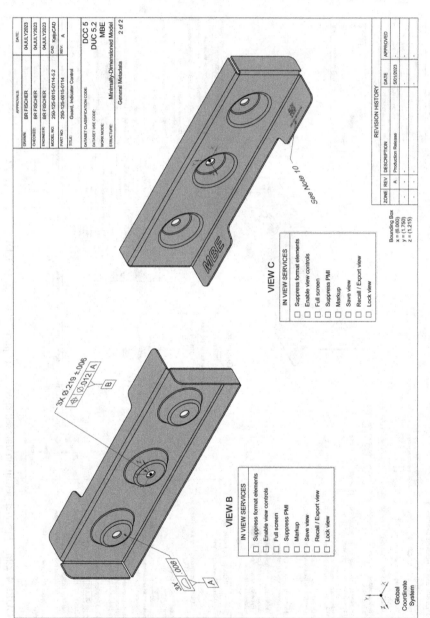

FIGURE 4.29b Product definition dataset: sheet metal part 2: DCC 5, MD, DUC 5.2, sheet 2/2.

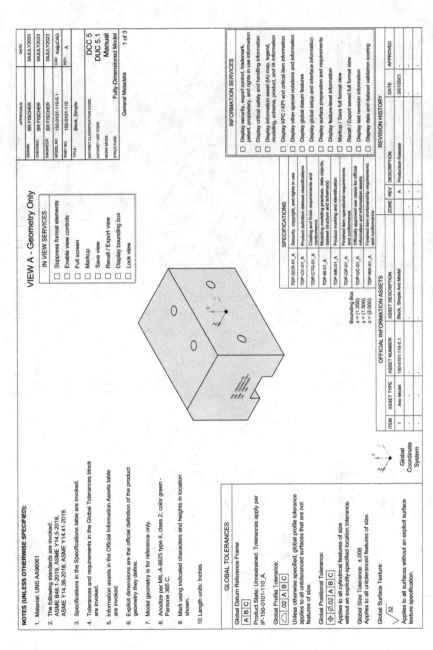

FIGURE 4.30a Product definition dataset: simple part 1, machined: DCC 5, FD, DUC 5.1, sheet 1/3.

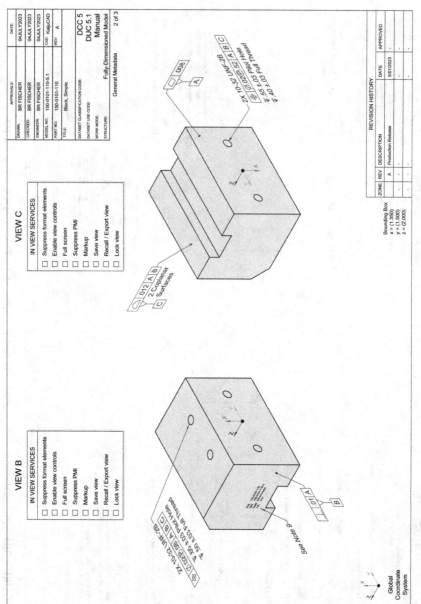

FIGURE 4.30b Product definition dataset: simple part 1, machined: DCC 5, FD, DUC 5.1, sheet 2/3.

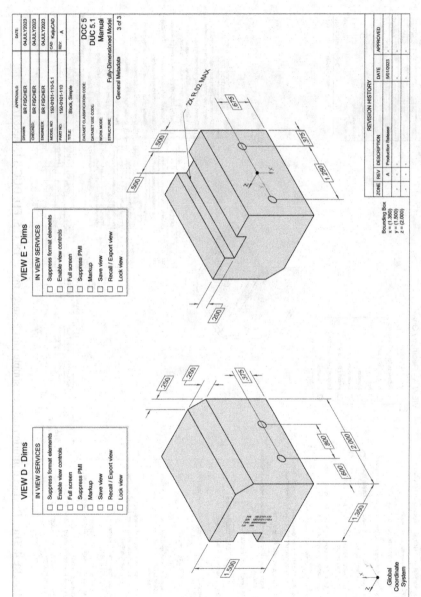

FIGURE 4.30c Product definition dataset: simple part 1, machined: DCC 5, FD, DUC 5.1, sheet 3/3.

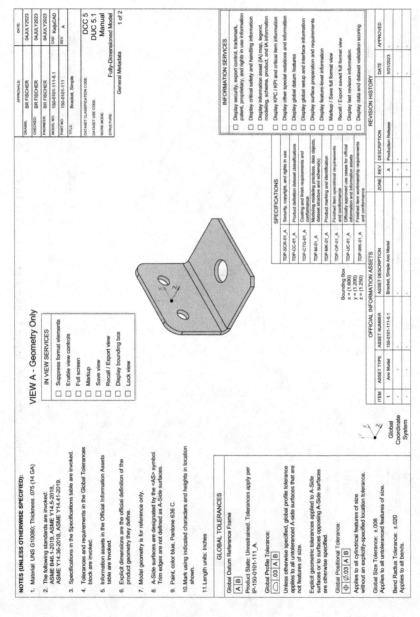

FIGURE 4.31a Product definition dataset: simple part 2, sheet metal: DCC 5, FD, DUC 5.1, sheet 1/2.

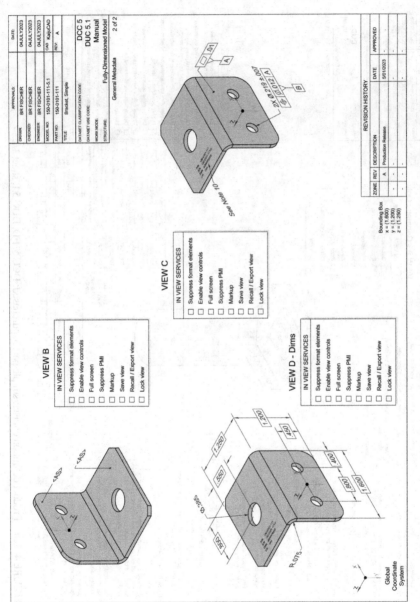

FIGURE 4.31b Product definition dataset: simple part 2, sheet metal: DCC 5, FD, DUC 5.1, sheet 2/2.

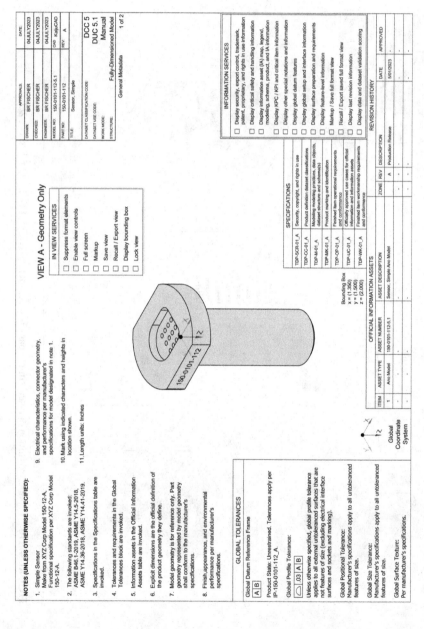

FIGURE 4.32a Product definition dataset: simple part 3, purchased: DCC 5, FD, DUC 5.1, sheet 1/2.

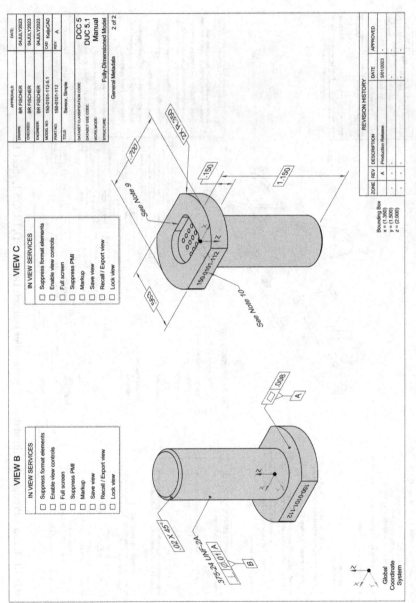

FIGURE 4.32b Product definition dataset: simple part 3, purchased: DCC 5, FD, DUC 5.1, sheet 2/2.

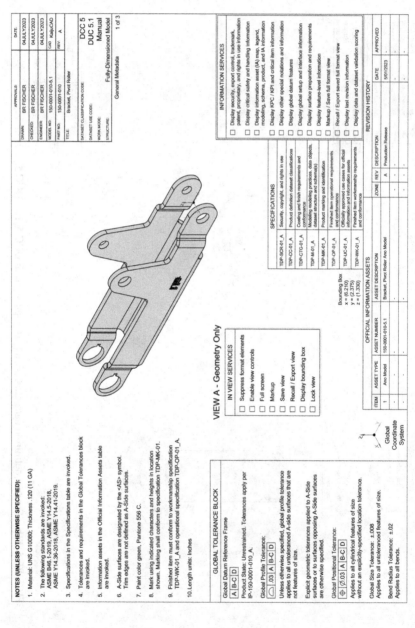

FIGURE 4.33a Product definition dataset: pivot roller bracket: DCC 5, FD, DUC 5.1, sheet 1/3.

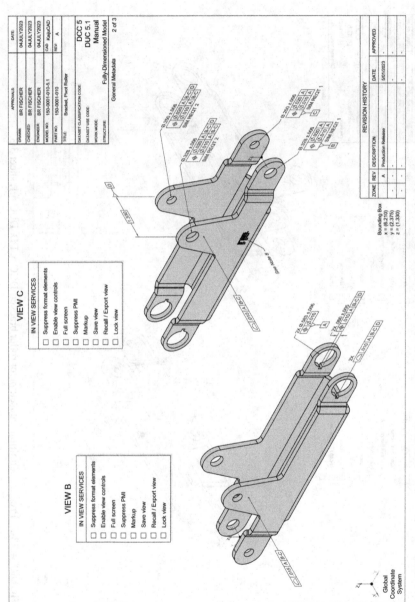

FIGURE 4.33b Product definition dataset: pivot roller bracket: DCC 5, FD, DUC 5.1, sheet 2/3.

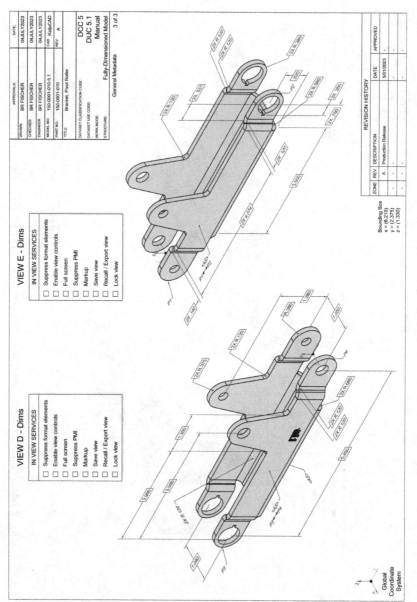

FIGURE 4.33c   Product definition dataset: pivot roller bracket: DCC 5, FD, DUC 5.1, sheet 3/3.

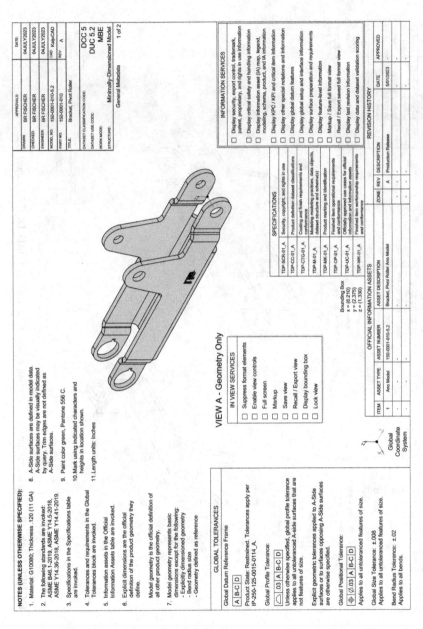

FIGURE 4.34a Product definition dataset: pivot roller bracket: DCC 5, MD, DUC 5.2, sheet 1/2.

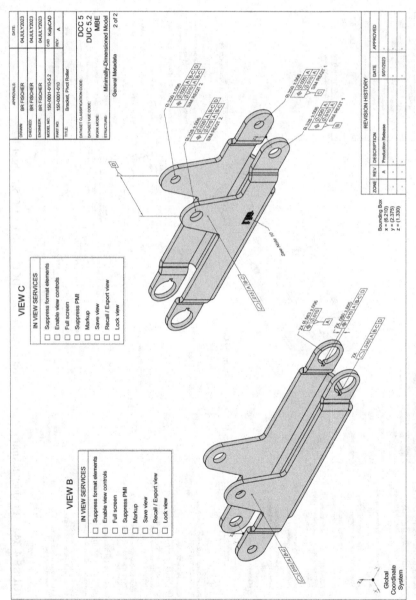

FIGURE 4.34b Product definition dataset: pivot roller bracket: DCC 5, MD, DUC 5.2, sheet 2/2.

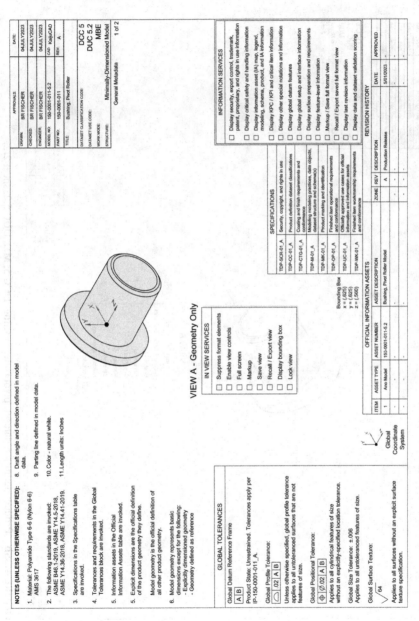

NOTES (UNLESS OTHERWISE SPECIFIED):

1. Material: Polyamide Type 6-6 (Nylon 6-6) AMS 3617

2. The following standards are invoked:
ASME B46.1-2019, ASME Y14.5-2018,
ASME Y14.36-2018, ASME Y14.41-2019.

3. Specifications in the Specifications table are invoked.

4. Tolerances and requirements in the Global Tolerances block are invoked.

5. Information assets in the Official Information Assets table are invoked.

5. Explicit dimensions are the official definition of the product geometry they define.

Model geometry is the official definition of all other product geometry.

6. Model geometry represents basic dimensions except for the following:
- Explicitly dimensioned geometry.
- Geometry defined as reference

8. Draft angle and direction defined in model data.

9. Parting line defined in model data.

10. Color - natural white.

11. Length units: Inches

GLOBAL TOLERANCES

Global Datum Reference Frame

A | B

Product State: Unrestrained. Tolerances apply per IP-150-0001-011_A.

Global Profile Tolerance:

⌓ | .02 | A | B

Unless otherwise specified, global profile tolerance applies to all untoleranced surfaces that are not features of size.

Global Positional Tolerance:

⊕ | ⌀.02 | A | B

Applies to all cylindrical features of size without an explicitly-specified location tolerance.

Global Size Tolerance: ±.006
Applies to all untoleranced features of size.

Global Surface Texture:

√⁶⁴

Applies to all surfaces without an explicit surface texture specification.

Global Coordinate System

VIEW A - Geometry Only

IN VIEW SERVICES
- ☐ Suppress format elements
- ☐ Enable view controls
- ☐ Full screen
- ☐ Markup
- ☐ Save view
- ☐ Recall / Export view
- ☐ Display bounding box
- ☐ Lock view

Bounding Box
x = (.625)
y = (.625)
z = (.560)

SPECIFICATIONS

TDP-SCR-01_A	Security, copyright, and rights in use
TDP-CC-01_A	Product definition dataset classifications
TDP-CTG-01_A	Coating and finish requirements and conformance
TDP-M-01_A	Modeling modeling practices, data objects, dataset structure and schema(s)
TDP-MK-01_A	Product marking and identification
TDP-OP-01_A	Finished item operational requirements and conformance
TDP-UC-01_A	Officially approved use cases for official information and information assets
TDP-WK-01_A	Finished item workmanship requirements and conformance

OFFICIAL INFORMATION ASSETS

ITEM	ASSET TYPE	ASSET NUMBER	ASSET DESCRIPTION
1	Ano Model	150-0001-011-5.2	Bushing, Pivot Roller Model
-	-	-	-
-	-	-	-

DRAWN:	BR FISCHER	DATE: 04/JUL/2023
CHECKED:	BR FISCHER	04/JUL/2023
ENGINEER:	BR FISCHER	04/JUL/2023

APPROVALS:

CAD KajuCAD
MODEL NO. 150-0001-011-5.2	REV. A
PART NO. 150-0001-011
TITLE Bushing, Pivot Roller
DATASET CLASSIFICATION CODE:	DCC 5
DATASET USE CODE:	DUC 5.2
WORK MODE:	MBE
	Minimally-Dimensioned Model
STRUCTURE:	General Metadata	1 of 2

INFORMATION SERVICES
- ☐ Display security, export control, trademark, patent, proprietary, and rights in use information
- ☐ Display critical safety and handling information
- ☐ Display information asset_(IA) map, legend, modeling, schema, product, and IA information
- ☐ Display KPC / KPI and critical item information
- ☐ Display other special notations and information
- ☐ Display global datum features
- ☐ Display global setup and interface information
- ☐ Display surface preparation and requirements
- ☐ Display feature-level information
- ☐ Markup / Save full format view
- ☐ Recall / Export saved full format view
- ☐ Display last revision information
- ☐ Display data and dataset validation scoring

REVISION HISTORY

ZONE	REV	DESCRIPTION	DATE	APPROVED
	A	Production Release	5/01/2023	
-	-	-	-	-
-	-	-	-	-

FIGURE 4.35a Product definition dataset: pivot roller bushing: DCC 5, MD, DUC 5.2, sheet 1/2.

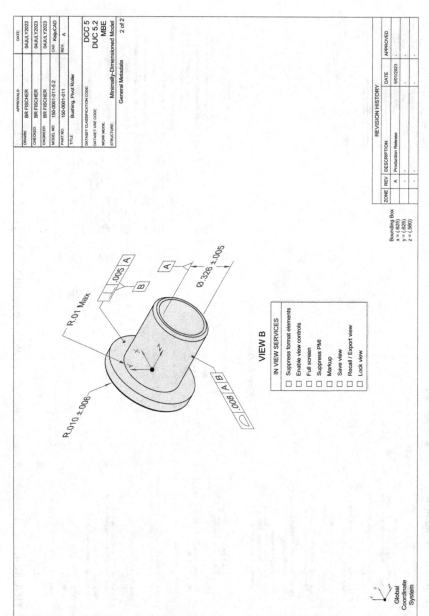

FIGURE 4.35b Product definition dataset: pivot roller bushing: DCC 5, MD, DUC 5.2, sheet 2/2.

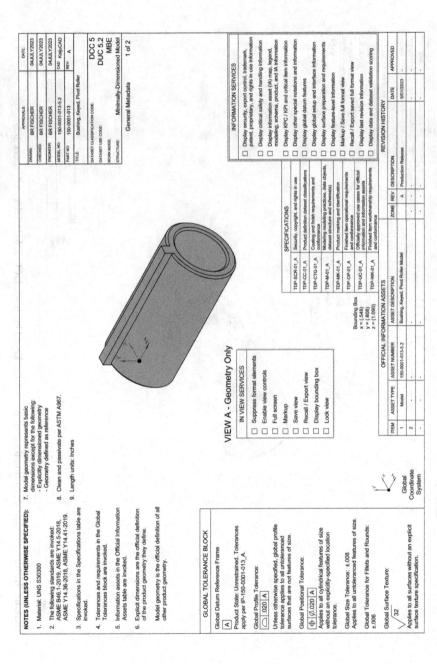

NOTES (UNLESS OTHERWISE SPECIFIED):

1. Material: UNS S30300

2. The following standards are invoked:
   ASME B46.1-2019, ASME Y14.5-2018,
   ASME Y14.36-2018, ASME Y14.41-2019.

3. Specifications in the Specifications table are invoked.

4. Tolerances and requirements in the Global Tolerances block are invoked.

5. Information assets in the Official Information Assets table are invoked.

6. Explicit dimensions are the official definition of the product geometry they define.

   Model geometry is the official definition of all other product geometry.

7. Model geometry represents basic dimensions except for the following:
   - Explicitly dimensioned geometry
   - Geometry defined as reference

8. Clean and passivate per ASTM A967.

9. Length units: Inches

GLOBAL TOLERANCE BLOCK

Global Datum Reference Frame

[A]

Product State: Unrestrained. Tolerances apply per IP-150-0001-013_A.

Global Profile Tolerance:

[⌓ .020 | A]

Unless otherwise specified, global profile tolerance applies to all untoleranced surfaces that are not features of size.

Global Positional Tolerance:

[⊕ | ⌀.020 | A]

Applies to all cylindrical features of size without an explicitly-specified location tolerance.

Global Size Tolerance: ±.008
Applies to all untoleranced features of size.

Global Tolerance for Fillets and Rounds:
±.006

Global Surface Texture:

√ 32

Applies to all surfaces without an explicit surface texture specification.

Global Coordinate System

VIEW A - Geometry Only

IN VIEW SERVICES

☐ Suppress format elements
☐ Enable view controls
☐ Full screen
☐ Markup
☐ Save view
☐ Recall / Export view
☐ Display bounding box
☐ Lock view

SPECIFICATIONS

TDP-SCR-01_A	Security, copyright, and rights in use
TDP-CC-01_A	Product definition dataset classifications
TDP-CTG-01_A	Coating and finish requirements and conformance
TDP-M-01_A	Modeling modeling practices, data objects, dataset structure and schema(s)
TDP-MK-01_A	Product marking and identification
TDP-OP-01_A	Finished item operational requirements and conformance
TDP-UC-01_A	Officially approved use cases for official information and information assets
TDP-WK-01_A	Finished item workmanship requirements and conformance

Bounding Box:
x = (.548)
y = (.608)
z = (1.090)

OFFICIAL INFORMATION ASSETS

ITEM	ASSET TYPE	ASSET NUMBER	ASSET DESCRIPTION
1	Model	150-0001-013-5.2	Bushing, Keyed, Pivot Roller Model
2	-	-	-

INFORMATION SERVICES

☐ Display security, export control, trademark, patent, proprietary, and rights in use information
☐ Display critical safety and handling information
☐ Display information asset (IA) map, legend, modeling, schema, product, and IA information
☐ Display KPC / KPI and critical item information
☐ Display other special notations and information
☐ Display global datum features
☐ Display global setup and interface information
☐ Display surface preparation and requirements
☐ Display feature-level information
☐ Markup / Save full format view
☐ Recall / Export saved full format view
☐ Display last revision information
☐ Display data and dataset validation scoring

	APPROVALS:	DATE:
DRAWN:	BR FISCHER	04JULY2023
CHECKED:	BR FISCHER	04JULY2023
ENGINEER:	BR FISCHER	04JULY2023
		CAD: KajuCAD
MODEL NO.	150-0001-013-5.2	REV: A
PART NO.	150-0001-013	
TITLE	Bushing, Keyed, Pivot Roller	
DATASET CLASSIFICATION CODE:		DCC 5
DATASET USE CODE:		DUC 5.2
WORK MODE:		MBE
STRUCTURE:	Minimally-Dimensioned Model	
	General Metadata	1 of 2

REVISION HISTORY

ZONE	REV	DESCRIPTION	DATE	APPROVED
	A	Production Release	5/01/2023	-
	-	-	-	-
	-	-	-	-

FIGURE 4.36a Product definition dataset: pivot roller keyed bushing: DCC 5, MD, DUC 5.2, sheet 1/2.

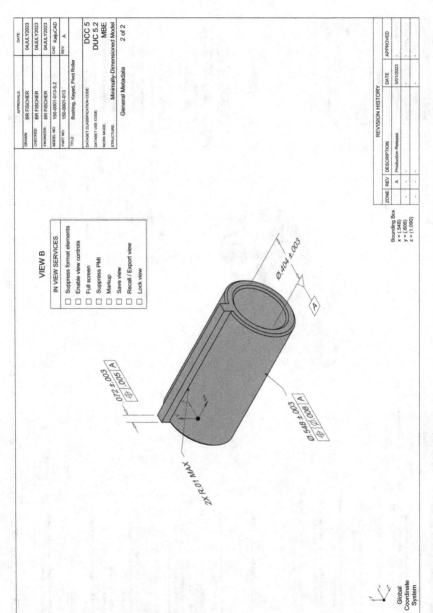

VIEW B

IN VIEW SERVICES
- ☐ Suppress format elements
- ☐ Enable view controls
- ☐ Full screen
- ☐ Suppress PMI
- ☐ Markup
- ☐ Save view
- ☐ Recall / Export view
- ☐ Lock view

Ø.404 ±.003

A

.072 ±.003

⊕ | .005 | A

Ø.548 ±.003

⊕ | Ø.008 | A

2X R.01 MAX

Bounding Box
x = (.548)
y = (.608)
z = (1.090)

Global
Coordinate
System

APPROVALS:		DATE:
DRAWN:	BR FISCHER	04JULY2023
CHECKED:	BR FISCHER	04JULY2023
ENGINEER:	BR FISCHER	04JULY2023
MODEL NO:	150-0001-013-5.2	CAD: KaijuCAD
PART NO:	150-0001-013	REV: A
TITLE:	Bushing, Keyed, Pivot Roller	
DATASET CLASSIFICATION CODE:		DCC 5
DATASET USE CODE:		DUC 5.2
WORK MODE:		MBE
STRUCTURE:	Minimally-Dimensioned Model	
	General Metadata	2 of 2

REVISION HISTORY				
ZONE	REV	DESCRIPTION	DATE	APPROVED
-	A	Production Release	5/01/2023	-
-	-	-	-	-
-	-	-	-	-

FIGURE 4.36b Product definition dataset: pivot roller keyed bushing: DCC 5, MD, DUC 5.2, sheet 2/2.

NOTES (UNLESS OTHERWISE SPECIFIED):

1. Material: ABS CS-72
2. The following standards are invoked: ASME B46.1-2019, ASME Y14.5-2018, ASME Y14.8-2022, ASME Y14.36-2018, ASME Y14.41-2019.
3. Specifications in the Specifications table are invoked.
4. Tolerances and requirements in the Global Tolerances block are invoked.
5. Information assets in the Official Information Assets table are invoked.
6. Explicit dimensions are the official definition of the product geometry they define.

   Model geometry is the official definition of all other product geometry.
7. Model geometry represents basic dimensions except for the following:
   - Explicitly dimensioned geometry
   - Geometry defined as reference
8. Draft angle and direction defined in model data.
9. Parting line defined in model data.
11. Color - gray, Pantone 877 C.
12. Mark using "150-0001-012" "Lot n", where "n" is applicable lot number, in location shown.
13. Length units: Inches

GLOBAL TOLERANCES

Global Datum Reference Frame

| A | B |

Product State: Unrestrained. Tolerances apply per IP-150-0001-012_A.

Global Profile Tolerance:

| ⌓ | .02 | A | B |

Unless otherwise specified, global profile tolerance applies to all untoleranced surfaces that are not features of size.

Global Positional Tolerance:

| ⊕ | ⌀.02 | A | B |

Applies to all cylindrical features of size without an explicitly-specified location tolerance.

Global Size Tolerance: ±.008
Applies to all untoleranced features of size.

Global Surface Texture:

√64

Applies to all surfaces without an explicit surface texture specification.

VIEW A - Geometry Only

IN VIEW SERVICES
☐ Suppress format elements
☐ Enable view controls
☐ Full screen
☐ Markup
☐ Save view
☐ Recall / Export view
☐ Display bounding box
☐ Lock view

Bounding Box
x = (1.875)
y = (1.875)
z = (.435)

Global Coordinate System

INFORMATION SERVICES
☐ Display security, export control, trademark, patent, proprietary, and rights in use information
☐ Display critical safety and handling information
☐ Display information asset (IA) map, legend, modeling, schema, product, and IA information
☐ Display KPC / KPI and critical item information
☐ Display other special notations and information
☐ Display global datum features
☐ Display global setup and interface information
☐ Display surface preparation and requirements
☐ Display feature-level information
☐ Markup / Save full format view
☐ Recall / Export saved full format view
☐ Display last revision information
☐ Display data and dataset validation scoring

SPECIFICATIONS

TDP-SCR-01_A	Security, copyright, and rights in use
TDP-CC-01_A	Product definition dataset classifications
TDP-CTG-01_A	Coating and finish requirements and conformance
TDP-M-01_A	Modeling modeling practices, data objects, dataset structure and schema(s)
TDP-MK-01_A	Product marking and identification
TDP-OP-01_A	Finished item operational requirements and conformance
TDP-UC-01_A	Officially approved user cases for official information and information assets
TDP-WK-01_A	Finished item workmanship requirements and conformance

OFFICIAL INFORMATION ASSETS

ITEM	ASSET TYPE	ASSET NUMBER	ASSET DESCRIPTION
1	Ano Model	150-0001-012-5.2	Wheel, Pivot Roller Model
-	-	-	-

	APPROVALS	DATE
DRAWN:	BR FISCHER	04JULY2023
CHECKED:	BR FISCHER	04JULY2023
ENGINEER:	BR FISCHER	04JULY2023
CAD: KaijuCAD
MODEL NO: 150-0001-012-5.2	REV: A
PART NO: 150-0001-012
TITLE: Wheel, Pivot Roller
DATASET CLASSIFICATION CODE: DCC 5
DATASET USE CODE: DUC 5.2
WORK MODE: MBE
STRUCTURE: Minimally-Dimensioned Model
General Metadata	1 of 2

REVISION HISTORY

ZONE	REV	DESCRIPTION	DATE	APPROVED
	A	Production Release	5/01/2023	
-	-	-		-
-	-	-		-

FIGURE 4.37a Product definition dataset: pivot roller wheel: DCC 5, MD, DUC 5.2, sheet 1/2.

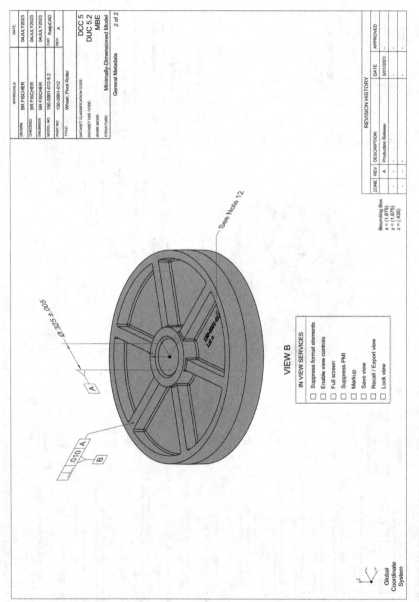

FIGURE 4.37b Product definition dataset: pivot roller wheel: DCC 5, MD, DUC 5.2, sheet 2/2.

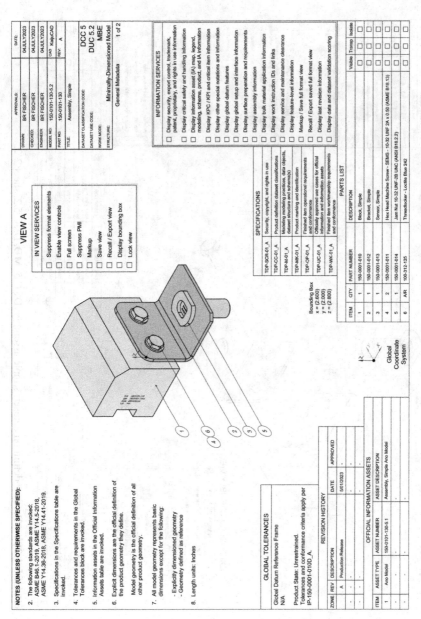

FIGURE 4.38a Product definition dataset: simple assembly: DCC 5, MD, DUC 5.2, sheet 1/2.

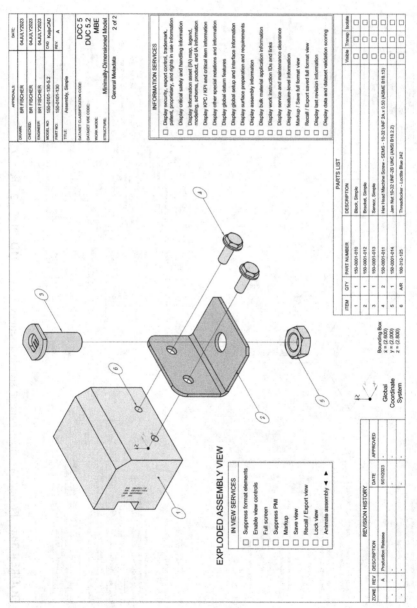

FIGURE 4.38b Product definition dataset: simple assembly: DCC 5, MD, DUC 5.2, sheet 2/2.

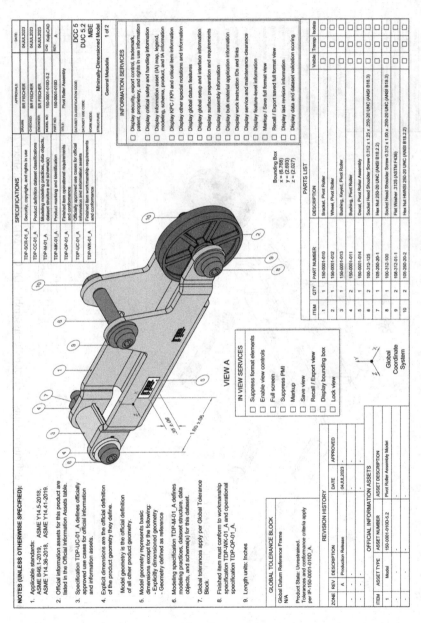

FIGURE 4.39a Product definition dataset: pivot roller assembly: DCC 5, MD, DUC 5.2, sheet 1/2.

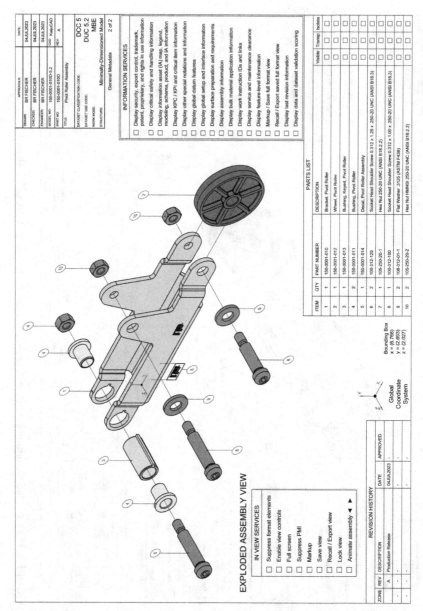

FIGURE 4.39b Product definition dataset: pivot roller assembly: DCC 5, MD, DUC 5.2, sheet 2/2.

	APPROVALS:	DATE:
DRAWN	BR FISCHER	04JUL2023
CHECKED	BR FISCHER	04JUL2023
ENGINEER	BR FISCHER	04JUL2023
MODEL NO:	150-0001-010D-5.2	CAD: KapcCAD
PART NO:	150-0001-010D	REV: A

TITLE: Pivot Roller Assembly

DATASET CLASSIFICATION CODE: DCC 5
DATASET USE CODE: DUC 5.2
WORK MODE: MBE
STRUCTURE: Minimally-Dimensioned Model

General Metadata 2 of 2

INFORMATION SERVICES

- [ ] Display security, export control, trademark, patent, proprietary, and rights in use information
- [ ] Display critical safety and handling information
- [ ] Display information asset (IA) map, legend, modeling, schema, product, and IA information
- [ ] Display KPC / KPI and critical item information
- [ ] Display other special notations and information
- [ ] Display global datum features
- [ ] Display global setup and interface information
- [ ] Display surface preparation and requirements
- [ ] Display assembly information
- [ ] Display bulk material application information
- [ ] Display work instruction IDs and links
- [ ] Display service and maintenance clearance
- [ ] Display feature-level information
- [ ] Markup / Save full format view
- [ ] Recall / Export saved full format view
- [ ] Display last revision information
- [ ] Display data and dataset validation scoring

PARTS LIST

ITEM	QTY	PART NUMBER	DESCRIPTION	Visible	Transp	Isolate
1	1	150-0001-010	Bracket, Pivot Roller			
2	1	150-0001-012	Wheel, Pivot Roller			
3	1	150-0001-013	Bushing, Keyed, Pivot Roller			
4	2	150-0001-011	Bushing, Pivot Roller			
5	1	150-0001-014	Decal, Pivot Roller Assembly			
6	2	100-312-125	Socket Head Shoulder Screw 0.312 x 1.25 x .250-20 UNC (ANSI B18.3)			
7	1	105-250-20-1	Hex Nut .250-20 UNC (ANSI B18.2.2)			
8	1	100-312-100	Socket Head Shoulder Screw 0.312 x 1.00 x .250-20 UNC (ANSI B18.3)			
9	2	106-312-01-1	Flat Washer .3125 (ASTM F436)			
10	2	105-250-20-2	Hex Nut HM/NSI .250-20 UNC (ANSI B18.2.2)			

EXPLODED ASSEMBLY VIEW

IN VIEW SERVICES

- [ ] Suppress format elements
- [ ] Enable view controls
- [ ] Full screen
- [ ] Suppress PMI
- [ ] Markup
- [ ] Save view
- [ ] Recall / Export view
- [ ] Lock view
- [ ] Animate assembly ▼ ▲

Global Coordinate System

Bounding Box
x = (6.766)
y = (2.603)
z = (2.027)

REVISION HISTORY

ZONE	REV	DESCRIPTION	DATE	APPROVED
	A	Production Release	04JUL2023	-
-	-	-	-	-
-	-	-	-	-

5 Fundamentals of MBE

ACRONYMS AND INITIALISMS

Many acronyms and initialisms used in this chapter are defined in Chapter 4. See Chapter 4 for definitions of other relevant acronyms and initialisms used in this chapter.

DAPVI   Documents, Annotation, PMI, and Visual Information
DAPVI are manual-use information assets and information. DAPVI are optimized for manual use.

LVOI   Lifecycle Value of Information
Refer to Formula 5.2 and corresponding text for a more detailed explanation of LVOI.

PrDCC   *Process definition Dataset Classification Code*
Alphanumeric designations that represent process definition dataset structure. In this text, PrDCCs 1-6 are defined; however, other DCCs could be defined if needed. Process definition dataset structures corresponding to PrDCCs range from document-only (e.g. PrDCC 1), document + model (e.g. PrDCC 3), model-only (e.g. PrDCC 5), to data-only (e.g. PrDCC 6).

Process definition Dataset Classification Codes are defined in and unique to this text.

PrDUC   *Process definition Dataset Use Code*
Alphanumeric designations that represent official use cases for process definition datasets and their constituent information assets. A PrDUC defines the official use for information and information assets and how information and information assets are supposed to be used. PrDUCs correspond to and augment PrDCCs. In this text, PrDUCs 1.1-6.1 are defined; however, many other PrDUCs could be generated. PrDUCs 1.1-6.1 represent common DCC use cases and can be used as a starting point.

Process Definition Dataset Use Codes are defined in and unique to this text.

VAT   Value Added Task
Tasks that add value to the organization. Enabling and empowering staff to focus more (or exclusively) on tasks that add value to the organization rather than on trivial low-value tasks is a core characteristic of an optimized Model-Based Enterprise (MBE).

Refer to the section in this chapter on VATs and Formula 5.1 for a more detailed explanation of VAT and how to calculate the net value of a task.

DOI: 10.1201/9781003203797-5   **229**

TERMS AND DEFINITIONS

Many terms used in this chapter are defined in Chapter 4. See Chapter 4 for definitions of other relevant terms used in this chapter.

NOMINAL PRODUCT

A perfect or ideal product. A nominal product exists in theory and may be represented by design information, a product definition dataset, a design information asset, etc.

The nominal product includes characteristics, typically with ideal values, such as product geometry, material, surface conditions, mechanical properties, electrical properties, components, subcomponents, and kinematic characteristics in the case of assemblies and systems, behavioral characteristics, chemical and elemental ratios in the case of compounds, etc. Depending on lifecycle maturity stage (e.g. conceptual, developmental, production-ready) and business decisions, the nominal product model may or may not completely define all aspects of the nominal product. However, a nominal product should be defined to a level as required for its lifecycle maturity stage.

NOMINAL PRODUCT MODEL

Information model that represents a nominal product.

Product Specification

Information that describes a product or requirements imposed on a product, such as its composition, allowable limits, boundary conditions, performance parameters, etc. In DCC 1–5 product definition datasets, product specifications are defined by text, symbols, code, mathematical formulas, manual-use information, direct-use information, simulation information, twins, or other methods. In a DCC 6 product definition, dataset product specifications are defined by direct-use information.

In discrete manufacturing, product specifications are specified in a product definition dataset and provide information about the nominal product, provide restrictions, targets, limits, boundary conditions, and other constraints and requirements on the nominal product. In some cases, product specifications define the nominal product, such as specifications for threaded holes, gears, sprockets, splines, welded joints, finish, surface texture, heat treating, etc.

Example 1: Specification Applies to Existing Nominal Product Information

A product definition dataset describes a mechanical part. The dataset includes a 3D model that represents nominal product geometry. Specifications such as explicit dimensions, GD&T, surface texture, etc. are applied to the model geometry. The specifications augment, refine, and add more information thereby defining requirements for actual product geometry. If the product specifications are designated as the official definition of the applicable product features and characteristics, then the specifications do not augment or refine the definition, the specifications are the official definition.

Example 2.1: Product Material Defined in Nominal Product Model

A product definition dataset of a mechanical part includes a 3D model that represents the nominal product geometry. The material is officially defined in the model. The model material attributes include a material standard (e.g. from ASTM), which indicates the ratios of elements in the material, allowable limits, physical properties, etc. The specification (e.g. the model material) defines both the nominal product and the allowable limits and boundary conditions.

Example 2.2: Product Material Defined by Specification

A product definition dataset of a mechanical part includes a 3D model that represents the nominal product geometry, but the material is not defined in the model. A default generic material is used in the model. Annotation (PMI), a note in this case, is the official definition of the material. The note references a material standard (e.g. from ASTM), which indicates the ratios of elements in the material, allowable limits, physical properties, etc. The specification (e.g. the note) defines both the nominal product and the allowable limits and boundary conditions.

SPECIFICATION MODEL

Information model that represents the specifications applied to a product.

Note that while most product definition datasets include a specification model, they do not contain the requirements the specifications impose upon the product. The specifications are included in the product definition dataset, but the meaning of the specifications is not included.

The meaning of the specifications and the requirements they impose on the product exist in the consumption model.

CONSUMPTION INFORMATION

Information extracted from received information to which additional information is added, usually to perform an activity or create derivative information. In a discrete manufacturing workflow context, consumption information includes a representation of the nominal product and product specifications, and the requirements, constraints, targets, and boundary conditions for an as-produced product. For example, in a manual workflow, a product definition dataset is used manually (read-interpret-determine implications-convert-transcribe, RIDICT), and the standards it references are used manually (RIDICT) to develop a manufacturing plan.

Consumption information exists separately from the source of the requirement. For example, a product definition dataset contains specifications that impose requirements upon a product, but the product definition dataset does not contain the requirements.

In most cases, requirements, constraints, targets, acceptable limits, and boundary conditions are not fully defined in a product definition dataset. Other information, such as standards and Standard Operating Procedures (SOPs) must be used to understand the measurand that the requirements, constraints, targets, and boundary conditions apply to, so conformance criteria can be fully understood.

Example 1: Consumption Information

A positional tolerance specification is applied to a hole. The tolerance is specified on an RFS basis. While it is clear the specification applies to the hole, the specification doesn't explain the requirements imposed by the specification. By referring to the ASME Y14.5 Standard, the recipient learns that the tolerance controls the axis of the hole, the axis must conform to a cylindrical tolerance zone of the indicated diameter, and the tolerance zone is basically-related to the indicated datum reference frame. The meaning of the specification and how it affects an as-produced product is consumption information.

In this example, consumption information defines the requirements that specifications impose upon a product, process, or other system of interest. Requirements include the characteristic(s) affected (e.g. which aspect of the product is controlled, constrained, or limited by the requirement), the applicable units of measurement, general rules for conformance to the requirement, the boundary conditions imposed by the requirement, and other information, such as the severity, criticality, or importance of the requirement.

Consumption also includes information that is not part of the product definition dataset, such as process parameters, tooling, machinery, process steps, and other information relevant to the use case. In the case of a manufacturing plan based on a product definition dataset, the information derived from the product definition dataset likely represents a small portion of the consumption information.

Consumption Model

Information model that represents consumption information.

Model-Based Enterprise (MBE)

Simple Definition

An organization focused on obtaining maximum value from its information throughout the organization, the supply chain, the product lifecycle, and the information lifecycle by creating and using MBD as much as possible.

Detailed Definition 1

An organization focused on creating and directly using MBD information as much as possible throughout the organization, the supply chain, the product lifecycle, and the information lifecycle, with the goals of maximizing the lifecycle value of information (LVOI) and minimizing the need to create and use manual-use information.

Detailed Definition 2

An organization that focuses on and prioritizes obtaining the greatest LVOI for the least cost and burden throughout the organization, the product lifecycle, and the information lifecycle by maximizing the creation and direct use of MBD information.

Detailed Definition 3

An organization that focuses on and prioritizes obtaining the greatest LVOI for the least cost and burden throughout the organization, the product lifecycle, and the information lifecycle by:

1. Using digital information directly in software and computation (i.e. automation) throughout the organization, the supply chain, the product lifecycle, and the information lifecycle.
2. Optimizing processes and methods to create and use direct-use MBD information throughout the organization, the supply chain, the product lifecycle, and the information lifecycle.
3. Reducing or eliminating the need for, use of, and preference for visually-presented manual-use information throughout the organization, the supply chain, the product lifecycle, and the information lifecycle.
4. Designating direct-use information assets and information as official instead of manual-use information assets and information throughout the organization, the supply chain, the product lifecycle, and the information lifecycle.

Alternate Definition

An organization focused on creating MBD and digital information and using that information directly as much as possible throughout the product lifecycle and information lifecycle, with the goal of maximizing the LVOI while incurring the least cost and burden from the information.

1. An MBE strives to create the least information needed in order to minimize lifecycle information costs and cognitive loads the information places on staff.
2. An MBE focuses on minimizing manual activities and document-based workflows as much as possible.
3. An MBE is formally structured to achieve these goals.
4. An MBE formally supports and encourages the behavior needed to achieve these goals.

MBE METHODS

Methods used to become and function as an MBE.

PROCESS DEFINITION DATASET

An information asset that defines a process using process definition data.

A process definition dataset defines processes used in an enterprise or its supply chain. Process definition datasets are often derivative datasets (e.g. a manufacturing plan, inspection plan, assembly work instructions, quality plan) created from information in a product definition dataset. Note that process definition datasets may be primary authored and not derivative datasets. Process definition datasets may contain primary-authored and secondary-authored information.

TRADITIONAL CONCEPT OF MBE

The original or traditional concept of MBE was based on the traditional concept of MBD, which is using 3D models instead of drawings, using 3D CAD or neutral model geometry instead of dimensioned drawing views. Design engineers would create and share 3D CAD model geometry with other departments, hoping that other departments would use the 3D geometry instead of other information that defined product geometry on 2D drawings, such as explicit dimensions. There were several steps to the original idea. Here is an abbreviated list:

1. Instead of defining products on 2D engineering drawings, design staff defined products using 3D information assets (product definition datasets and TDPs that contain 3D data).

   MBD was considered as mainly something that design engineers did using 3D CAD (MBD was and continues to be mostly thought of in a CAD context – this is insufficient and limits the potential of MBD from the very first step).

   MBD information assets created using this traditional paradigm are primarily intended for manual use.

2. The information assets containing 3D data are released as official information assets. 3D product definition datasets are packaged using 2D and 3D PDF or other format for release.

3. Lifecycle activities are supposed to use the information assets containing 3D data.

4. The primary use case for these information assets is visualization (a manual use case).

5. Thus, activities in such an environment that use the information provided do so manually. They use the RIDICT sequence for using visually presented information shown in Figure 5.1 to obtain information needed to create another manually-created information asset (e.g. a process definition dataset) for a lifecycle use case, such as a material requisition, a manufacturing plan, an inspection plan, etc.

   Documents in = documents out.

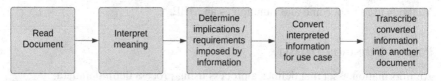

FIGURE 5.1 Read-interpret-determine implications-convert-transcribe (RIDICT) sequence of visually-presented information.

Note that 3D information assets intended for manual use or that are used manually are equivalent to documents.

6. In most cases, the lifecycle information assets (e.g. process definition datasets) that result from using 3D product definition datasets and TDPs consist of one or more documents that are intended for manual use. The lifecycle information asset is a disconnected derivative dataset (DDD), as it is not logically and digitally associated with the information and information asset from which it was derived.

The information value chain defined above is broken.

The main drivers for adopting these traditional MBD methods were design engineering. Their reasons were often self-serving and done at the expense of lifecycle user's/other departments' productivity. I call this a "push implementation". I introduce the concepts of "push" and "pull" in Chapter 2.

In general, the reasons traditional MBD methods were implemented followed something like the following trajectory:

- Design engineers didn't/don't like making drawings, as the engineers and their managers were not trained to make drawings in school.
- Engineers were/are not particularly good at making drawings, as they were not trained as drafters.
- Engineers had/have a lot of other things to think about that were/ are prioritized by their managers, such as design, analysis, regulatory compliance, etc.

Often, engineers and their managers were/are not rewarded for creating good drawings. Metrics and KPIs were/are aligned with meeting schedule. Checkers, which are now largely extinct, had the job and the skills to do the job of ensuring drawings had high enough quality and adequately served their purpose. Design engineers and their managers often viewed checkers as bottlenecks and impediments to meeting their objectives rather than as enablers of their objectives. (RIP checkers.)

Thus, MBD methods were often adopted for self-serving siloed reasons. However, the reason people believed that doing this was a good idea was because of their perceived environment and perceived priorities. The responsibility for scenarios like this lies in equal parts with the organization's upper management, middle management, and the design staff. Note that at each level, the actors are responding to their perceived environment and implicit or explicit reward structure. Everyone is trying to do the right thing, but they are not. What they should be doing, why they should be doing it, and the consequences of not doing it are unclear to them.

Old Dogs, New Tricks

To recap, MBD was and continues to be considered a CAD topic, mostly driven by annotated models, where 3D PDFs are released instead of 2D drawings, and MBD information is expected to be used manually (e.g. visualization). MBE was and continues to be considered as an environment where 3D PDFs are used manually instead of using 2D drawings manually. These ideas permeate most MBD and MBE implementations regardless of whether the implementation leads to worthwhile business value or not. In the abbreviated list of steps above, the end result is that the organization is using information assets that include 3D data, but they are using the information assets manually. The most common way the information assets are used follows the RIDICT sequence shown in Figure 5.1, with its inherent cost, burden, risk, and inefficiency. Organizations have changed the media but not the method.

In most MBE implementations, organizations have changed the media but not the method.

I have spent a lot of time thinking about historical parallels to the transition from document-based to model-based workflows, from manual-use as the primary way information is used to direct-use as the primary way information is used. There are many examples. In cases where a technological transformation was successful and sticky enough to overcome inertia against change, the result was a process that was much easier, faster, and/or cheaper than what it replaced. *What we used to do was automated so well, we not only didn't need to do it anymore, we didn't even need to think about it anymore.* We no longer needed to understand and deal with it. The burden and the need for the skills to deal with it were eliminated, which freed us to focus on more meaningful and valuable activities. The funny thing is that even with such hindsight, people still tend to resist change, and often resist it strongly.

In cases where a technological transformation was successful and sticky enough to overcome inertia against change, the result was a process that was much easier, faster, and/or cheaper than what it replaced. What we used to do was automated so well, we not only didn't need to do it anymore, we didn't even need to think about it anymore.

Note that the idea that we don't have to know or think about something anymore may, in some cases, not be true or as valuable as it seems. Some experts don't see this as a good thing; they decry it as detrimental. In *"Shaping the Future of the Fourth Industrial Revolution"* by Klaus Schwab, Mr. Schwab states that there are risks if individuals become reliant upon new technologies and automation if they encourage the loss of important skills. That seems to be in conflict with what Arthur C. Clarke wrote in *"3001: The Final Odyssey"*, in which what were once important skills become useless or no longer valuable when technology and progress obviate the need for those skills (e.g. manual drafting skills). This is from a passage in *"3001: The Final Odyssey"* where after 1000 years drifting in space, Dr. Frank Poole is resuscitated and finds himself on a starship where the crew does not know how to maintain the ship's systems. Frank learns that they don't need to memorize or practice the skills because they have technology to acquire the skills instantaneously on an as-needed basis. The contention I draw from this is that many of the skills that we have today, that we need to have today, are, in fact, non-value added and really only important because we're inefficient and still using outdated holdover methods from earlier document-based times.

Important note – I am not advocating that we no longer need to be technically competent and that we do not need to know technical details. We still need expertise and to be proficient in many things. Engineering and technical disciplines remain critical. What I am advocating is that we don't need to be experts in every topic that used to be important. Some things are overcome by events; they are obviated by technology and new ways of doing things. We need to recognize the total cost incurred for being able to do something that we no longer need to do and then decide if we should still be doing it. I have had a long career and seen many skills that were once important become commoditized and later rendered moot by broadly adopted technology.

Many of the skills that we have today, that we need to have today, are, in fact, non-value added and really only important because we're working inefficiently and still using outdated holdover methods from earlier document-based workflows.

VALUE ADDED TASKS (VATS)

I provided a definition for VAT at the beginning of this chapter. As I stated, enabling and empowering staff to focus more (or exclusively) on tasks that add value to the organization rather than on trivial low-value tasks is a core characteristic of an optimized MBE.

Considering value of a task as a function of skills and the costs associated with those skills, our goal should be to achieve the greatest value from commonly-available skills, emphasizing low-skill high-value tasks and deemphasizing high-skill low-value tasks. We can measure the value with a simple formula.

Formula 5.1 – Net Value of Task (NVoT)

5.1 $NVoT = VoT - CoT$

Where
    VoT = Value of Task (for the organization) = OBE
    OBE = Organizational Benefits Enabled
    OBE = Enabled activities (e.g. analysis, simulation, costing, estimating, pro-
        curement, production, assembly, quality, inspection, installation,
        logistics, sustainment, service, maintenance, retirement), automa-
        tion, efficiency, productivity, increasing value, visibility, usability,
        and usefulness of IP and information, reduced cost, time, informa-
        tion required to achieve goals, need for manual use information, risk
        (scarcity of skills, resources, potential for error, redundant informa-
        tion, etc.), utilization of CAPEX and OPEX, …

    CoT = Cost of Task (for the organization) = OCI
    OCI = Organizational Costs Incurred
    OCI = Cost of task in terms of currency, time, schedule, risk (scarcity of
        skills, resources, potential for error, redundant information, etc.),
        development, maintenance, and validation of information and infor-
        mation assets, development, maintenance, and validation of skills,
        overhead, staff, resources, facilities, CAPEX, OPEX, logistics, morale,
        churn, reputation …

Formula 5.1a – Lifecycle Net Value of Task (LNVoT)

5.1a $LNVoT = LVoT - LCoT$

Where
    LVoT = Lifecycle Value of Task (for the organization)
    LCoT = Lifecycle Cost of Task (for the organization)

A good Lean exercise for an organization is to evaluate the value achieved from selected tasks using the formula above. Note that skills, how common they are, how much it costs for an organization to hire people with those skills, and how much it costs to train staff that do not have the skills or have inadequate skills, the cost of maintaining those skills, and the cost of verifying that the staff has the required skills are usually not considered at all. These costs are overlooked and baked into "just how it is" or some other description of the status quo. The skills or lack thereof are part of the noise of the status quo, and part of the inertia we must overcome to become a value-driven organization.

As you read the remainder of this text, you should be thinking about how to apply the ideas in the text to your organization and answer the questions, "how do we want to do business?" and "how can we increase the value we obtain from information while decreasing how much it costs?" Again, cost may be in terms of money (e.g. profit or loss), it may be in terms of time (e.g. cycle time, time to market, etc.), or it may be in terms of resources required or consumed (e.g. facilities, staff, skills, software, etc.).

LIFECYCLE VALUE OF INFORMATION (LVOI)

I describe cases where we don't need to do something anymore, or don't need to think about it anymore, as automation. Automation can be a loaded term that scares people and conjures images of robots and automation in factories. Factory automation is a good example of automation. I spent the first ten years of my design career designing equipment for factories that automated tasks in packaging, material handling, food processing, and nuclear material processing and handling. In this book, however, I am thinking more about automation in the context of information. Creating information, managing and maintaining it, releasing it, using it, revising it, protecting it, building on it, learning from it, retiring it, and the overall lifecycle business value obtained from the information versus the cost of creating/managing/maintaining/releasing/understanding/using/revising/protecting/building on/learning from/retiring it/.... We can imagine a simple formula:

Formula 5.2 – Lifecycle Value of Information (LVOI) as a Function of Lifecycle Cost of Information

$$5.2 \qquad LVOI = LVEBI - (CoC + CoX)$$

Optimally, our goal should be to maximize LVOI (maximize the result of formula 5.2).

Where:
LVOI	Lifecycle Value of Information
LVEBI	Lifecycle Value Enabled by Information
CoC	Cost of Creating Information
CoX	Cost of "X", where X is managing, maintaining, releasing, understanding, using, revising, protecting, building on, learning from, wading through, misunderstanding, interpreting, ignoring, and retiring information, repeating the RIDICT sequence over and over throughout the information lifecycle...

The formulas above yield a monetary value in applicable currency.

If we prefer to represent the value cost function as a ratio, as a number, we can rearrange the formula as shown in Formula 5.2_{alt}

$$5.2_{alt} \quad LVOI = LVEBI / (CoC + CoX)$$

Regardless of which approach we use, our goal should be to determine a baseline LVOI for current operation and compare LVOI for the work methods we want to adopt.

Note: Maximizing the value of information versus its cost (the net value obtained from information) is a core idea of this book. These formulas treat LVOI and LVEBI as positive attributes (beneficial to our business goals) and CoC and CoX as negative attributes (detrimental to our business goals).

The traditional ideas of MBD and MBE are examples of people trying to adapt a new technology to an old paradigm. They force their current work methods and thinking upon scenarios where technology drives a completely new paradigm. This can be because of a lack of will, commitment, and imagination, or even just a heavy dose of stubbornness and resistance to change. In many ways, current implementations of MBE that focus on manual use cases miss the point. See Figure 5.2 for a historical comparison to creating 3D digital product definition datasets for manual use.

The problem in the scenario shown in Figure 5.2, if it is not obvious, is that the automobile was treated as a replacement for the buggy, not the whole system; the automobile was not recognized as a completely new form of transportation. This is analogous to most current MBD and MBE implementations. The use cases for MBD information assets are constrained by the imagination of the users, stuck in the past trying to support inefficient no-longer-necessary manual use cases. Software vendors also contribute to this problem, as they push

FIGURE 5.2   Missing the point: thinking about new work methods using old paradigms – if we transitioned from horse and buggy to automobiles like companies transition to MBE.

software that primarily supports legacy manual use cases. Developers working on standards for legacy methods such as drawing standards also contribute to this problem.

VALUE-DRIVEN MBE

If we want to succeed with MBE and obtain real business value from MBD, we need to change how we think about information, we need to change how we use information, and we need to change how we work. If we don't, we are wasting time and money, fooling ourselves thinking we're progressing, and causing a lot of grief in the workforce as people have to deal with change that doesn't add much value.

We need to change how we think about information, how we use infor-mation, and how we work. If we don't, we are wasting time, money, and resources, and decreasing goodwill as people grapple with change that doesn't add much value.

In the context of MBE, our goal is automation and maximizing the value that may be obtained and the value that is obtained from our information. Our goal with MBD is to create information and information assets that facilitate the automation of lifecycle tasks and activities. A simple example is how 3D solid modeling CAD software allows CAD users to drag and drop orthographic views on a 2D drawing. Engineers and designers no longer need to learn, pos-sess, and maintain the skills needed to manually create 2D orthographic draw-ing views. While I learned these skills in the 1970s and 1980s, today I can drag and drop views like every other mechanical 3D CAD user. The value of some of the skills that used to be needed to define mechanical products has been automated away. And we can say *goodbye* and *good riddance* to those skills, at least for production work.

A similar leap occurred with stress analysis. Finite Element Analysis (FEA) and other 3D-data-based analyses have largely replaced manual methods of the past. With the advent of 3D digital geometry models, people realized they could use the models directly in lifecycle activities. FEA is a good example because it is relatively simple to execute, and we can use CAD model geometry or a derivative of it as a representation of product geometry in the analysis. Here is a simplified MBD CAD-to-FEA use case:

1. CAD model geometry is used to represent product geometry directly in FEA software. If needed, native CAD geometry is converted to another data format required by the FEA software. A CAD design engineer is the primary author for the CAD model geometry.

2. An analytical engineer uses the CAD model geometry or its derivative directly in the analysis software. The analytical engineer may process the geometry further (e.g. the geometry is simplified, a mesh is wrapped around the geometry, the geometry may be converted to another data format, etc.).

The analytical engineer acts as a secondary author, as they create one or more derivative datasets/information assets based on/derived from the CAD model geometry representing the product. The analytical engineer creates new content, such as:

a. The simplified model and the mesh
b. The final results of the analysis
c. Records of the results
d. Recommendations and disposition of the design.

3. In some cases the results may be used directly to modify the design, but usually the results are presented in a report, which is used manually. A design engineer reads the report (using the RIDICT sequence), there is some back and forth between the design engineer and the analyst, and the design engineer manually modifies the design accordingly.

4. Documentation of the initial design, the analysis report, the iterative discussion, and the final design are created and stored. Documents of record. Documents, documents, documents....

A few good things and a few inefficient things are listed in the steps above. Using the CAD model geometry or its derivative as the basis for the analysis is a good example of MBE, as the 3D digital data is used directly in a lifecycle activity (analysis). In this case, we automated away the need to recreate the initial product geometry in the analysis software.

In a drawing-based workflow, say one using DCC 1 or DCC 2, where drawings are official information assets, the product geometry would be fully defined by explicit dimensions. The analyst would have to recreate the product geometry from scratch, as they'd only receive a 2D drawing with dimensions. Such a process not only takes a long time but may be fraught with errors.

In steps 1 and 2 above, MBD methods facilitate automation, thereby negating the need to recreate a product geometry model for analysis. Recreating geometry that already officially exists in the workflow is unnecessary, inefficient, error-prone, slow, costly, risky, and thus a bad idea. As should be obvious at this point in the book, I believe we must use existing MBD information directly wherever possible. Designating manual-use information as the official definition in design or any lifecycle-stage information asset removes the potential for automation and forces lifecycle users to recreate whatever the manual-use information represents. In the example above, steps 1 and 2 deal with product geometry. If the official definition of the product geometry is a fully-dimensioned drawing or annotated model, that means that the official definition of the product geometry is explicit dimensions, and each lifecycle user must use the explicit dimensions as the basis for manually rebuilding product geometry models for their lifecycle activity.

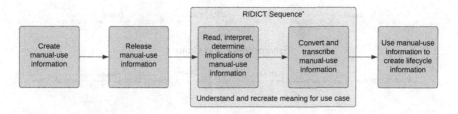

* In many activities, users must recreate information because they received manual-use information. The portion of their process that is based on manual-use information cannot be linked to the information they received as input because the information is not in a directly-usable form, or the information has not been designated, validated, and approved for direct use.

  A common example is where explicit dimensions are used as the official definition of product geometry in a product definition data set and a lifecycle activity requires a model of the product geometry. In such cases, the lifecycle user must recreate the product geometry for use in their process from scratch (e.g. as basis for G-code for CAM, as basis for a CMM program, as representation of product geometry used in analysis, etc.)

FIGURE 5.3   Vicious cycle of manual-use information.

See Figure 5.3 for a generalized view of what happens when we create and release manual-use information.

MBE FOR MAXIMUM VALUE

If we want to maximize the value of information, we need to maximize the accessibility and usefulness of the information. We need to be able to use the information in all applicable contexts, and we need the information to be as useful as possible. We need rich information, information that represents its content with sufficient depth and granularity so that the information is modeled such that lifecycle activities can use the information directly with minimal human intervention. We need information that is optimized for direct use. We need to be able to bypass the RIDICT sequence (RDICT needs to become extinct). We need information assets that are not only structured for direct use but also so additional information and feedback can be appended, attached, and linked to the information elements within the information asset. We need information that can be built upon, not just viewed and manually acted upon. We need to operate in a manner that allows us to obtain maximum value from information and to minimize the need for official manual-use information. Lastly, we need a culture and environment that support doing these things, which empower us to increase the maximum business value of information and decrease the lifecycle cost of our information.

EFFECT OF BUSINESS ENVIRONMENT AND CULTURE ON MBE

Figure 5.4 shows the cyclical nature of business culture and environment – how tradition and our expectations drive our values, which drive our metrics and our behavior, which in turn drive how we work, the tools we have available, the skills

FIGURE 5.4   Environmental and cultural cycle – why we work a certain way.

we need, our capabilities, and the information we consume and create and how we consume and create it (our inputs and outputs). Note that tradition, our expectations, and our values drive what we think we should be doing, which information we think we need, and in which form. And our culture, environment, leadership, and what we are rewarded for and what we are penalized for, drive our values at work. If we want to truly benefit from MBE, we need a culture, environment, and leadership that truly values benefiting from MBE. We don't just need to see a shiny corporate poster stating "We ♥ MBE". We need leaders that talk-the-talk, walk-the-walk, own it, support it, and get the obstacles out of the way. Or the leaders need to get out of the way.

For MBE to succeed, we need leaders that talk-the-talk, walk-the-walk, own it, support it, and get the obstacles out of the way. Or the leaders need to get out of the way.

The self-reinforcing cycle shown in Figure 5.4 can work in our favor, too. Once we have functioning optimized MBE systems in place, and once the value from them is realized, it will be extremely difficult to regress back to earlier less efficient methods. I often challenge people attending my MBE workshops by asking them to consider a future scenario where a company has operated as an optimized MBE for some time, and someone suggests that the organization should revert back to using document-based information in manual activities. This idea would be met with laughter and considered ridiculous.

There is a middle ground where the self-reinforcing cycle plays a negative role. Consider organizations that have adopted MBD and MBE principles but continue to prioritize document-based work methods, creating information and information assets for manual-use, and using that information manually. The organizations use MBD methods to create information and information assets optimized for manual use, they purchase tools and systems that support manual use cases, they designate the manual-use information as official information, they use the MBD information manually, and their environment and culture ossifies around manual use. Their status quo, their expectations, values, behavior, metrics, etc. have migrated toward model-based information and work methods, but they believe that this not-so-efficient and not-very-productive middle ground is where they should be. They believe that they've adopted a modern approach, and they come to accept it as an improvement over their prior work methods, but they are stuck working in a way in which their information has little value, drives little value to the organization, and the cost of creating, managing, maintaining, and using the information is very high. If we apply Formula 5.2 to this scenario, we can quantify that these work methods yield little value for the organization. Many companies that have adopted MBD and MBE are stuck in this middle ground.

To be clear, using model-based methods to define manual-use information and using that MBD information manually is better than using traditional documents and document-based methods. There is an improvement. However, the improvement is limited to improving visualization in manual use cases. MBD information optimized for manual use is easier for a person to understand (RIDICT), and given the right tools (hardware, software, platforms) and business environment (formal support and ownership by management), it takes less time and there should be fewer mistakes, fewer RFIs (requests for information), and less scrap and rework from these manual use cases. However –this is an important point – the benefits of moving from traditional document-based manual work methods to model-based manual work methods are strictly improved visualization. In such an environment, the official information is manual-use information and the official manual-use information must be used manually in lifecycle activities. Thus, the benefits are limited to the improvements of manually using dynamic model-based manual-use information instead of traditional static document-based information. How much benefit can we obtain by improving visualization? Short answer – I don't know. But I have an estimate. I assume we could achieve a 5–10% lifecycle benefit by migrating from traditional document-based work methods to model-based manual-use information. What are we improving if we improve visualization? I assume we are improving how quickly we understand information, the context that the information is provided in, how well and how thoroughly we understand the information, how easy it is for us to recognize that the information is incomplete, how the information compares to an earlier revision (e.g. what changed, this is a real benefit), and the interrelationships between information elements in the information asset (e.g. specifications associated to model geometry, query and highlight, etc.). I am not suggesting workflows where model-based manual-use information is designated as the official information and used as such are optimal. I included many examples of DCC 5 information assets designated as DUC 5.2 in the previous chapter. I highlight those examples as they tend to be the safe place for organizations to start – model geometry officially represents product geometry and is used directly in lifecycle processes, but visually-presented manual-use information officially represents specifications and metadata and is used manually in lifecycle processes. This middle ground feels less foreign to many organizations. It is okay to start with this approach if a plan is formally in place to migrate further toward creating and using information directly in software. Define the middle ground as a step in the right direction, but formally indicate that it is a short-term step. There will still be a lot of low-hanging fruit, a lot of value left on the table when using the middle-ground approach. The potential to achieve 10X cost-to-value improvements will occur when we move to creating and optimizing information for direct-use and using the information directly.

The value obtained by organizations that use MBD information manually is limited to improved visualization in manual use cases. Organizations realize little benefit when implementing MBE in such an inefficient manner.

Figure 5.5 shows how a document-based environment drives our expectations, what we do, and what we think is important and necessary. Similar to Figure 5.4, it highlights why cultural inertia is so difficult to overcome. Our environment, our culture, our expectations, our values, what we do, and what we are expected to do are driven by what have been doing – what we do tomorrow is driven by what we did today, and what we did today was driven by what we did the day before, which was ultimately driven by what our predecessors did 20 or more years ago. Figure 5.6 shows the same diagram as Figure 5.5 but emphasizes manual-use information instead of documents. As discussed earlier in this chapter, if we become an MBE that emphasizes creating manual-use information and using it manually, we are not much better off than when we were a document-based organization.

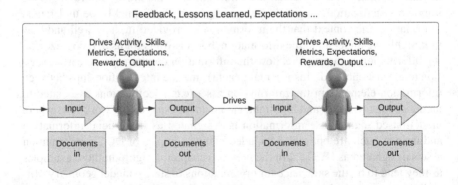

Self Perpetuating Document / Manual Activity Cycle

FIGURE 5.5   Self-perpetuating document/manual activity cycle.

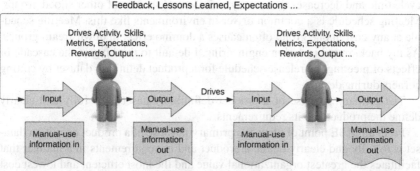

Self Perpetuating Manual-use Information / Manual Activity Cycle

FIGURE 5.6 Self-perpetuating manual-use information/manual activity cycle.

MAXIMIZING VALUE OF INFORMATION

If our goal is to use MBD and become an MBE to increase efficiency, reduce errors, increase quality, decrease cycle time and development time, decrease cost, decrease risk, decrease time to market, etc., then we should develop information that facilitates automation, which increases the value of information and decreases the amount of information we have to create, manage, and deal with to successfully achieve our objective. We should be trying to make it so the information we create satisfies its needs, has maximum value, and places the least burden on lifecycle activities. To say it differently, we need to create information that fully serves its purpose and optimally allows lifecycle users to use that information as efficiently as possible. Core components of this idea are:

1. Focus on creating information that can be used directly in software and computation.
2. Create an environment where all lifecycle activities have the tools and support needed to use information directly in software and computation.
3. Champion goals for creating information and information assets that allow manual activities to be eliminated or decreased as much as possible.
4. Reward organizations and individuals for creating information and information assets that maximize direct use and minimize the burden and cost they impose on lifecycle activities.
5. Reward organizations and individuals for using information and information assets directly and efficiently.

Item 4 is interesting, as it addresses a core problem that plagues industry. Organizations tend to be siloed, their management is siloed, their goals and metrics are siloed, and the people managing and working in these siloes are rewarded for achieving siloed goals. They are rewarded for achieving things that apparently benefit the siloed group, but achieving the metric increases the burden, cost, and cycle time and decreases the quality, morale, and success of other siloed groups. Meeting schedule is a common driver in environments like this. Meeting schedule at any cost in one group often causes a domino effect in downstream groups. As my background is design engineering, I default to thinking of the cascade of effects of meeting the release schedule for a product definition dataset by cutting corners during design.

The primary purpose of a product definition dataset is to fully and clearly define the product and its requirements.

From an MBE point of view, the primary purpose of a product definition dataset is to fully and clearly define a product and its requirements in a manner that facilitates the greatest organizational value and the most efficient and lowest cost use of that information in all its uses. Thus, in a model-based environment, an abbreviated list of the primary metrics for a product definition dataset should be:

1. How completely the product definition dataset defines the product and its requirements.
2. The usefulness of that information for the organization and relevant life-cycle activities.
3. The cognitive load the information places on the workforce (how easy it is to understand the information, skill levels required to understand the information).
4. The burden the information places on the workforce (how easy it is to use the information, skill levels required to use the information).
5. How well the information facilitates direct-use in other activities.
6. How rich the information is.
7. How well the information enables direct-use and automation and reduction of manual activities driven by the information.
8. How much benefit the information drive during its lifecycle and the product lifecycle.
9. How much savings in time, overhead, burden, cognitive load, skill requirements, staffing, effort, time to market, and cost of all sorts does the information facilitate?
      In summary:
10. How much bang do we get from the bucks it costs to create, understand, own, use, maintain, and protect information during its lifecycle across the extended enterprise?
      Note that we also want to minimize:
11. How much content the information includes that is superfluous to achieving business objectives.

From an MBE point of view, the primary purpose of a product definition dataset is to fully and clearly define a product and its requirements in a manner that facilitates the greatest organizational value and the most efficient and lowest cost use of that information in all its uses.

These criteria can be quantified using Formula 5.2, which calculates the lifetime value of information (LVOI) as a function of lifecycle cost of information. If the items in the list above are treated as metrics, and if we can simply apply a score to how well we achieve the goals they represent, we want to achieve the highest aggregate score possible for items 1–9. As item 9 states, we want to get most value from our information while minimizing the cost the information incurs across the organization. Item 10 is a measure of the content included in the information that is superfluous to achieve the information's business objectives. Thus, we want to minimize item 10. This is an important point of clarification: if our goal is to be efficient and achieve maximum business value from information across the organization and throughout the information's lifecycle, and if we have or can implement the capability to use the information directly, then any content in that information that is intended to officially support manual use is superfluous. To say this differently, if we create information intended to be used manually, then the information includes content to support manual use that incurs extra cost and burden – it costs a lot in terms of money, time, and skills to create, manage, use, and own manual-use information. A lot of the cost and burden are hidden, as they are buried in how we do business today, our expectations, the status quo, etc., and we don't even recognize the cost and burden exists.

As an example, consider the cost of acquiring, training, maintaining, managing, and retaining a workforce that understands how to create, understand, and use engineering drawings (and by extension annotated models). These skills are becoming less and less common in new hires and recent college graduates (RCGs). The training regimen to become adequately proficient in these skills is often longer than a year (it is often much longer). I assume most people don't want anyone to know that they are uncomfortable doing something that their management thinks they should be able to do or taught them to do in on-the-job training. So not only are there direct costs to giving new hires the skills to work with engineering drawings and annotated models, but there is also a prolonged period of reduced productivity from the employees while they are in the learning period and risks from them not admitting their skills are deficient. We could add another item to the list of product definition dataset criteria that measures how long it takes for a new hire to be able to create or use information to perform a task in their job.

If we have the potential to eliminate much of the burden, cost, and skills required to work in a document-based environment where information is primarily intended for manual use, we should embrace that opportunity.

THE PURPOSE OF MBE

The central premise of this book, and the premise and promise of MBE, is to achieve the greatest business value from our information while minimizing the lifecycle cost of the information. Thus, the purpose of MBE is to maximize the LVOI, specifically the lifecycle value of *our* information and intellectual property.

MBE requires taking a Lean approach to information. We need to recognize and understand that there is a significant cost to the information we create and use in our organizations. We need to recognize and accept that there are many extra costs and burdens associated with how we do business today, with what we think is *the right way* to do business. We need to recognize and understand that the costs and burdens of how we currently create, use, manage, and maintain information are not sustainable or scalable, and that how we currently create, use, manage, and maintain information are not green. We need to recognize and understand that the costs and burdens of how we currently create, use, manage, and maintain information are not compatible with the skills, values, expectations, and priorities of new hires, the incoming workforce, and what they learn in school.

Properly implemented, MBE aligns how we work, how we create, use, manage, and maintain information, which information we prioritize, and which information we think we need with modern society. Nearly every new hire has a smart phone in their pocket. Nearly every new hire is proficient with their smart phone and expects our work systems and methods to be as streamlined, easy to use, and straightforward as the apps on their phone. By contrast, the way our organizations operate, our systems, our approach to information, what we think is important, our organizational culture and values are not streamlined, easy to use, or straightforward. Our environment is cluttered, disjointed, and burdened by trivia required by legacy power structures, be they political, tribal, or personal. Our environments are often chaotic, and what we do and what we are rewarded for often conflict with our corporate mission statements.

- We want to delight our customers and provide joy!
- Quality is our number one goal.
- Safety is our number one goal.

In capitalist economies, the things listed above, while important (not so sure about "joy"), our priority is bringing value to our stakeholders and making a profit. This idea has expanded recently to broaden the definition of stakeholders, but the goal of a business is to maximize profit and return on investment, while not harming anyone, breaking any rules, laws, etc. In the jargon of this text, our goal is to maximize value of what we own, what we do, and how we do it, for the organization. In the context of MBE, we apply these principles and guardrails to our information, the information we create, use, manage, and maintain. We want the most bang for the buck from our information.

One of the challenges of the mission statement synopses above is that there is a ghost lurking in the shadows in every organization, a priority that wins almost every time, and it is not listed in the synopses. The priority is meeting or beating schedule. I am not saying that meeting or beating schedule is unimportant or a bad thing. It is obviously important. However, when schedule is prioritized over everything else, as it often is, organizations often throw everything else out the window and under the bus to meet schedule when it appears they will be late.

For public companies, meeting quarterly profits or quarterly expectations is another important factor that is often detrimental in practice.

There is a multi-stage problem driven by these priorities and the environment and culture they engender. I'll describe it as follows:

1. Management formally establishes values, establishes priorities, guardrails, targets, goals, etc.
2. The workforce learns the values, priorities, guardrails, targets, goals, etc.
3. Work proceeds to achieve those goals.
4. Problems are encountered as work proceeds.
5. Workers circumvent approved work methods and processes to achieve targets.
6. Management explicitly or implicitly rewards the circumvention.
7. The workforce learns that what really matters is a little different than what was formally laid out in step 1.
8. The way we work and train our workforce adjusts to the reality of what is really deemed as important (what we are rewarded for) versus what it says on paper.
9. We develop a culture of informality, where getting the job done happens regardless of what it says on paper.

Remember the stories I told in Chapter 2 about out-of-control informal processes?

1. I was working as a GD&T mentor on a large vehicle project, and I learned that another group was nearly done designing the inspection fixture for the chassis assembly even though we weren't done developing the GD&T.
2. I was teaching a tolerance analysis course to a group of manufacturing engineers at a large aerospace company, I asked them what they did at their company, and they said, "We fix design engineering's mistakes", and they didn't think their answer was a sign of a problem.

These are examples of informal out-of-control processes. Informal processes are used, expected, and rewarded in many work environments. Doing what it takes to get the job done, even if that requires circumventing the rules, becomes an operational priority and reality. In these environments, adopting something like MBE is challenging. We find these problems during the discovery and development phases of MBE implementation projects. It is fairly straightforward to replace manual document-based workflows and processes with MBE processes. We map how work is currently being done and restructure it using MBD and MBE methods. We replace manual-use information and processes that use manual-use information with direct-use information and automated processes where possible. It becomes difficult to do this when the way companies are supposed to work (the way their procedures and formal processes say they work) differs from how they actually work.

There are several methods for assessing an organization's MBE readiness, maturity, or capability. In my company, I use my own version, which focuses on things I deem as most important. In addition to technical characteristics such as hardware, software, platforms, data formats, etc., I apply significant importance and weight to how companies intend to use information, how they actually use information, their culture, and specific to this conversational thread, how closely how they actually work and their values match their formally-defined rules, procedures, and values. Specifically, how much are informal work methods used, how important are they in their environment and culture, and how much do they recognize that they operate that way? This requires more than one answer, as we must ask different levels of people in the company these questions: upper management, middle management, workers, repeat for supply chain and other stakeholders...

Getting back to the lead-in to several paragraphs ago, that MBE requires a Lean approach, that we must apply Lean principles to our information and the information lifecycle, we should minimize the amount of information needed to accomplish our business goals. We should recognize that every piece of information (information element) has a cost associated with it, and that every information element imposes a burden to the organization. We should ensure that each information element provides value for each relevant use case and activity, we should ensure that the value provided exceeds the burden and cost the information imposes on the use case and activity, and we should work to maximize the lifetime value of each information element (its LVOI). The way we do this is to create rich, complete, high-quality secure information that is built and optimized for direct use and use it directly as much as possible. Direct-use information must be used as official information wherever possible, and transitioning away from official document-based workflows using manual-use information must be done wherever possible. The ability to do these things is limited by many factors, technical, cultural, and legal or regulatory in some cases. However, if we don't press software and platform vendors, standards development organizations, government agencies, industrial leadership, corporate management, the workforce, and educators to work toward these goals today, it will take a long time to get there.

To realize value from this Lean MBE approach to information means that we must first align how our organization operates and our formally approved rules and principles. This is not unique to MBE – it applies to adoption of any quality improvement or similar new business method. We can, of course, choose to skip this painful step and dive right in to adopting, developing, and implementing MBE methods, but we will not realize the expected benefits. This is where a lot of companies are today. They skipped the hard part and focused on the technical part. They started using MBD and adopted some form of MBE. They may have done so through ignorance or with full awareness that they skipped a step. They may have had no choice. If the boss says to adopt MBE, you respond by explaining the reality of an informal environment and culture, and you are told to do what you can within the context of your silo anyway, you end up with a partial implementation and partial benefits. If we don't address how formally people work today, and thus how closely they adhere to the organization's procedures and rules, there is little likelihood that they'll adhere to new procedures and rules.

Upper-level, preferably C-level support is needed for digital transformation to succeed. People need to understand that the initiative isn't merely the buzzword of the month to be treated with lip service, but it is actually something that is important to the organization. I recommend the book "Why Digital Transformations Fail". More information can be found in the bibliography. Note that I have read many books, reports, papers, and articles and over the last ten years to help broaden my understanding of digital transformation, information science, the Fourth Industrial Revolution, the history of manufacturing, process mapping, and related topics.

DOCUMENTS, ANNOTATION, PMI, AND VISUAL INFORMATION (DAPVI)

Documents, annotation, PMI, and visual information (DAPVI) play a significant role in how companies have worked in the past. DAPVI continues to play a significant role in how we work today and how we create, record (document), use, maintain, and store our intellectual property (IP). We use DAPVI in manual processes, and we think we must create, use, maintain, record, and store our IP in documents because we have always worked that way. What we did last year and how we did it drives last month, drives last week, drives yesterday, drives today, which drives tomorrow. As stated above, creating, using, and owning manual-use information comes with significant cost and burden. Every piece of manual-use information has a cost. Every piece of manual-use information clutters the information landscape and our ability to differentiate between important or critical information and trivial or less important information. Confronted with three hundred information elements on an engineering drawing or annotated model, we often struggle to determine which of those information elements matter and which are trivial. There is a lot of noise in manual-use information. (You may feel that way about what I've written in this book!)

Historically, in product definition datasets, we have used DAPVI as the official definition of product definition data. Today, the discrete manufacturing industry, the industry this book targets, still mostly operates that way. We define the nominal (perfect) product using explicitly-dimensioned views and we define specifications with presented information, both of which are intended and optimized to be used manually. We do our best to make sure that what we create is easy to read and follows the often peculiar and nuanced rules for drafting and written language.

We accept document-based workflows, creating, using, and prioritizing manual-use information because it is what we are used to doing, regardless of whether it is easy, if the rules are straightforward and consistent, and if we are able to achieve satisfactory quality, productivity, and results. We do what we do because we think it's what we should do. And without a compelling alternative, it is what we should do. Today, however, we have a compelling alternative as described in this book. We can choose to deprioritize document-based workflows and creating, using and prioritizing manual-use information. We can opt for value-focused MBE methods. We can choose to become an optimized MBE that focuses on and prioritizes creating and using direct-use information directly to achieve maximum LVOI. We can change. If you have read this far, you know there is an alternative to approaches that prioritize manual workflows and manual-use information. All it takes is courage and a champion to realize the potential of these ideas. Whether we decide to fully embrace direct-use information and workflows throughout the organization or do so on a case-by-case basis, we can reduce our reliance on DAPVI and its associated cost and burden. Once we realize and quantify the value of doing this, our efforts and ROI will snowball.

The Chain of DAPVI

Our legacy of creating and using DAPVI has created a chain or series of linked activities, goals, metrics, expectations, behavior, etc. Refer again to Figures 5.4–5.6.

Using DAPVI means people must create DAPVI. Creating DAPVI requires skills, training, and time, which come with an opportunity cost. Time spent achieving, maintaining, and verifying DAPVI-related skills could be better spent on something more directly valuable to the organization. (Note that I have made a living and authored many books about DAPVI, yet I suggest the burden of using traditional DAPVI may not be worth it and there is a better way.) Because we create DAPVI, people must use it. Using DAPVI puts a tremendous burden on the workforce, as incoming workers lack DAPVI-related skills. There are many steps required to create DAPVI with adequate quality for use in an enterprise and supply chain. The rules and best practices for DAPVI are complex, nuanced, and sometimes contradictory (see content about checkers earlier in the text). DAPVI rules and best practices have evolved over many years, providing a broad range of nuanced capabilities. Proficiency in these complex nuanced methods requires rote memorization and extreme specialization to master if they can be mastered at all. This drives a significant expense for an organization to achieve and manage even a very low capability.

Many organizations possess low skill levels and proficiency for creating and using DAPVI and they don't have a clear understanding of the ensuing consequences. They understand the consequences, the problems created by the deficiencies, but they don't understand the true root cause why the problems occurred.

As stated earlier, degradation of the workforce through the demise of designers and drafters, who specialized in DAPVI, has led to a state where there is inadequate skill, priority, and time alloted to produce high-quality DAPVI. Few people in industry today have mastered or want to master DAPVI-related skills.

Given the low level of mastery of DAPVI, people create a lot of extraneous or unnecessary DAPVI – they create a lot of noise. There is a lot of trivial information on drawings and annotated models and in documents. For DAPVI, visual format of information is critical. How the information looks and how similar it looks to similar information is important, as it must correspond to people's expectations and experience. Not following this practice affects their ability to understand the information. Visual format and its rules and best practices are challenging. This is where the mastery (or lack of mastery) comes into play. So much effort, focus, emphasis, and experience are needed to become proficient with DAPVI, that we get a lot of DAPVI that is low-value noise. It is extraneous and only gets in our way as we try to ferret out the important bits of information. This extraneous DAPVI has a tremendous cost that exponentially increases across the enterprise and through the product lifecycle.

Using the information lifecycle idea, every information element incurs a cost in every step and stage in the information's lifecycle. There is a cost when information is created and each time it is processed, translated, read/viewed, interpreted, converted, transcribed, stored, released, revised, etc. This is true for each and every information element, every word, every line in a diagram, every Boolean solid in a model, every character in a formula, etc. Of course, the information element also has potential value in all of these activities and use cases. Using Formula 5.2 and careful analysis we can assign values and costs to these, and determine if we come out ahead, if we obtain greater value than cost from our information elements. We must consider the maintenance cost of our information. We must consider the context of our information. We must consider how efficiently the information may be used and the burden using it places on our organization and our workforce. We must keep the ever-increasing misalignment between DAPVI-related skills and the smartphone-wielding digital natives in our incoming young workforce that expect things to be as easy as swiping app icons on their phones. As you know, most of what you do at work is not that easy. It should be.

VALUE OF MBE

UNDERSTANDING COST AND BURDEN OF OPERATIONS AND ACTIVITIES

To establish a baseline, we need to understand the cost and burden imposed by how we currently do business. A simple way to do this is by mapping current operations, activities, tasks, etc., and breaking down the costs and burdens they impose on the organization. It is important to recognize that there are hidden costs and burdens for all the things we do at work and all the information we create and use. We must be vigilant and avoid complacency of thinking of how we work as "just the way it is" and glossing over the overall cost and burden imposed.

For example, we can think about an information lifecycle task such as obtaining information to determine what needs to be inspected and the limits of acceptability for each characteristic to be evaluated. Continuing the example, let's assume that today an inspector obtains that information from an engineering drawing and converts the information into criteria for inspection (e.g. measurand, boundary conditions, importance, criticality, etc.) using the RIDICT process. As this is a manual document-based process, the inspector needs to have certain knowledge, skills, and experience to perform the task. Generally speaking, the inspector must be able to read and understand engineering drawings. With regard to specifications included on the drawing, the inspector must be able to recognize which information represents specifications, understand the meaning of each specification, understand the target set for each specification (what the specification applies to), understand the requirements imposed on the product by each specification, and understand the criticality of the specification and the imposed requirements. In addition, the inspector must understand what should be evaluated or measured to determine conformance to the specification, how that would be accomplished (what tools, staff, machinery, software, hardware, procedures or operations, etc. would be used), how the activity best fits into their workflow, how to document the requirements, how to document the inspection process, how to document the workflow, how to report the results, what type of information needs to be included in the inspection report (generally and for each characteristic evaluated), who should receive the results, what information the recipient needs to do their job, etc. While this is a long list, it is abbreviated (and probably insufficient, as I am not an inspector). While there are many steps, and it is possible that the responsibilities may be distributed to multiple people, the abbreviated list of steps starts with the inspector being able to read a drawing.

What is the cost of having a staff member able to read engineering drawings? How much does that cost us? Do we just take it for granted? What is the replacement cost? (How much would it cost to hire someone with equivalent skills needed to satisfactorily perform the activity? This includes how easy or difficult it would be to hire someone with comparable skills, how long it would take, the loss in productivity, cost of HR and management, overtime for other staff to temporarily fill the gap, etc. incurred during the hiring process.) How sure can we be that the staff member or their replacement has the necessary skills? What does it cost to determine this? Is there a cost to maintain those skills? Do staff members like to read drawings? Does having to deal with drawings affect their morale and retention rate? Do they do a good job, are they thorough, do they hurry through the task, how often do they miss things, how often do they misinterpret drawings, will they tell us if this occurs…? Can we tell if this occurs…? What do these things cost the organization in terms of time and money? How do these things affect our quality, schedule, morale, bottom line, training regimen, and the number of people needed to perform, manage, and assess these activities? How well do the skills of the incoming workforce match our staffing requirements? How much and how long do we have to train new hires with little or no relevant experience to be proficient in our environment for the tasks they were hired to perform? What are the costs (opportunity costs) in terms of time, money, quality, throughput, and productivity incurred during the training period? Creating and reading engineering drawings are archaic processes that require archaic skills. There are significant costs to doing things using archaic methods where technology presents a more streamlined readily-available alternative.

What is the cost of having a staff member able to read engineering drawings? How much does that cost the organization?

It may be difficult to think about this scenario in such detail, as it represents how we work today and generally we expect new hires to possess the skills listed. Much of what I describe above are what I call *hidden costs* and *hidden burdens*. We take them for granted, as they are part of the status quo. A significant problem with many MBD and MBE implementations is that they are primarily performed as technical activities by technical professionals. MBD and MBE are treated as technical topics and are implemented to achieve technical results; people develop technical solutions to technical problems. This is insufficient.

We must take a broad, business-oriented, organization-wide and lifecycle-spanning approach to MBD and MBE, what their value propositions are, and how best to achieve value from their implementation. It cannot just be replacing drawings with models, for as I have seen in some implementations, the OEM replaces drawings with annotated models but ends up using the annotated models in almost the same way they used drawings. Most activities are performed using visual information even though an MBD product definition dataset was created. Formulas 5.3–5.6 below provide methods to quantify the costs outlined in Formula 5.2 with more detailed information for each step in workflows of interest. Note that the cost and value of our operations should be modeled; we should create a cost model and a value model to justify and quantify the value of our digital transformations. Likewise, we should treat versions of these models as twins, allowing us to understand, manage, and optimize our transformation over time (Footnote 5.1).

FOOTNOTE

5.1 In the book "Why Digital Transformations Fail", Tony Saldanha explains that digital transformation is not a discrete event that occurs once. It isn't just one and done. Our transformations and the benefits we realize from them are dynamic and change over time. We need to apply continuous improvement principles to maintain and increase the value obtained from transformation.

Formula 5.3 – Lifecycle Cost of Information (CoX) (expanded)

5.3 $\text{CoX} = (k_1 q_1 r_1 \text{CoU}_{Op1} + k_2 q_2 r_2 \text{CoU}_{Op2} + k_3 q_3 r_3 \text{CoU}_{Op3} + k_4 q_4 r_4 \text{CoU}_{Op4}$

$+ k_5 q_5 r_5 \text{CoU}_{Op5} + ...)$

CoX: Cost of "X", where X is managing, maintaining, releasing, understanding, using, revising, protecting, building on, learning from, wading through, misunderstanding, interpreting, ignoring, retiring, etc. information, repeating the RIDICT sequence over and over throughout the information lifecycle...

CoU_{Opn}:   Cost of Use for Operation n (lifecycle cost incurred from how information is used in operation n)

k_n Coefficient: Importance/subsequent impact of operation (weighting)

q_n Coefficient: Occurrence: the number of times an operation occurs in the lifecycle (quantity). Occurrence represents the effect of parallel occurrences of a process (e.g. a process that includes steps or use cases occurring in multiple work cells or places), simultaneously or asynchronously (e.g. multiple inspectors inspecting instances of a part, assembly, or system at the same time).

r_n Coefficient: Recurrence: the number of times operation recurs in the lifecycle (cyclic activities). Recurrence represents the effect of serial recurrences of process, where an activity occurs more than once at different stages of the product and information lifecycles, generally as part of a change/revision process (e.g. changing, checking, and reviewing a drawing or annotated model for each revision; incorporating changes into a process or other information asset from the revised drawing or annotated model for each revision, recurring steps in processes that occur more than once in a process, such as many semiconductor manufacturing and inspection processes (e.g. photolithography, deposition, plasma etch, planar surfacing, acid etch, heat treatment, microscopy, etc.)).

Notes about formulas:

1. I don't provide guidance or examples of the values that could be used for any of the variables and coefficients in this book. There are many ways to evaluate processes and how information is used, its effectiveness and cost, the burden the information places on the users, and its overall value in its lifecycle (the information lifecycle).

2. For simplicity, q_n and r_n may be combined if desired. However, I recommend treating these as separate variables as each represents a different aspect of the lifecycle and provides a clearer understanding of the effect of occurrence and recurrence of activities. When mapping the information lifecycle, it is likely that occurrence will vary over time for recurring processes. Using the example from the occurrence definition above, multiple inspectors may be used for the initial first article/PPAP inspection, but in later recurrences, more or less people may perform the activity, thereby reducing the quantity value for the recurrence.

3. A simple way to model the effect of quantify (q_n) and recurrence (r_n) is to use whole numbers representing the number of times an activity and the use case occurs and recurs. In cases that only happen once, value of these coefficients should be set to 1.

Formula 5.4 – Lifecycle Value of Information (LVOI) as a Function of Lifecycle Cost of Information (expanded version of Formula 5.2)

$$5.4 \quad LVOI = LVEBI - (CoC + (k_1 q_1 r_1 CoU_{Op1} + k_2 q_2 r_2 CoU_{Op2}$$

$$+ k_3 q_3 r_3 CoU_{Op3} + k_4 q_4 r_4 CoU_{Op4} + k_5 q_5 r_5 CoU_{Op5} + \ldots))$$

Formula 5.2 is repeated here with CoX replaced by the series from Formula 5.3 representing the cost of using information for each applicable operation in the lifecycle.

Formula 5.5 – Change in Lifecycle Value of Information (LVOI)

5.5 $\Delta\,LVOI = Cur\,LVOI - New\,LVOI$

$$= Cur\left(LVEBI - (CoC + CoX)\right) - New\left(LVEBI - (CoC + CoX)\right)$$

Where:

$\Delta\,LVOI$ Change in Lifecycle Value of Information (change in lifecycle value of information by replacing current processes with new processes)

Cur   Current (for currently-used processes)

New   New (for new processes)

Formula 5.5 quantifies the change in lifetime value of information and how it is used between a current operating process and a new operating process.

Formula 5.6 – Lifecycle Value Enabled by Information (LVEBI): Step-by-Step

5.6 $LVEBI = (k_1q_1r_1VoU_{Op1} + k_2q_2r_2VoU_{Op2} + k_3q_3r_3VoU_{Op3}$

$$+ k_4q_4r_4VoU_{Op4} + k_5q_5r_5VoU_{Op5} + \ldots)$$

Formula 5.6 quantifies the lifecycle value that information enables.

Where:

LVEBI   Lifecycle Value Enabled by Information

VoU_{Opn} Value of Use for Operation n (lifecycle value obtained from how information is used in operation n)

Note: CoU_{Opn} and VoU_{Opn} provide different ways to quantify and understand the effects of how information is used in an operation on the information lifecycle.

Formula 5.7 – Change in Lifecycle Value Enabled by Information ΔLVEBI

5.7 $\Delta\,LVEBI = Cur\,LVEBI - New\,LVEBI$

Formula 5.7 quantifies the change in lifetime value of information and how it is used enabled by migrating from a current operating process to a new operating process. Note this formula may be used to calculate theoretical (possible) value or actual value enabled.

Where:

$\Delta\,LVEBI$ Change in Lifecycle Value Enabled by Information (The change in lifecycle value enabled by information from replacing current processes with new processes)

Cur LVEBI   Lifecycle Value Enabled by Information in Current processes

New LVEBI   Lifecycle Value Enabled by Information in New processes

Note that Formulas 5.2–5.7 treat LVOI and LVEBI as positive attributes (beneficial to our business goals) and CoC and CoX as negative attributes (detrimental to our business goals).

Formula 5.8 – Lifecycle Cost Driven by Information (LCDBI)

5.8 $\quad$ LCDBI $= (k_1q_1r_1CoC + k_1q_1r_1CDBRI_{Op1} + k_2q_2r_2CDBRI_{Op2}$

$\qquad\qquad + k_3q_3r_3CDBRI_{Op3} + k_4q_4r_4CDBRI_{Op4} + k_5q_5r_5CDBRI_{Op5} + ...)$

Where:
LCDBI $\quad$ Lifecycle Cost Driven by Information (The cost of creating, using, maintaining, and managing information over the information lifecycle)
CDBRI $\quad$ Cost Driven by Received Information (The cost driven by type, structure, and format of information, the necessity of information, how the information was created, adequacy of the tools and systems used/available to create the information, adequacy of the tools and systems used/available to use the information, the intended use of information, how well the information is optimized for the intended use case, and the value obtained or lost from the intended use case (savings versus opportunity cost))

Note that the coefficients for quantity (q_n) and recurrence (r_n) applied to CoC in Formula 5.8 represent situations where information or an information asset is recreated, remodeled, etc. In cases that only happen once, value of these coefficients should be set to 1.

Formula 5.8 treats the creation of information, which includes how it was created, and more importantly, the method of use it was intended and optimized for, as the primary driver of lifecycle cost. This approach allows us to understand and isolate the lifecycle cost and burden resulting from creating information for a certain use. For example, how creating information for manual-use drives cost, inefficiency, and burden for all lifecycle activities that use the information.

Formula 5.9 – Lifecycle Cost Driven by How Information is Used (LCDBHIU)

5.9 $\quad$ LCDBHIU $= (k_1q_1r_1CoUI_{Op1} + k_2q_2r_2CoUI_{Op2} + k_3q_3r_3CoUI_{Op3}$

$\qquad\qquad + k_4q_4r_4CoUI_{Op4} + k_5q_5r_5CoUI_{Op5} + ...)$

Where:
LCDBHIU $\quad$ Lifecycle Cost Driven by How Information is Used (The cost incurred by using information over the information lifecycle)
CoUI$_{Opn}$ $\quad$ Cost of Using Information in Operation n (The cost incurred from using received information in lifecycle operations and activities. Cost is primarily driven by how information is used, how well the information suits the use case (suitability of information), information quality, and the efficiency and productivity of the use

case. The cost is also driven by the intended use of information (direct use or manual use), how well the information is optimized for the intended use case, and the value obtained or lost from the intended use case (savings versus opportunity cost). Cost is also affected by the type, structure, and format of information, whether information is needed, whether information obfuscates important information, and the adequacy of tools and systems used/available to use the information.

Note that Formula 5.8 focuses on information creation (which information was created, its intended use case, its content, structure, format) as the primary driver of lifecycle cost, and Formula 5.9 focuses on how information is used as the primary driver of lifecycle cost. I make these distinctions because it is common for information to be created to satisfy one use case or method and for the information to be used differently than intended or expected. Also, it is common that information is created with a mishmash of goals, some intended for manual use, and some intended for direct use, and none of it satisfying either use case very well.

Formulas 5.2–5.9 present different ways to predict and evaluate the value and costs of our information (our intellectual property). The formulas allow us to understand the tradeoffs we make or drive when we create information for certain use cases and methods. These formulas are just a starting point, as there are many other ways to think about and model our information-related costs and value chain.

Formulas 5.10–5.12 are used to calculate the savings possible or obtained by transforming from an initial state to a new state.

Formula 5.10 – Lifecycle Savings from Information (LSFI)

5.10 $LSFI = (k_1 q_1 r_1 SICI + k_1 q_1 r_1 SIUI_{Op1} + k_2 q_2 r_2 SIUI_{Op2} + k_3 q_3 r_3 SIUI_{Op3}$
$$+ k_4 q_4 r_4 SIUI_{Op4} + k_5 q_5 r_5 SIUI_{Op5} + \ldots)$$

Where:

LSFI   Lifecycle Savings from Information (The lifecycle savings from creating and using information enable over the product and information lifecycles. The content, structure, format, and intended use of the information (the information itself) and how the information is used in lifecycle activities affect the results.)

SICI   Savings Incurred Creating Information (The savings incurred from transforming the information creation process, content, format, etc. from an initial state to a transformed state)

$SIUI_{Opn}$ Savings Incurred by Using Information in each Operation (The savings incurred by using the information in an operation in the product and information lifecycles).

The savings in the cost of creating information is calculated in Formula 5.11, and the savings from using the information for each operation is calculated by Formula 5.12.

Formula 5.11 – Savings Incurred Creating Information (SICI)

5.11 $SICI = \left(CoC_{BT} - CoC_{AT}\right)$

Where:

SICI   Savings Incurred Creating Information (The savings incurred from transforming the information creation process, content, format, etc. from an initial state to a transformed state)

CoC_{BT} Cost of Creating Information Before Transformation (The cost of creating information in an initial state)

CoC_{AT} Cost of Creating Information After Transformation (The cost of creating information using in a transformed state)

Formula 5.12 – Savings Incurred Using Information (SIUI)

5.12 $SIUI_{Opn} = \left(CoU_{BT} - CoU_{AT}\right)$

$SIUI_{Opn}$ Savings Incurred Using Information in Operation n (The savings incurred from transforming which information is used and how it is used in lifecycle activities (from an initial state to a transformed state). Note that lifecycle operations also include activities performed by design engineering staff, such as revising product definition datasets and TDPs, responding to RFIs, attending meetings, and other information-related activities.)

CoU_{BT} Cost of Using Information Before Transformation (The cost of using information in an initial state)

CoU_{AT} Cost of Using Information After Transformation (The cost of using information in a transformed process)

To repeat an earlier idea, cost in these formulas may be in terms of currency, or it may be in terms of time, burden, skill requirements, or other metric, whatever unit is being studied and used for measuring, evaluating, or justifying current methods or a transformation.

Note that coefficients and variables defined in previous formulas are not always redefined in subsequent formulas (e.g. k_n, q_n, r_n, etc.).

Figure 5.7 shows hypothetical savings from switching how an information asset is created and used from a document-based to a mixed workflow, such as from DCC 5 + DUC 5.1 to DCC 5 + DUC 5.2. Examples of these product definition datasets are shown in Chapter 4, Table 4.3, and Figures 4.33 and 4.34. Note that the values shown in Figure 5.7 are hypothetical and do not represent the product definition datasets shown in Figures 4.33 and 4.34. I created this spreadsheet and figure in 2018, whereas I created the figures more recently. While the values shown are hypothetical, I felt that the potential savings shown in the diagram were conservative. Generally speaking, the larger costs in the lower third of the diagram are for activities that are repetitive ($r_n > 1$), activities that recur during the product and information lifecycles.

The hypothetical savings in the diagram represent a small change. The initial state uses a document-based manual-use workflow (documents as original information assets used manually). The transformed state uses a mixed-use workflow

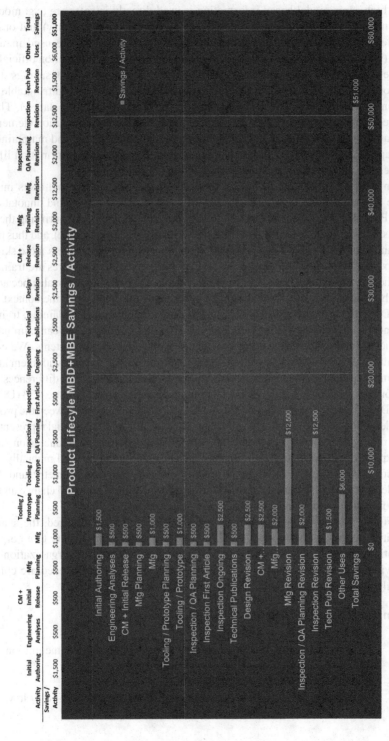

FIGURE 5.7 MBE lifecycle savings estimates.

in which some model-based information is used directly in software (e.g. model geometry is used directly and officially represents product geometry) and other model-based information is used manually (e.g. specifications are used manually (viewed, read, selected, queried, etc.) and the presented information officially represents the specifications). Thus, in this scenario, there are a lot of savings still left on the table. Note that much of the recurring savings is directly attributable to having less explicitly-defined information in the product definition dataset. This corresponds to the idea that every piece of annotation, every annotation element, has a lifecycle cost. Superfluous annotation or annotation created and maintained to support low-value activities adds more cost than value to the information life-cycle, and it drives more cost than value to the organization.

In the case of product definition annotation and PMI, authors and users must have specialized and highly developed skills to create and understand annotation and PMI. There is a tremendous cost to obtaining, maintaining, and verifying these skills. The skills are scarce in new hires (the skills are a scarce resource), and thus the organization must either pay a premium to obtain new hires that possess these skills or spend a lot of time and money and lose productivity while new hires are trained on the job. I call these "low-value high cost" skills. I say they are low value because much of the skill relates to dealing with superfluous information. In the context of MBD, MBE, and product definition data, creating, maintaining, explaining, trying to understand, and using explicit dimensions is a good example. We can eliminate explicit dimensions today with commercially available tools and systems. We can eliminate the cost associated with the skills and time needed to work with them and the mistakes and misinterpretations so common in manual use. The differences in annotation are depicted in the product definition datasets shown in Figure 4.33 (DCC 5 + DUC 5.1) and Figure 4.34 (DCC 5 + DUC 5.2). The difference between the product definition datasets is the level of explicit dimensioning and the official representation of product geometry. In Figure 4.33, the official definition of product geometry is explicit dimensions, and those dimensions are supposed to be used manually. In Figure 4.34, the official definition of product geometry is model geometry, and the model geometry is supposed to be used directly in software. This small change alone can drive tremendous savings in time, cost, skill requirements, etc.

Figure 5.8 shows the relationship between the skill level required, the availability or commonality/scarcity of skills, and the value of an activity to an organization. Our goal should be to achieve the greatest value for the organization in activities where a low level of skills is required, and the skills needed are commonly available.

Scenario 1: Worst Case (High skill, low value)

- High level of skills required.
- Skills are not commonly possessed by new hires. Skills are uncommon and must be taught using on-the-job training (OJT).
- Skills have low value to the organization.
- Actual value achieved by the organization from these skills is very low. (Cost and burden far outweigh the value realized.)

Scenario 2: Middle Ground (moderate skill, moderate value)

- Moderate level of skills required.
- Skills are possessed unevenly by new hires. Complete skill set required is uncommon and gaps must be filled by training new hires using OJT.
- Skills have moderate value to the organization.
- Actual value achieved by the organization from these skills is low, as the skills are not really that important but still require OJT and time to apply the skills, use the information created, manage and maintain the information created, etc. (Cost and burden outweigh the value realized.)

Scenario 3: Almost Best Case (low skill, high value)

- Moderate level of skills required.
- Skills are commonly possessed by new hires. Little to no OJT is required.
- Skills have high value to the organization.
- Actual value achieved by the organization from these skills is high, as the skills are important and require little or no OJT and time to apply the skills, understand, use the information created, manage and maintain the information created, etc. (Value realized far outweighs the cost and burden.)

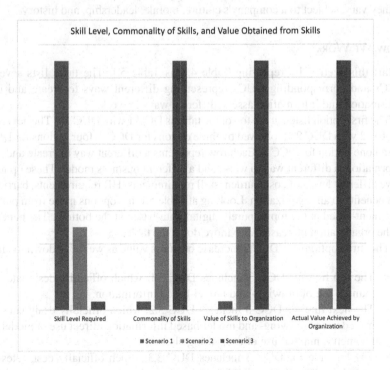

FIGURE 5.8 Skill level, commonality of skills, and value obtained from skills.

Scenario 4:   Best Case (not shown in Figure 5.8) (creation and use of high-value information is automated, skill requirements are low).

- The best case is where the skills to create, understand, use, manage, and maintain information are no longer required because information is used directly in software (automation).
- The need for skills to manually create, understand, use, manage, and maintain information is replaced by skills to create, understand, use, manage, and maintain systems that generate and use the information (e.g. programming, generative design, digital twins of processes (e.g. analysis, manufacturing, inspection, etc.)).

Often, companies want to take a stepwise approach to digital transformation toward becoming an MBE. As discussed earlier, it might be a good idea to transform in small steps. Or it may be that the transformation stalls on one of these small steps because the value obtained relative to the cost of the change and the perceived discomfort experienced by staff and suppliers from the change appear to insufficiently justify the change.

I should add another formula or two that quantify the pain and discomfort that will be incurred by a change. We consider potential pain, discomfort, and cultural issues in our corporate MBE readiness and capability evaluations and assessments, as they vary, subject to a company's culture, morale, leadership, and history.

How We Work

I start this section by repeating Table 4.3 as Table 5.1. The table lists several DCCs and corresponding DUCs representing different ways to create and use information and information assets, different ways to work.

The first option listed at the top of the table is DCC 1 with DUC 1.1. The next row is DCC 2 with DUC 2.1, followed by three options for DCC 3, four options for DCC 5, and one option for DCC 6. Each row represents a different way to create and use information, a different way to work, and a different business model. These options drive different levels of cost, burden, skill requirements, HR requirements, barriers, and benefits to an organization. Looking at Table 5.1, the options move from purely document-based at the top to purely digital-data-based at the bottom. The benefits to the organization increase as we move down the table.

The three options for DCC 3 increase business value as we move down the list:

- The top line for DCC 3 includes DUC 3.1, which officially designates manual use of drawing- and model-based information.
- The next line for DCC 3 includes DUC 3.2, which officially designates mixed use of drawing- and model-based information (direct use of model geometry, manual use of everything else).
- The last line for DCC 3 includes DUC 3.3, which officially designates mixed use of model-based information (some direct use, some manual use).

TABLE 5.1
Product Definition Dataset Classification Codes (DCCs) and Dataset Use Codes (DUCs) (repeat of Table 4.3)

| Information Asset Designation | | | | Official Information Asset - Official Product Definition Dataset Content | | | | | | | | | |
| | | | | Geometry | | Specifications | | Product Metadata | | File Metadata | | Ancillary Info | |
DCC	DUC	Schema Level	Use Type	Official Information	P/R₁	Official Information	P/R₂	Official Information	P/R₃	Official Information	P/R₄	Official Information	P/Rₙ
1	1.1	1.1	Manual	Drawing - Dims	P	Drawing - Ano	P	Drawing - Ano	P	PDM/PLM**	P	Manual use	P
2	2.1	2.1	Manual	Drawing - Dims	P	Drawing - Ano	P	Drawing - Ano	P	PDM/PLM**	P	Manual use	P
3	3.1	3.1	Manual	Model - Dims	P	Drawing - PMI	P	Drawing - Ano	P	PDM/PLM**	P	Manual use	P
3	3.2	3.2	Mixed	Model - Geometry	R	Drawing - PMI	P	Drawing - Ano	P	PDM/PLM**	P	Manual use	P
3	3.3	3.3	Mixed	Model - Geometry	R	Model - PMI	P	Model - PMI	P	PDM/PLM**	P	Manual use	P
5	5.1	5.1	Manual	Model - Dims	P	Model - PMI	P	Model - PMI	P	PDM/PLM**	P	Manual use	P
5	5.2	5.2	Mixed	Model - Geometry	R	Model - PMI	R	Model - PMI	P	PDM/PLM**	P	Manual use	P
5	5.3	5.3	Mixed	Model - Geometry	R	Model - PMI	R	Model - PMI	R	PDM/PLM**	P	Manual use	P
5	5.4	5.4	Direct	Model - Geometry	R	Model - PMI	R	Model - PMI	R	PDM/PLM**	R	Direct use	R
6	6.1*	6.1	Direct	Direct-use geometric definition	R	Direct-use specifications	R	Direct-use metadata	R	PDM/PLM**	R	Direct-use information	R

Notes

Official information and P/R columns for geometry, specifications, metadata, and ancillary info designate the source of official information in the product definition dataset that is to be used in lifecycle activities.

P/R   Indicates if presented or represented information is the official information

P   Presented information is official

R   Represented information is official

*   DUC 6.1   Information is created solely for direct use. Corresponding presented information may or may not be defined. Information may be in pre-compiled form, such as ASCII or source code (which are human readable), or it may be compiled to facilitate direct use, such as object code (which is generally not human readable).

**   While PDM or PLM are likely sources for the information, it is not mandatory that they are the sources. Information in linked digital platforms and datasets may drive or be driven by another system. How this is done is a business decision.

The four options for DCC 5 increase business value as we move down the list:

- The top line for DCC 5 includes DUC 5.1, which officially designates manual use of model-based information.
- The next line for DCC 5 includes DUC 5.2, which officially designates mixed use of model-based information (direct use of model geometry, manual use of everything else).
- The next line for DCC 5 includes DUC 5.3, which officially designates mixed use of model-based information (direct use of model geometry and specifications, manual use of everything else).
- The last line for DCC 5 includes DUC 5.4, which officially designates direct use of model-based information for everything.

The last line for DCC 6 includes DUC 6.1, which officially designates direct use of digital/model-based information for everything. The information is optimized for direct use and may or may not include information that supports manual use. If manual-use information is provided, it is for reference only.

Legacy approaches provide comfort for experienced workers, as experienced workers possess skills to function in document-based manual workflows, and they generally accept document-based manual workflows as how we work and how we should work. By contrast, legacy methods pose challenges for new hires, as they lack many skills needed for document-based manual workflows, they don't expect to need those skills, they don't really want to work the same way their parents (or grandparents) did, they are digital natives, they are used to information automation, and they are used to not having to deal with or think about the burden imposed by document-based manual workflows (keep in mind, they may not even know how to write a paper check, have checks, or even think that dealing with a piece of paper is necessary when they can do banking immediately using their smartphone...). Quite frankly, many young people don't accept the idea that document-based manual workflows and the skills needed to work in them are valuable. At least initially. Over time they may be worn down, square pegs pounded into round holes, learning the skills needed and coming to accept the drudgery of working in document-based manual workflows.

It doesn't have to be that way.

Recently I've had a lot of discussions about AI, what it enables, its risks, how it is and will continue to be used in the workplace, and how it will change our information landscape. I'm on board with it, and I assume it'll grow and evolve whether we like it or not. General AI is a work in progress, but it is improving and showing significant value. However, there is a trend that I've mentioned several places in the text about using new technology to do the wrong thing, using technology to do what we do today in almost the same way we do it today.

New technology often provides a way to completely bypass what we used to do – often, new technology makes some skills irrelevant and less valuable. A lot of AI buzz in the workplace today is about AI assistants, where text-based AI tools allow us to create a report, write a paper, draw a figure, or some other visual asset, thereby saving us time and the need for creative skills and capabilities to create the asset. The problem I see in this scenario is that we're using AI to create an asset optimized for manual use. This is like the visual dashboard I described in Chapter 3. I saw many presentations at a conference in which the pinnacle use case touted for a very powerful software platform was creating information dashboards for visualization, for manual use – for me, those presentations showed a lack of imagination and an understanding of the potential offered by their software platform. In many current AI scenarios, powerful AI models with increasing capabilities are used to create information for visualization. We need to be using our powerful software platforms to reduce the amount of information we need to view. We need to develop more software that allows us to use information directly, thereby freeing us to focus on higher-value activities.

These are not baseless hypothetical ideas. There are many historical precedents. Looking back at the First Industrial Revolution, then the Second Industrial Revolution, then the Third Industrial Revolution, now somewhere in the fourth or fifth depending on your preferred spin, history is full of examples where technology and automation have freed us to and focus on problems and pursue activities that provide more value for the individual and society. The move from agriculture to living in cities, to using steam power, electrical power, from horses and oxen to cars, trucks, and tractors, from heating by fire to central heating, from ice boxes to refrigerators... – in each case people's lives were improved and they were able to spend their time focusing on more valuable activities and solving new problems. The ideas I list here are not original. I am adapting what I have read in many books and articles about history, technology, and transformation to what I see in industry. There are also potentially detrimental consequences of technology. Technological innovation has consequences, in terms of our lives, our planet, our values, our expectations, our quality of life, etc.

We can adopt new technologies in a way that makes things better, not worse. This isn't a sociology book, but there are a lot of bright spots. I am an optimist. I am an entrepreneur. Entrepreneurs are optimists by nature, otherwise we wouldn't take the risks associated with trying to create something new and do it on our own. I write this book as an optimist. If being optimistic is a choice, and I believe it generally is, then we should choose optimism.

MBE is lean. MBE is green. Creating documents, using documents, working manually, are inefficient, costly, slow, burdensome, wasteful, and out of touch with the times. In addition to the cost and burden in terms of time, skills and training, time to market, etc. that I've already described, document-based manual workflows require a lot of energy. It costs money for every unit of time an organization operates. There's a burden associated with time. Energy is required to keep the lights on (as a metaphor). Energy is also required to create, release, distribute, understand, use, manage, revise, maintain, and store information. As stated earlier, every piece of information and every information asset has a lifecycle cost. Some of that cost can be attributed to energy. Broadly, MBE requires less information and thus less energy as we progress down the list in Table 5.1. MBE can be very green indeed. Or, if our goal is to use MBD manually, such as DCC 3 + DUC 3.1 or DCC 5 + DUC 5.1, we will probably use more energy than we did before migrating to inefficient implementations of MBE.

MBE is green.

According to the International Energy Agency, industry accounted for roughly one-third of global energy consumption in 2023 (Footnote 5.2). Industry uses a lot of energy. Note that this industrial energy use data includes industries and sectors other than the target topics of this text (e.g. concrete production, although concrete production is related to the target industries by virtue of the facilities and building OEMS and their suppliers inhabit). MBE's green potential is significant. Everything we can do to decrease our energy consumption footprint and the energy expended to create, manage, use, and maintain information matters. But someone might counter with "But it's just a bunch of bits…!" In a digital environment, information is stored digitally, and it is just a bunch of bits. But, in traditional workflows, it takes people, a lot of them, to create, manage, use, and maintain those bits. By definition, manual-use information is hands-on throughout its lifecycle. Figure 5.9 shows industrial energy consumption by fuel from 2000 to 2030. The units on the vertical axis are exajoules. The information and chart are taken from the context of CO_2 emissions. Other data available depict heat generation and costs related to industrial energy consumption. Industrial energy consumption has consequences.

The more efficient we make our operations, which includes how much information we create and use and how we use the information, the greener our operations and the greater our productivity.

MBD AND INFORMATION FLOW IN AN MBE

As clarified in this text, Model-Based Definition (MBD) may be created in many activities for many different purposes throughout the enterprise, supply chain, and product lifecycle. As clarified in this text, information has a lifecycle all its own, the information lifecycle. To be an optimized MBE, to obtain the maximum value from creating MBD and transforming to an MBE, we need to create direct-use MBD

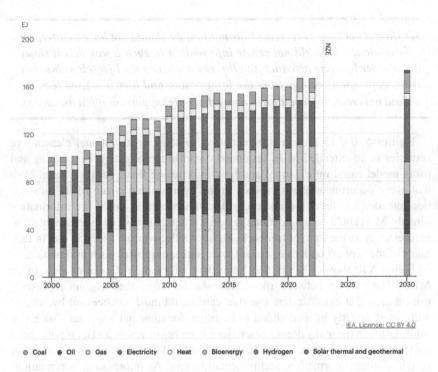

○ Coal   ● Oil   ○ Gas   ● Electricity   ○ Heat   ◎ Bioenergy   ● Hydrogen   ● Solar thermal and geothermal

FIGURE 5.9   Industrial energy consumption by fuel in the net zero scenario, 2000–2030 (Footnote 5.2).

FOOTNOTE

5.2 Source: IEA, Industrial energy consumption by fuel in the Net Zero Scenario, 2000-2030, IEA, Paris https://www.iea.org/data-and-statistics/charts/industrial-energy-consumption-by-fuel-in-the-net-zero-scenario-2000-2030, IEA. License: CC BY 4.0

information for all activities and use the information directly in those activities. The best-case scenario is where we create MBD information in the earliest lifecycle stages and use that information as the basis for information created in later lifecycle activities. Preferably, we only create information once. We create that information so that it rigorously represents its underlying subject and so it can be used directly in software, and we use that information in every applicable lifecycle process. We do not want to recreate information, we do not want to have to interpret information, we do not want to create information in such a way that it slows down other lifecycle activities and thereby decreases the lifecycle value that the organization realizes from the information and how it is used (e.g. we do not want to create information in a way that only supports manual use cases).

We should not have to recreate information, we should not have to interpret information, we should not create information in such a way that it slows down other lifecycle activities and thereby decreases the lifecycle value that the organization realizes from the information and how it is used (e.g. we should not create information in a way that only supports manual use cases).

Figures 5.10–5.13 present different ways to show organizational elements or activities in an extended MBE (extended organization or enterprise creating and using model-based information) and how information flows between them. Note that every department/activity shown in the figures could create MBD information intended for direct use and every department/activity could use information directly. Most could do so intentionally, while others, such as end users, might do so inadvertently, in the form of feedback and data collected from lifecycle use. In fact, many of the "opt in" options in our software and apps provide such information.

Figure 5.10 shows a simplified Venn diagram of the information flow in an MBE. The overlap between the domains/activities in the diagram represents information that exists in one use that can/should/must be accessed by, shared with, used directly by, and added to by other domains and activities. Access to information in the many domains/activities in an organization and its supply chain will vary. The diagram in Figure 5.10 shows interrelationships of organizations or activities, their information, and information flow. As information interrelationships and flow can be complex, and vary by organization, the diagram only hints at all the possible relationships, responsibilities, and roles.

Figures 5.11 and 5.12 present more detailed views of information flow and the many organizations/activities involved in, affecting, contributing to,

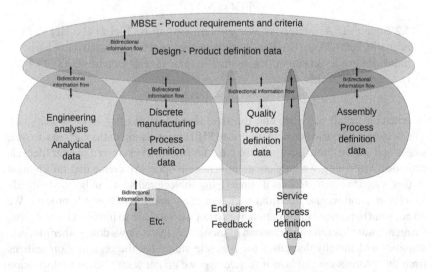

FIGURE 5.10   Information flow in an MBE (simplified) 1.

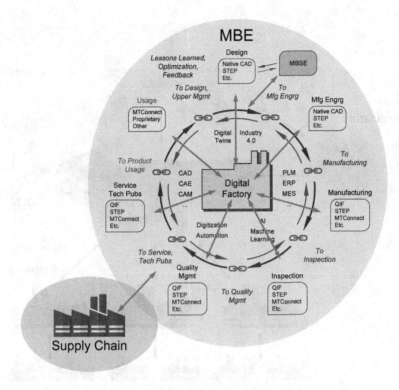

FIGURE 5.11 Information flow in an MBE (simplified) 2.

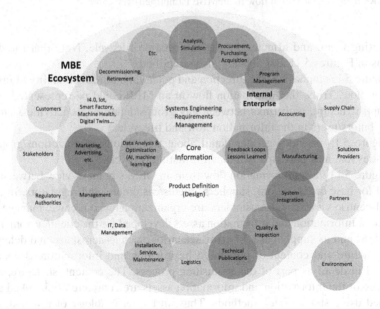

FIGURE 5.12 Information flow in an MBE (simplified) 3.

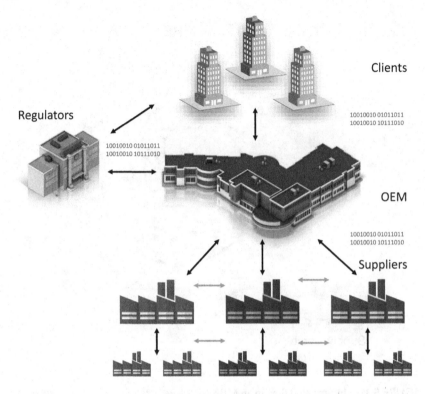

FIGURE 5.13 Information flow in an MBE (simplified) 4.

benefiting from, and affected by the information lifecycle. Note that the diagrams in Figures 5.11 and 5.12 are also simplified.

Figure 5.11 emphasizes the direction and potential multidirectionality of information flow. Truly, the information flow in an MBE could be represented with a graph structure, with links between any groups/departments/activities/actors/contributors/users in the information flow and lifecycle.

Figure 5.12 provides greater detail about the various organizations/groups/activities that may participate in the information lifecycle.

Figure 5.13 shows information flow from an OEM's point of view between an OEM (or base organization), suppliers, clients/customers, and regulators. In this OEM-centric view, clients/customers are organizations who receive, manage, or purchase information and information assets developed for the client/customer by the OEM and its supply base. In many scenarios, such as aerospace and defense, the structure and content of purchased information and information assets are defined in detail as part of the acquisition process. The content, structure, and purpose of the information and information assets are contractually invoked and defined using standardized methods. Thus, in the terminology of this text, the contractually invoked requirements are binding on the developing party (e.g. the

OEM). Conforming to these requirements drives the development and confor-mance to standardized model organizational structures (schemas) as discussed in earlier chapters.

In a document-based environment, information is often shared unidirection-ally. It flows from domains/activities at the start of lifecycle to domains/activi-ties later in the lifecycle, in a waterfall fashion. In this traditional information model, information value and IP are buried in documents, which impede obtain-ing greater value from information in every use case. Information in a traditional document-based manual-use workflow is not optimized for obtaining greater value or adding value to information developed in other domains/activities. The information just flows along with the documents like a piece of driftwood in a river, and at each step people use the RIDICT sequence to get relevant informa-tion and translate it into a form suitable for their manual activity.

PROCESS DEFINITION DATASETS

In Chapter 4, we established the importance of standardized product definition datasets and how they may affect productivity and quality. Product definition datasets are one of many types of information assets. In this section, we will address a broad category of information assets, which is process definition data-sets. While most organizations routinely categorize and treat product definition datasets rigorously, process definition datasets are often treated casually. Process definition datasets typically don't receive the same attention, rigor, and respect given to product definition datasets. This is a mistake.

Whether they realize it or not, most OEMS and their supply chain create and own far more process definition datasets than product definition datasets. There is tremendous value in process definition datasets, in the intellectual property they contain, and the investment required (e.g. in terms of CAPEX, OPEX, or skills, time, labor, capital, etc.) to create, manage, maintain, and use them. Process defi-nition datasets are a crucial part of our workflows, allowing us to operate produc-tively and respond rapidly to changing market demands.

Thus, we must recognize the value of process definition datasets and treat them rigorously, and we must define standardized official process definition data-set structures and use cases just like we do for product definition datasets. This information should be defined in organizational standards and SOPs much like we do for product definition datasets.

While many organizations routinely treat product definition datasets rigor-ously, process definition datasets are often treated casually. This is a mis-take. Most OEMS and their supply chain create and own far more process definition datasets than product definition datasets. Process definition data-sets should be treated formally and with the same rigor as product definition datasets. There is tremendous value and IP in process definition datasets.

Process definition datasets should be created using MBD methods and they should be developed and optimized for direct use wherever possible. Ideally, as indicated in the preceding section, direct-use information created for other life-cycle activities would be used as the basis for information in process definition datasets. A likely scenario is that a product definition dataset is used as the basis for a process definition dataset (the process definition dataset would be built from the product definition dataset). Even though defining a product and defining a process are different activities with different requirements, a lot of the content in a product definition dataset is used in developing processes to manufacture and validate the product. Thus, product definition information would be included or used as the basis for process definition information – product definition data would be modified to suit the needs of the process and the use cases it supports. Ideally, the product definition dataset and the process definition dataset derived from it would be created and optimized for direct use, and related information elements in the datasets would be linked.

To reiterate, MBD is not limited to the design activity. MBD methods can and should be used throughout the organization and the information lifecycle.

Process definition datasets are usually derived from and based on information in product definition datasets. Thus, in one way or another, process definition datasets are derivative datasets. In environments that have a formal systems engineering discipline, process definition datasets are also based on information from systems engineering. (Note that while some organizations have formal systems engineering disciplines, many organizations do not. The role of requirements management is handled differently. How companies deal with this is a business decision.) Figure 5.14 depicts how information commonly flows to and from a process definition dataset. Inputs to the process are shown in the box on the left side of the figure, which are the product definition dataset, systems engineering, and other information sources (e.g. procurement, facilities, safety, etc.), the process definition dataset(s) is shown in the center box, and the output released from the process definition activity is shown in the box on the right side of the figure.

Table 5.2 includes Dataset Classification Codes (DCCs) for Process Definition Datasets (PrDCCs) and Dataset Use Codes for Process Definition (PrDUCs). The

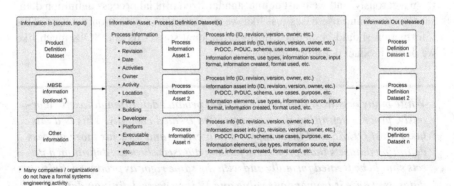

FIGURE 5.14 Product definition dataset to process definition dataset information flow.

TABLE 5.2

Process Definition Dataset Classification Codes (PrDCCs) and Dataset Use Codes (PrDUCs) – Source of Information Corresponding to Information Elements Listed in Table 5.1

| Information Asset Designation | | | | Official Information Asset - Official Process Definition Dataset Content | | | | | | | | | |
| PrDCC | PrDUC | Schema Level | Use Type | Geometry | | Specifications | | Product Metadata | | File Metadata | | Ancillary Info | |
				Official Source	P/R₁	Official Source	P/R₂	Official Source	P/R₃	Official Source	P/R₄	Official Source	P/Rₙ
1	1.1	1.1	Manual	Drawing - Dims	P	Drawing - Ano	P	Drawing - Ano	P	PLM, ERP, …	P	Manual use	P
2	2.1	2.1	Manual	Drawing - Dims	P	Drawing - Ano	P	Drawing - Ano	P	PLM, ERP, …	P	Manual use	P
3	3.1	3.1	Manual	Model - Dims	P	Drawing - PMI	P	Drawing - Ano	P	PLM, ERP, …	P	Manual use	P
3	3.2	3.2	Mixed	Model - Geometry	R	Drawing - PMI	R	Drawing - Ano	P	PLM, ERP, …	P	Manual use	P
3	3.3	3.3	Mixed	Model - Geometry	R	Model - PMI	R	Model - PMI	P	PLM, ERP, …	P	Manual use	P
5	5.1	5.1	Manual	Model - Dims	P	Model - PMI	P	Model - PMI	P	PLM, ERP, …	P	Manual use	P
5	5.2	5.2	Mixed	Model - Geometry	R	Model - PMI	P	Model - PMI	P	PLM, ERP, …	P	Manual use	P
5	5.3	5.3	Mixed	Model - Geometry	R	Model - PMI	R	Model - PMI	P	PLM, ERP, …	P	Manual use	P
5	5.4	5.4	Direct	Model - Geometry	R	Model - PMI	R	Model - PMI	R	PLM, ERP, …	P	Direct use	R
6	6.1	6.1	Direct	Direct use geometric definition	R	Direct use specifications	R	Direct use metadata	R	PLM, ERP, …	R	Direct use	R

Notes

P/R Indicates if presented or represented information is the official information

P Presented information is official

R Represented information is official

PrDCCs and PrDUCs are unique to this text, as the applicable ASME and ISO standards only define Dataset Classification Codes 1-5 (DCC 1-5) for *product* definition datasets. In addition to DCCs, PrDCCs, DUCs, and PrDUCs, organizational structures/schemas should also be standardized and used for process definition datasets. Companies need DUCs and PrDUCs to formally define how their information should be used, and these codes are published for the first time in this text.

Process definition datasets contain different or additional information elements than product definition datasets. If a product definition dataset is incorporated into a process definition dataset, then the information elements defined for the product definition dataset also occur in the process definition dataset as instantiated derivative and/or reference information. However, it is likely that process definition datasets will include information elements that are not part of a product definition dataset, such as programs (e.g. source code, object code for execution), instructions (e.g. work instructions, G-code, M-code, C/C++, Python, etc.), tool definitions, tool behavior (e.g. G-code), sequential information, setup and environmental information, target data, process control variables and limits, etc. Table 5.2 is structured like Table 5.1 and has the same information elements, except it specifies the source data (e.g. from a product definition dataset) and how the information is supposed to be used (its official use case). Note that some process definition datasets will include these information elements, and some will not. In processes used to create and evaluate products, such as manufacturing, inspection, and assembly, instances of product geometry and product specifications may exist, be referenced into, or recreated in the process definition dataset. For example, product geometry may be used to define toolpaths and tool offsets for machining processes, and specifications may be linked to requirements defined by the specification, such as defining the size limits for a produced hole from the size dimension and tolerances on a drawing or annotated model.

Often SOPs and allowable practices for an MBE are initially defined manually, with initial information entered manually into a form or dialog box. A sample data entry form for this information is shown in Figure 5.15. This form lists allowable PrDCC + PrDUC combinations similar to those listed in Table 5.2. Common choices for the cells in each column are shown as Allowable Content below the form in Figure 5.15. Note that the number and type of information elements are extensible. Alternate formats are shown in later figures that include more information elements.

Information elements listed in Figure 5.15:

Information Element 1	Product geometry
Information Element 2	Process geometry
Information Element 3	Design limits and requirements
Information Element 4	Process information 1
Information Element n	Process information n

In Figure 5.16, information representing process definition dataset types and content was entered into a spreadsheet via the form shown in Figure 5.15.

Process Information
Process _____
Activity _____
Rev _____
Date _____
Process Owner _____

Product Information
Product _____
Product Definition Dataset _____
Rev _____
Product Definition Dataset Owner _____

PrDCC _____
PrDUC _____
Schema _____

PrDCC	PrDUC	Schema Level	Use Type	Information Element 1	Information Source 1	P/R₁	Information Element 2	Information Source 2	P/R₂	Information Element 3	Information Source 3	P/R₃	Information Element 4	Information Source 4	P/R₄	Information Element n	Information Source n	P/R n
1	1.1	1.1	Manual	Product geometry	Document	P	Process geometry	Document	P	Design limits and requirements	Document	P	Process information 1	Document	P	Process information n	Document	P
2	2.1	2.1	Manual		Document	P		Document	P		Document	P		Document	P		Document	P
3	3.1	3.1	Manual		Model	P		Model	P		Model	P		Model	P		Model	P
3	3.2	3.2	Mixed		Model, code	R		Model	P		Model	P		Model	P		Model	P
3	3.3	3.3	Mixed		Model, code	P		Model, code	R		Model	P		Model	P		Model	P
5	5.1	5.1	Manual		Model	P		Model	P		Model	P		Model	P		Model	P
5	5.2	5.2	Mixed		Model, code	R		Model	P		Model	P		Model	P		Model	P
5	5.3	5.3	Mixed		Model, code	R		Model, code	R		Model	P		Model	P		Model	P
5	5.4	5.4	Direct		Model, code	R		Model, code	R		Model, code	R		Model, code	R		Model, code	R
6	6.1	6.1	Direct		Direct-use information	R		Direct-use information	R		Direct-use information	R		Direct-use information	R		Direct-use information	R

Allowable Content

	PrDUC	Schema Level	Use Type	Information Element 1	Information Source 1	P/R₁	Information Element 2	Information Source 2	P/R₂	Information Element 3	Information Source 3	P/R₃	Information Element 4	Information Source 4	P/R₄	Information Element n	Information Source n	P/R n
1	1.1, 1.2, ... 1.n	Same value as PrDUC	Manual	May be any product- or process-related information element	Document	P	May be any product- or process-related information element	Document	P	May be any product- or process-related information element	Document	P	May be any product- or process-related information element	Document	P	May be any product- or process-related information element	Document	P
2	2.1, 2.2, ... 2.n		Mixed		Model	R		Model	R		Model	R		Model	R		Model	R
3	3.1, 3.2, ... 3.n		Direct		Model, code			Model, code			Model, code			Model, code			Model, code	
4	4.1, 4.2, ... 4.n				Code			Code			Code			Code			Code	
5	5.1, 5.2, ... 5.n				Direct-use info			Direct-use info			Direct-use info			Direct-use info			Direct-use info	
6	6.1, 6.2, ... 6.n																	

FIGURE 5.15 Process definition dataset type, content, and structure data entry form.

Official Process Definition Dataset Classes, Schema, Uses, and Constituent Information Elements

PrDCC	PrDUC	Schema Level	Use Type	Information Element 1	Information Source 1	P/R$_1$	Information Element 2	Information Source 2	P/R$_2$	Information Element 3	Information Source 3	P/R$_3$	Information Element 4	Information Source 4	P/R$_4$	Information Element n	Information Source n	P/R$_n$
1	1.1	1.1	Manual		Document	P		Document	P		Document	P		Document	P		Document	P
2	2.1	2.1	Manual		Document	P		Document	P		Document	P		Document	P		Document	P
3	3.1	3.1	Manual	Product geometry	Model	P	Process geometry	Model	P	Design limits and requirements	Model	P	Process information 1	Model	P	Process information n	Model	P
3	3.2	3.2	Mixed		Model, code	R		Model	P		Model	P		Model	P		Model	P
3	3.3	3.3	Mixed		Model, code	R		Model, code	R		Model	P		Model	P		Model	P
5	5.1	5.1	Manual		Model	P		Model	P		Model	P		Model	P		Model	P
5	5.2	5.2	Mixed		Model, code	R		Model	R		Model	P		Model	P		Model	P
5	5.3	5.3	Mixed		Model, code	R		Model, code	R		Model	P		Model	P		Model	P
5	5.4	5.4	Direct		Model, code	R		Model, code	R		Model, code	R		Model, code	R		Model, code	R
6	6.1	6.1	Direct		Direct-use information	R		Direct-use information	R		Direct-use information	R		Direct-use information	R		Direct-use information	R

Allowable Content

PrDCC	PrDUC	Schema Level	Use Type	Information Element 1	Information Source 1	P/R$_1$	Information Element 2	Information Source 2	P/R$_2$	Information Element 3	Information Source 3	P/R$_3$	Information Element 4	Information Source 4	P/R$_4$	Information Element n	Information Source n	P/R$_n$
1	1.1, 1.2, ... 1.n	Same value as PrDUC	Manual	May be any product- or process-related information element	Document	P	May be any product- or process-related information element	Document	P	May be any product- or process-related information element	Document	P	May be any product- or process-related information element	Document	P	May be any product- or process-related information element	Document	P
2	2.1, 2.2, ... 2.n		Mixed		Model	R		Model, code	R		Model, code	R		Model, code	R		Model	R
3	3.1, 3.2, ... 3.n		Direct		Model, code			Code			Code			Code			Code	
4	4.1, 4.2, ... 4.n				Code			Direct-use info			Direct-use info			Direct-use info			Direct-use info	
5	5.1, 5.2, ... 5.n				Direct-use info													
6	6.1, 6.2, ... 6.n																	

FIGURE 5.16   Official process definition dataset table.

The information entered defines the official process definition dataset types and content supported, recommended, and allowed by the organization. Figure 5.16 is extracted from Figure 5.15.

In Figures 5.15 and 5.16, process information elements 1-n are shown in columns followed by columns indicating the source of the information. It is important to indicate the source of process definition information because process definition datasets are usually derived from product definition datasets and other information.

Representing Process Definition Dataset Types with Direct-Use Information

Per the best practices recommended in this text, processes, standards, and SOPs should be defined using MBD methods, as model-based information, and preferably as direct-use information. Thus, while it is convenient to create a table defining officially-approved process definition dataset types, structure, and content, we should create more than just a table. A table presents information graphically and is structured and optimized for manual use. We want information that is structured and optimized for direct use. So, we can use the tabular data to generate machine-readable/computer-sensible/whichever-term-you-prefer information. In this work-flow example, we used the tabular data to generate the HTML shown in Figure 5.17. Note that the HTML shown in Figure 5.17 is incomplete, as rows for PrDUC 3.1, 3.2, 3.3, 5.1, 5.2, 5.3, 5.4, 6.1 are not included. The HTML is shortened such that it only contains a representative sample. Note that while the code shown in Figure 5.17 is truncated, Table 5.3 was generated from HTML that represents all table rows.

The HTML in Figure 5.17 was created using generative AI, thereby automating the programming task. Indeed, generative and narrow AI trained on our organization's specific information and information assets will come to dominate most of our authoring activities. People will still drive the process and refine the results, but the process of authoring will be via AI. Note that the skills needed to work this way are different than the skills we currently need to create IP, which in turn are different than the skills needed before the widespread adoption of computers, and more recently, the skills needed to drive 3D CAD and analytical software.

The table shown in Table 5.3 is a presented version of the HTML in Figure 5.17.

Alternate Process Definition Dataset Data Entry Form

An alternate data entry form is provided in Figure 5.18 to close this section. Allowable content is shown below each column. The form in Figure 5.18 is arranged with information elements by row rather than by column. Some information elements shown in each figure are not included in other figures. Portions of the two approaches can be combined as desired.

Note: There is no limit to the number of information elements (rows) that may be defined in this form. Only 11 rows are included in the form in Figure 5.18, so the figure is easier to read in the small area of the pages in this book.

```
<!DOCTYPE html>
<html lang="en">
<head>
 <meta charset="UTF-8">
 <meta name="viewport" content="width=device-width,
initial-scale=1.0">
 <title>Table Representation</title>
 <style>
  table, th, td {
   border: 1px solid black;
   border-collapse: collapse;
   padding: 8px;
   text-align: center;
  }
  table {
   width: 100%;
  }
  th {
   background-color: #f2f2f2;
  }
 </style>
</head>
<body>
 <table>
  <thead>
   <tr>
    <th>PrDCC</th>
    <th>PrDUC</th>
    <th>Schema Level</th>
    <th>Use Type</th>
    <th>Information Element 1</th>
    <th>Information Source 1</th>
    <th>P/R 1</th>
    <th>Information Element 2</th>
    <th>Information Source 2</th>
    <th>P/R 2</th>
    <th>Information Element 3</th>
    <th>Information Source 3</th>
    <th>P/R 3</th>
    <th>Information Element 4</th>
    <th>Information Source 4</th>
    <th>P/R 4</th>
    <th>Information Element n</th>
    <th>Information Source n</th>
    <th>P/R n</th>
   </tr>
  </thead>
  <tbody>
   <!-- First Row -->
   <tr>
```

FIGURE 5.17 Official process definition dataset table represented as HTML.

```
     <td>1</td>
     <td>1.1</td>
     <td>1.1</td>
     <td>Manual</td>
     <td>Product geometry</td>
     <td>Document</td>
     <td>P</td>
     <td>Process geometry</td>
     <td>Document</td>
     <td>P</td>
     <td>Design limits and requirements</td>
     <td>Document</td>
     <td>P</td>
     <td>Process information 1</td>
     <td>Document</td>
     <td>P</td>
     <td>Process information n</td>
     <td>Document</td>
     <td>P</td>
    </tr>
    <!-- Second Row -->
    <tr>
     <td>2</td>
     <td>2.1</td>
     <td>2.1</td>
     <td>Manual</td>
     <td>Product geometry</td>
     <td>Document</td>
     <td>P</td>
     <td>Process geometry</td>
     <td>Document</td>
     <td>P</td>
     <td>Design limits and requirements</td>
     <td>Document</td>
     <td>P</td>
     <td>Process information 1</td>
     <td>Document</td>
     <td>P</td>
     <td>Process information n</td>
     <td>Document</td>
     <td>P</td>
    </tr>
   </tbody>
  </table>
 </body>
</html>
<! - The HTML above is incomplete, as rows for PrDUC 3.1, 3.2,
3.3, 5.1, 5.2, 5.3, 5.4, 6.1 are not included. Table 5.3 is
generated from the data that represents all table rows. ->
```

FIGURE 5.17   (Continued)

TABLE 5.3
Process Definition Dataset Table Generated from HTML

PrDCC	PrDUC	Schema Level	Use Type	Information Element 1	Information Source 1	P/R_1	Information Element 2	Information Source 2	P/R_2	Information Element 3	Information Source 3	P/R_3	Information Element 4	Information Source 4	P/R_4	Information Element n	Information Source n	P/R_n
1	1.1	1.1	Manual	Product geometry	Document	P	Process geometry	Document	P	Design limits and requirements	Document	P	Process information 1	Document	P	Process information n	Document	P
2	2.1	2.1	Manual	Product geometry	Document	P	Process geometry	Document	P	Design limits and requirements	Document	P	Process information 1	Document	P	Process information n	Document	P
3	3.1	3.1	Manual	Product geometry	Model	P	Process geometry	Model	P	Design limits and requirements	Model	P	Process information 1	Model	P	Process information n	Model	P
3	3.2	3.2	Mixed	Product geometry	Model code	R	Process geometry	Model	P	Design limits and requirements	Model	P	Process information 1	Model	P	Process information n	Model	P
3	3.3	3.3	Mixed	Product geometry	Model code	R	Process geometry	Model code	R	Design limits and requirements	Model	P	Process information 1	Model	P	Process information n	Model	P
5	5.1	5.1	Manual	Product geometry	Model	P	Process geometry	Model	P	Design limits and requirements	Model	P	Process information 1	Model	P	Process information n	Model	P
5	5.2	5.2	Mixed	Product geometry	Model code	R	Process geometry	Model	P	Design limits and requirements	Model	P	Process information 1	Model	P	Process information n	Model	P
5	5.3	5.3	Mixed	Product geometry	Model code	R	Process geometry	Model code	R	Design limits and requirements	Model	P	Process information 1	Model	P	Process information n	Model	P
5	5.4	5.4	Direct	Product geometry	Model code	R	Process geometry	Model code	R	Design limits and requirements	Model code	R	Process information 1	Model code	R	Process information n	Direct-use information	R
6	6.1	6.1	Direct	Product geometry	Direct-use information	R	Process geometry	Direct-use information	R	Design limits and requirements	Direct-use information	R	Process information 1	Direct-use information	R	Process information n	Direct-use information	R

Process Definition Dataset ID

Process _____	Activity ID _____
Rev _____	PrDCC _____
Date _____	PrDUC _____
Process owner _____	Schema _____

Rev _____	
Product _____	
Product Definition Dataset _____	
Rev _____	
Product Definition Dataset Owner _____	

Information Element No	Information Element	Owner	Use Type	Information In (source, input)	Input Format	Information Out (output)	Output Format	Asset ID	Asset Name or ID number	Asset Rev	Asset Version
1										A/R	A/R
2											
3											
4											
5											
6											
7											
8											
9											
10											
n											

Allowable Content

	Name of owner		Information In	Input Format	Information Out	Output Format	
1 May be any product- or process-related information element		Manual	A/R	.usd, varies	A/R	.usd, varies	A/R
2		Mixed	Animation	.mp4, ...	AI	.mp4, ...	
3		Direct	Audio	.mp3, ...	Animation	.mp3, ...	
4		Source code	Boundary conditions	.csv	Audio	.csv	
5		Executable	Document	.pdf, ...	Boundary conditions	.pdf, ...	
6			Drawing	.dwg, ...	Document	.dwg, ...	
			Environmental info	.usd, varies	Drawing	.usd, varies	
			Graphic diagram	.png	Environmental info	.png	
			Model	.stl, .stp, ...	Graphic diagram	.stl, .stp, ...	
			Model data	.stl, .stp, ...	Model	.stl, .stp, ...	
			Process state	.nc, ...	Model data	.nc, ...	
			Process sequence	.nc, ...	Process sequence	.nc, ...	
			Program	.exe, ...	Process state	.exe, ...	
			Sensor data	.xml, ...	Program	.xml, ...	
			Set up	-	Sensor data	-	
			Simulation	-	Set up	-	
			SPC data	.csv, ...	Simulation	.csv, ...	
			Tool behavior	-	SPC data	-	
			Tooling	-	Tool behavior	-	
			Twin	-	Tooling	-	
			V/R	.usd, varies	Twin	.usd, varies	
			Video	.mp4, ...	V/R	.mp4, ...	
			Work instructions	.pdf, .exe, ...	Video	.pdf, .exe, ...	
			Other		Work instructions		
					Other		

FIGURE 5.18 Process definition dataset type, content, and structure data entry form (alternate).

PARTS LISTS AND BILLS OF MATERIALS IN AN MBE

Parts Lists (PLs) and Bills of Materials (BOMs) are essential parts of product definition and enterprise activity. From the design phase through procurement, purchasing, manufacturing planning, quality, manufacturing, tooling, assembly, inspection, prototyping, testing, installation, service, end use and client use cases, information in PLs and BOMs are essential to enterprise operation. In some respects, a PL and a BOM are the same thing, existing at different lifecycle phases. Indeed, these terms were developed by different organizations, and in many cases, the terms are used and mean different things to different organizations. *Parts List* is defined in ASME Y14.34 Associated Lists, whereas *Bill of Materials* is a broadly used term that is commonly used in industry, literature, and business software. In many companies, I have found that people use the terms interchangeably, or one organization calls the information asset a "parts list" and another organization calls it a "bill of materials".

Based on usage, ownership, and where/when in the product and information lifecycles the information appears, the terms do not necessarily mean the same thing. In the simplest sense, parts lists evolved from engineering drawings, as they were and continue to be defined on engineering drawings and related documentation. This thinking has also been extended to annotated models. Generally, people who work in configuration management, purchasing and information management, software development, or in large OEMs tend to talk about BOMs. Whether we call them PLs or BOMs, their role is similar and important for MBE.

In general, a PL is developed during the design process and released as part of the product definition dataset. The PL is defined in CAD, PDM, PLM, or a combination of these systems. The information in the product definition dataset is converted into a TDP, and the TDP is used by people outside of the design organization. This is not a universal workflow but is common. The PL is used as the basis for an engineering bill of materials (EBOM) as defined in Chapter 4. The EBOM is converted to a manufacturing bill of materials (MBOM), also defined in Chapter 4. Often, the EBOM is owned and managed by one group (e.g. department A) related to design engineering, and the MBOM is owned and managed by another group (e.g. department B). Information in the EBOM is from the point of view of the product definition released by design (i.e. the constituent items that make up the product), and information in the MBOM is from the point of view of manufacturing (i.e. items that make up the product and additional materials needed to execute manufacturing process, more specific quantities of bulk materials such as adhesive, paint, etc., and also rearranging items in the EBOM for more efficient manufacturing workflow). See Figure 5.19 for a simplified PL-to-BOM workflow. There are many variations on this theme. I don't profess to have it right here or to represent every company's processes. Suffice it to say, companies deal with this information in different ways and by now should have evolved into doing so formally. I remember the first time I dealt with computer-based inventory management and bill of material systems in 1989 or so. Things have evolved a lot!

In many organizations, the process of creating PLs and BOMs and how the information in PLs and BOMs is distributed, managed, and propagated through the lifecycle is a good example of MBD and MBE. Information in the PL is linked to and derived from product model data and file metadata, EBOMs are derived

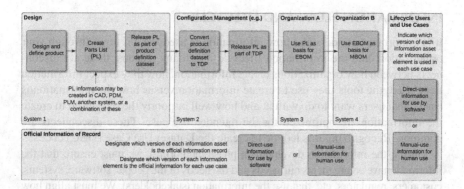

FIGURE 5.19 Parts list to BOM information lifecycle example.

from PLs and inventory data in PDM or PLM, and MBOMs are derived from EBOMs and inventory data bases, often in ERP software. Some organizations may have digital links between these systems and the information in the PL, EBOM, and MBOM, while other organizations may segregate these information assets, keeping them in silos, with a wall between siloes that requires manual information management and data entry between each version of the information. Note that the same information may exist in all three of these information elements (the PL, EBOM, and MBOM). Forward-thinking organizations will have these information assets linked and the means to manage and coordinate their content automated or semiautomated. As the PL, EBOM, MBOM, and other BOMs serve different purposes and roles at different times in the product and information lifecycles, there may be different versions of these information assets and different effectivities.

Today, many organizations have complex distributed environments for creating and managing, maintaining their various BOMs. Often, software and platforms used have evolved over time, with lots of custom programming to get to their current state. It isn't always clean or pretty. And in a lot of cases, getting to their current state has occurred over many years and has been costly.

But the most important point I want to make here is that BOMs, the information they contain, their distribution, and how the information is used are digitalized. The workflow is digital and many use cases are direct. Yes, we still read tables onscreen and print them, but a lot of the use cases of the information are automated. Much of this workflow is already model-based (e.g. driven by and populated with direct-use information).

Notes: The workflow shown in Figure 5.19 is hypothetical. Four systems are shown, but an actual implementation may include more or less systems. One or more CAD, PDM, PLM, ERP, other systems, or a combination of these could be used to develop, manage, distribute, and maintain the information. Thus, the source of the official information for each use case must also be designated.

INFORMATION FLOW AND LIFECYCLE

It is important to understand how information is used, how it flows through your organization, and the information lifecycle. For many reasons, there are disconnects between how organizations create information, how they expect information to be used, the tools they use to create information versus how the information is used, what users want to do with it, and how well or poorly the tools used to create the information are optimized for the intended use case. These disconnects are common. Recognizing this led me to create the Dataset Use Case (DUC) codes.

During the information development phase (authoring), we must ensure that the information we create meets our needs and the needs of the people, software, systems, customers, regulators, etc. that use the information (stakeholders). We must align how we expect information will be used with how it is actually used. We must ensure that information content, structure, format, and type optimally convey the necessary information for its purpose and how it is used. We must ensure that we are efficient and that we provide the right information in the right form to achieve maximum lifecycle value.

One problem is that the way we work today uses authoring tools that have evolved from our document-based past. Modeling tools evolved from drafting tools, 2D CAD evolved from board drafting, BOMs and PLM evolved from spreadsheet and database software, which evolved from typed and handwritten spreadsheets and tabular documents, and on and on. Referring back to Figure 1.1 in Chapter 1, its documents all the way down.

In the National Institute of Standards and Testing (NIST) PMI test model projects (circa 2011–2014) we found that commercially available, state-of-the-art CAD software had different levels of capability to annotate models. CAD annotation tools were optimized for creating graphical information, information for manual use, to be read and viewed. A lot of work was done to improve capabilities of authoring software. Much of the work was focused on breadth rather than depth, increasing support for applying more types of annotation and PMI to 3D models. In retrospect, I think that focusing on annotating models, while important at the time, added to the overall focus on using model-based information manually. Today, CAD annotation tools and most companies' work methods are still optimized for creating graphical annotation rather than direct-use information. Software developers respond to their users' requests, so if software users ask software developers to provide capabilities that support outdated inefficient workflows, the software developers oblige. Looking back at Figure 5.2, I envision a parallel to this of drivers asking car manufacturers to provide a way to connect horses to pull their cars as the engine sits unused under the hood.

NIST has supported MBD, MBE, information modeling standards, and many related activities for many years. NIST has done a lot for industry and helped pave the way for the capabilities we have today and will have tomorrow. NIST has and continues to help this topic and direct-use capabilities evolve, through support for research projects, standards development (e.g. STEP, QIF), and other activities.

Our MBD software tools are optimized for adding presented information to models – our software tools are optimized for using models manually. This is great if our goal is to use models manually. (It's difficult for me to write that last sentence with a straight face.) CAD software tools aren't initially optimized to create information for direct use. There's a good reason: industry wasn't and isn't asking for it, or there isn't enough of industry asking for it for the software developers to prioritize direct-use authoring capability. Product definition standards such as ASME Y14, ISO TC 10, and TC 213 also play a role in this, as they are derived from legacy drafting standards, which are mainly about presenting information for manual use. Major CAD developers are obliged to support engineering drawing standards and MBD standards like ASME Y14.41 and ISO 16792, which are also mainly about presenting information for manual use. There is also the issue of data structure supported by the software and data formats. Data formats like native CAD, STEP, JT, and more recently QIF, support direct use, but the interfaces for creating the information and the ecosystem for structuring direct-use information are weak and underdeveloped. To be clear, given that we are still using software optimized for manual use cases, and our standards are optimized for manual use cases, and people in industry still think that they should use information manually, it is very difficult to realign everyone in this ecosystem to prioritize direct use.

In the early days of MBD and MBE, we added fuel to the inertial fire (so to speak), as our goals for research were a mishmash of evaluating authoring, translation, and validation, and for manual use, direct use, and whether software could create and maintain certain PMI elements or not. We uncovered a lot of issues, and in the end, the work has provided great benchmarks for developers and users. Many early MBx R&D projects were focused on archaic ideas of industry's needs. Indeed, if we ask people in industry today about their needs in this area, they may still have archaic ideas. Such is the nature of cultural inertia and the effect of the status quo.

I created the following figures to help people working on Y14 standards understand what we need to be thinking of when developing standards that have evolved from drafting standards. Drafting standards are inadequate for industrial automation and creating direct-use information and guiding developers of software that could create or use direct-use information. I didn't get a lot of traction with the figures. Maybe too many old dogs and people stuck in the status quo. I'm reminded of the discussion I described earlier with my friend who said, "We need to retire …" Note that the issue of standards developers being stuck in the past, in continuing to standardize past practices, is not unique to ASME. I have worked a lot with ISO GPS standards development and in many ways the problem is more pronounced with ISO GPS standards. There are many people involved in Y14 standards development who use 3D CAD and participate in development of the ASME Y14.41 and other model-based standards. There is a culture and understanding of 3D modeling and digital data, although the understanding is insufficient. In ISO GPS, there didn't seem to be much thinking about 3D CAD or exposure to 3D CAD. By comparison, working with ISO TC 184/SC 4 in STEP, QIF, and related standards efforts, everyone is thinking about direct use of information. It's a very different environment. The focus is on rigorous digital information models. However, STEP, QIF, etc. do not exist in a vacuum, and they are usually not the initial data format in which information is authored. These data standards are burdened by supporting the methods in PMI-defining standards from ASME, AWS, ISO, and others – the data modeling standards are affected by the manual-use methods defined in these legacy drafting standards. I will address this in greater detail in Chapter 6. Generally, native 3D CAD is the original (native) format. Thus, we face challenges. CAD vendors work to please their biggest and most important clients, who in most cases are asking for refinements and additions for creating manual-use information. They are asking software vendors to prioritize legacy capabilities for legacy use cases. It's difficult to translate information suitable for direct-use into STEP or QIF if the information is not initially structured for direct-use in native CAD. Note that I am not criticizing CAD vendors. Software vendors try to keep their customers happy. If we ask for the wrong things, we will get the wrong things.

Figures 5.20–5.24 depict how information flows, how it is used, and who or what uses the information in a simplified information lifecycle. There is a lot going on in these figures. Pay particular attention to the arrowed lines in the figures, as the lines represent information flow from one process to the next, from source to user. The source of the information is at the end of the line without an arrowhead, and the recipient/user of the information is at the end of the line with an arrowhead. Note that in the figures *VAT* stands for *Value Added Task.*

Note that all the workflows presented in the figures include computers, software, networks, platforms, and other operational technologies (OT) used by modern organizations. All the workflows operate in environments fully supported by IT as well. The expense for the IT and OT in terms of CAPEX and OPEX exists in every workflow example. The difference between the examples is how information is

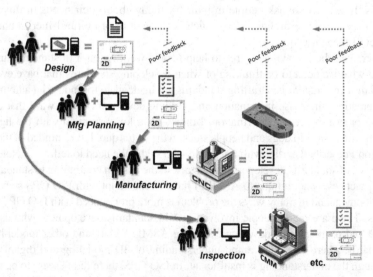

FIGURE 5.20 Information flow in the information lifecycle – 1 document-based manual-use workflow.

structured, how it is intended to be used, and how it is used. The last two examples in Figures 5.23 and 5.24 add the additional processing power enabled by cloud-based processes. Properly implemented, the expense for the workflows decreases as we progress from Figure 5.20 to Figure 5.24. We do less and get more in return.

The simplified workflows shown in Figures 5.20–5.24 include four activities/functions: design, manufacturing planning, manufacturing, and inspection. The workflows are represented as serial activities. I assume it is easy to extrapolate and apply the logic shown to other activities.

Design → Manufacturing Planning → Manufacturing → Inspection

These are some of the most important figures in this book.

INFORMATION FLOW 1: DOCUMENT-BASED MANUAL-USE WORKFLOW

In this workflow, design creates a product definition dataset consisting of documents (e.g. DCC 2 + DUC 2.1), which is converted into a TDP, the TDP is released, manufacturing planning uses the TDP documents to create their own documents, manufacturing uses the manufacturing plan documents to develop their own documents, they make the product, and inspection uses the TDP documents and manufacturing documents to create more documents so they can evaluate the as-produced product. Design creates documents, as they expect that the information will be used manually, and other activities oblige.

Design
- Receives an idea, request, or information describing a product
- The information is used manually (RIDICT) to formulate a problem statement
- Solves the problem by creating a design
  - Design engineers use computers and software to generate the design
- Creates/authors a document-based product definition dataset of the design
- A document-based product definition dataset is created
- The product definition dataset is converted into a TDP
- A TDP is released as official information for lifecycle activities to use.

Manufacturing Planning
- Receives the drawing and documents in the TDP
- Uses the documents manually (RIDICT)
- Develops the manufacturing plan and creates/authors and releases more documentation.

Manufacturing
- Receives manufacturing planning's documents
- Uses the documents manually (RIDICT)
- Manufactures the part per the manufacturing plan and creates/authors and releases more documentation.

Inspection

- Receives the drawing and documents in the TDP and the manufacturing documents
- Uses the documents to develop an inspection plan document
- Develops an inspection report to document the inspection results
- Receives the as-produced product
- Inspects the product to determine conformance to its requirements
- Enters the inspection results in the inspection report document.

Feedback and Lessons Learned

- Document-based
- Manual-use information I/O
- Unlikely, no direct links possible
- Inefficient if it occurs at all.

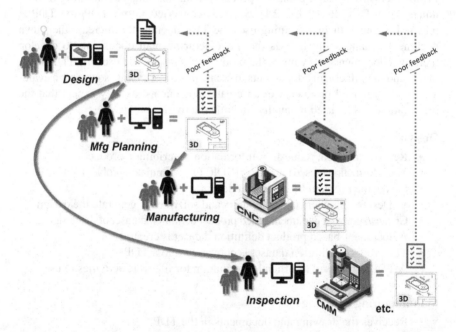

FIGURE 5.21 Information flow in the information lifecycle – 2 model-based manual-use workflow.

INFORMATION FLOW 2: MODEL-BASED MANUAL-USE WORKFLOW

In this workflow, design creates a product definition dataset consisting of a model and documents (e.g. DCC 3 + DUC 3.1), which is converted into a TDP, the TDP is released, manufacturing planning uses the TDP model and documents manually to create their own models and documents, manufacturing uses the manufacturing plan model and documents manually to develop their own models and documents, they make the product, and inspection uses the TDP (DCC 3 + DUC 3.1) and manufacturing documents to create more documents so they can evaluate the as-produced product.

Design
- Receives an idea, request, or information describing a product
- The information is used to formulate a problem statement
- Solves the problem by creating a design
  - Design engineers use computers and software to generate the design
- Creates/authors a model-based product definition dataset of the design
  - Creates a 3D CAD model of the product
  - Documents the product and its specifications on a 2D drawing
- A mixed-media product definition dataset is created
- A TDP is released as official information for lifecycle activities to use.

Manufacturing Planning
- Receives the model, drawing, and documents in the TDP
- Uses the model, drawing, and documents in the TDP manually (RIDICT)
- Develops the manufacturing plan and creates/authors and releases their own models and documents.

Manufacturing
- Receives manufacturing planning's documents
- Uses the documents manually (RIDICT)
- Manufactures the part per the manufacturing plan and creates/authors and releases their own models and documents.

Inspection
- Receives the models, drawings, and documents in the TDP and the manufacturing models and documents
- Uses the models, drawings, and documents to develop inspection plan models and documents
- Develops an inspection report to document the inspection results
- Receives the as-produced product
- Inspects the product to determine conformance to its requirements
- Enters the inspection results in the inspection report document.

Feedback and Lessons Learned
- Model-based
- Manual-use information I/O
- Unlikely, no direct links
- Inefficient if it occurs at all.

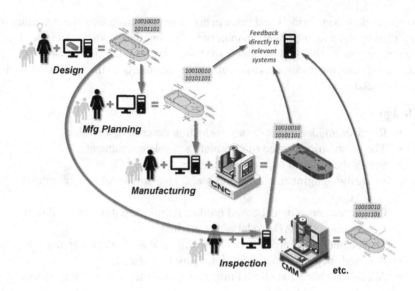

FIGURE 5.22 Information flow in the information lifecycle – 3 model-based direct-use workflow.

INFORMATION FLOW 3: MODEL-BASED DIRECT-USE WORKFLOW

In this workflow, design creates a product definition dataset consisting of an annotated model and ancillary information (e.g. DCC 5 + DUC 5.4), which is converted into a TDP, the TDP is released, manufacturing planning uses the TDP model and information directly to create their own models, manufacturing uses the manufacturing plan models directly to develop their own models, they make the product, and inspection uses the TDP (DCC 5 + DUC 5.4) and manufacturing models so they can evaluate the as-produced product.

Design
- Receives an idea, request, or information describing a product
- The information is used to formulate a problem statement
- Solves the problem by creating a design
  - Design engineers use computers and software to generate the design
- Creates/authors a model-based product definition dataset of the design
  - Creates a 3D CAD model of the product
  - The product and its specifications are defined in the model
- A model-based product definition dataset is created
- A TDP is released as official information for lifecycle activities to use.

Manufacturing Planning
- Receives the model in the TDP
- Uses the model in the TDP directly

- Develops the manufacturing plan and creates/authors and releases their own models.

Manufacturing
- Receives manufacturing planning's models
- Uses the models directly
- Manufactures the part per the manufacturing plan and creates/authors and releases their own models.

Inspection
- Receives the model in the TDP and the manufacturing models
- Uses the models to develop inspection plan models
- Uses an inspection record to contain the inspection results
- Receives the as-produced product
- Inspects the product to determine conformance to its requirements
- The inspection results are entered in the inspection record.

Feedback and Lessons Learned
- Model-based
- Direct-use information I/O
- Likely, direct links possible
- Enabled, even at the feature level.

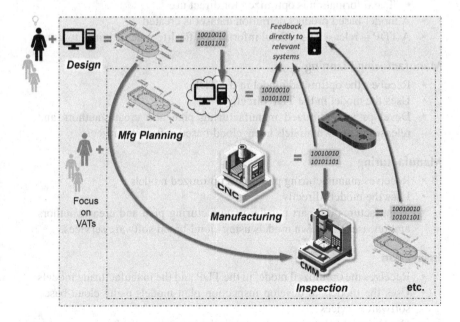

FIGURE 5.23 Information flow in the information lifecycle – 4 model-based direct-creation and direct-use workflow – cloud – phase 1.

Information Flow 4: Model-Based Direct-Use Workflow – Cloud – Phase 1

In this workflow, design creates a product definition dataset consisting of an annotated model and ancillary information optimized for direct use (e.g. DCC 5 + DUC 5.4), which is converted into a TDP, the TDP is released, manufacturing planning uses the TDP model and information directly to create their own models optimized for direct use, manufacturing uses the manufacturing plan models directly to develop their own models optimized for direct use, they make the product, and inspection uses the TDP (DCC 5 + DUC 5.4) and manufacturing models so they can evaluate the as-produced product. The entire process takes advantage of cloud-based services to increase productivity, quality, and throughput.

Design
- Receives an idea, request, or information describing a product
- The information is used to formulate a problem statement
- Solves the problem by creating a design
  - Design engineers use cloud-based software services to generate the design
- Creates/authors a model-based product definition dataset of the design
  - Creates a 3D CAD model of the product
  - The product and its specifications are defined in the model
  - The information is optimized for direct use
- A model-based product definition dataset is created
- A TDP is released as official information for lifecycle activities to use.

Manufacturing Planning
- Receives the optimized model in the TDP
- Uses the model in the TDP directly
- Develops an optimized manufacturing plan and creates/authors and releases their own models using cloud-based software services.

Manufacturing
- Receives manufacturing planning's optimized models
- Uses the models directly
- Manufactures the part per the manufacturing plan and creates/authors and releases their own models using cloud-based software services.

Inspection
- Receives the optimized model in the TDP and the manufacturing models
- Uses the models to develop inspection plan models using cloud-based software services
- Uses an inspection record to contain the inspection results
- Receives the as-produced product

- Inspects the product to determine conformance to its requirements
- Inspection results are automatically entered in the inspection record.

Feedback and Lessons Learned
- Model-based
- Direct-use information I/O
- Automatic, direct links
- Efficient and enabled, even at the feature level.

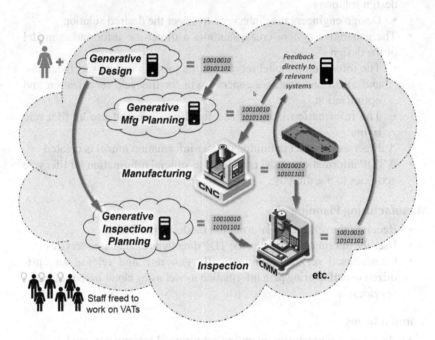

FIGURE 5.24 Information flow in the information lifecycle – 5 model-based direct-creation and direct-use workflow – cloud – phase 2.

INFORMATION FLOW 5: MODEL-BASED DIRECT-USE WORKFLOW – CLOUD – PHASE 2

In this workflow, design uses generative and simulation software to create an optimized product definition dataset consisting of an optimized direct-use information model (e.g. DCC 6 + DUC 6.1), the generation system converts the product definition dataset into a TDP information model, the TDP is released, manufacturing planning uses the TDP information model directly to create their own optimized direct-use information model, manufacturing uses the manufacturing planning information model directly to develop their own optimized direct-use information model, they make the product, and inspection uses the TDP information model (DCC 6 + DUC 6.1) and manufacturing information model directly to evaluate the as-produced product. The entire process uses cloud-based generative, simulation, and twin software services to optimize productivity, quality, throughput, and ROI.

Design

- Receives an idea, request, or information describing a product
- The information is entered into generative simulation software to formulate a problem statement and solve the problem
- A generative system solves the problem and provides a family of optional design solutions
  - Design engineers may intervene to select the desired solution
- The generative system creates/authors a direct-use information model of the design
  - The information model represents the product and its specifications, and also functions as a generic twin for lifecycle serialization and optimization
  - The information is optimized for direct use and use as lifecycle twins
- A direct-use product definition dataset information model is created
- A TDP information model is released as official information for lifecycle activities to use directly.

Manufacturing Planning

- Receives the optimized information model in the TDP
- Uses the information model in the TDP directly in cloud-based services
- Cloud-based services automatically generate and release an optimized manufacturing plan information model using cloud-based software services.

Manufacturing

- Receives manufacturing planning's optimized information model
- Uses the information model directly in cloud-based services
- Cloud-based services automatically generate and release an optimized manufacturing information model using cloud-based software services.
- Manufactures the part per the manufacturing plan.

Inspection

- Receives the optimized information model in the TDP and the manufacturing information model
- Uses cloud-based services to automatically generate an optimized inspection plan information model using cloud-based software services
- Uses an automatically generated inspection record to contain the inspection results
- Receives the as-produced product
- Inspects the product to determine conformance to its requirements
- Inspection results are automatically entered in the inspection record.

Feedback and Lessons Learned
- Model-based
- Direct-use information I/O
- Automatic, direct links
- Product and process optimization is built in, even at the feature level.

LIFECYCLE PRODUCT MODELS

As discussed earlier in this chapter, it is useful to separate product definition information into several models. First, we have the product definition dataset, the information asset that defines a product and its specifications. A product definition dataset should completely define the product and its specifications explicitly or by reference. However, a product definition dataset cannot be used alone to produce the product it represents or to verify that an as-produced product conforms to the requirements defined by its specifications. We need another model that defines the implications of those models and the physical and performance requirements they impose upon the product, which is the consumption model.

The nominal product model contains and defines the nominal, perfect, ideal product. Often this is a geometric model but may also include materials, properties, rules and logic for and components of electrical and electronic systems, systems containing or transporting fluids, etc., products that may be represented by a ladder diagram, schematic, process and instrumentation diagram (P&ID), or another logical diagram. The nominal model represents the theoretically-perfect product.

The specification model contains and defines the specifications applied to the product. In manual workflows, specifications are defined by text, symbols, and other graphical methods. While a product definition dataset information model includes a specification model, it does not contain the requirements those specifications impose upon the product – a product definition dataset does not contain a requirements model. The specifications are included in the product definition dataset, but the meaning of the specifications is not included. The meaning is left to the recipient of the product definition dataset. This is a source of problems and failure modes. In manual workflows, the meaning of specifications and the requirements they impose upon a product are left to interpretation. The people who created the product definition dataset have their interpretation of what the specifications require, and the people who received the product definition dataset have their interpretation of what the specifications require, and these two interpretations often differ.

In a manual process, specifications are read, understood or interpreted, broken down (RIDICT), and other information such as standards and SOPs (RIDICT again) and experience are used to determine the requirements that the specifications impose upon the product. In a direct-use process, the specifications are parsed, and code and programmed-in rule sets determine the requirements imposed. These are oversimplified descriptions of the processes, but my point is that the meaning and implications of a specification exist outside of the information asset that contains the specification. In the case of product definition

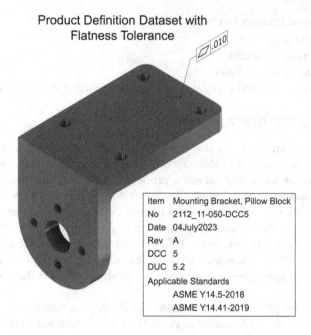

FIGURE 5.25 Product definition dataset containing a specification.

specifications, such as GD&T, the meaning of a specification requires understanding relevant portions of the ASME Y14.5 Standard or ISO GPS standards. The ASME Y14.5 Standard explains what the GD&T specification controls (which geometry is controlled by a specification), and the standard explains the requirements the specification imposes upon applicable geometry. See Figures 5.25–5.27 for examples.

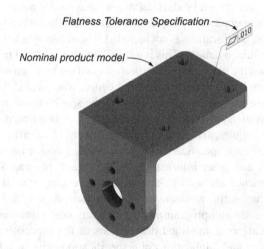

FIGURE 5.26 Specification model for flatness tolerance.

Figure 5.25 shows a very simple DCC 5 + DUC 5.2 product definition dataset, an annotated model. The nominal product model is defined by model geometry, and the specifications are defined by PMI (e.g. text, numbers, and graphical elements applied in the model).

This simple product definition dataset contains one specification, a feature control frame directed to the top surface by a leader. The specification is a flatness tolerance, which applies to the top surface. The specification alone does not indicate what it means – without additional information, we don't know the requirements or restrictions the specification imposes on the product geometry. All we know is that the specification is a flatness tolerance per ASME Y14.5-2018 and the tolerance applies to the top surface.

The requirements imposed by the flatness specification are shown in Figure 5.27. In a manual workflow, to understand the specification, we must read and understand the ASME Y14.5 standard (or the ISO 1101 standard if this was an ISO GPS-based product definition dataset) to determine which aspect of the product geometry is controlled by the flatness tolerance and to determine the conformance criteria the specification defines. In this case, the flatness tolerance defines a tolerance zone bounded by two parallel planes .010 inches apart, and the entire as-produced surface must fit on or between these parallel planes. Note that the as-produced surface is imperfect and the entire surface must conform to the tolerance zone. GD&T 101.

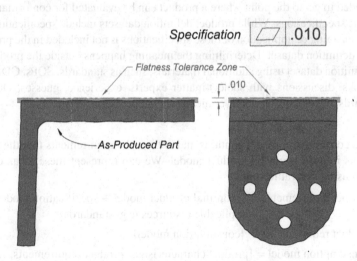

Specification ⟦ ▱ ⟧ .010

The flatness tolerance defines a tolerance zone that applies to the
indicated surface

Imposed Requirement:
The as-produced surface must fit on or between two parallel planes
.010" apart

FIGURE 5.27 Consumption model for flatness tolerance.

The meaning of a specification and the implications it imposes on the product must be understood by many parties in product and information lifecycles. For example, the design engineer should understand the requirements imposed on the product by specifications they define, the manufacturer should understand the requirements imposed on the product by the specification, and the inspector should understand the requirements imposed on the product by the specification. Hopefully, these actors share a common understanding. If not, there may be conflict when products are evaluated.

In a direct-use workflow, the idea of information being *models*, that the nominal product, specifications, and consumption information are *models* makes sense. Establishment of the consumption model can be direct, driven by ingesting a product definition dataset into manufacturing, inspection, or other software, as the software contains additional rules and information that build the requirements from the product and specification models. Note that software like this is commercially-available today. Coordinating and standardizing the meaning and implications of specifications across different lifecycle activities is easier and more consistent when using such methods. This is a critical business advantage of using MBD and becoming an MBE.

I developed the concepts of specification model and consumption model while working with the teams developing STEP standards. A few times I heard people imply that the information in a product definition dataset represented as a STEP file included the requirements. Given my background in design generally and GD&T specifically, I realized that there are a few more things needed to get to the point where a product can be evaluated for conformance to its specifications. While product definition datasets include specifications, the meaning or implication of those specifications is not included in the product definition dataset. Determining the meaning happens outside the product definition dataset using additional materials, such as standards, SOPs, GD&T books, discussions with subject-matter experts, experience, guesses, Ouija boards, eight balls, or whatever means are deemed appropriate.

The consumption model contains and defines the requirements that the specifications impose upon the product model. We can represent these ideas of this section using several formulas:

Product requirements $= f$ (nominal product model + specification model + applicable resources (e.g. standards))

Product requirements $\in$ {consumption model}

Consumption model = {product characteristics, product requirements, ...}

Process-specific consumption model = {product characteristics, product requirements, process information, process constraints, etc., ...}

A final comment about consumption models. Given that MBD occurs throughout the product lifecycle and information lifecycle, that many different organizations and activities use and generate MBD information, and most downstream activities base their information assets on received information, such as a product definition dataset, there are a lot more process definition datasets than product definition datasets. And most process definition datasets include consumption information or consumption models.

WHERE IS MBD CREATED AND USED IN PRODUCT AND INFORMATION LIFECYCLES?

We create and use information throughout the product lifecycle and the information lifecycle. As stated previously, using MBD methods to create model-based information is not just something that design engineers do. In an MBE, we want as many organizations and activities as possible to create MBD information and use it directly, thereby obtaining the maximum value from our efforts and our information. In this text, that is the simplified explanation of what an MBE is.

Figure 5.28 shows a high-level view of some of the many activities and organizations where MBD methods are used/should be used to create model-based information. The diagram does not depict a complete organization or enterprise, as most complete organizations have many divisions, departments, groups, teams, cells, and activities, and the diagram would be much too large to show in the small format of this text. (Note that we use diagrams like these to help clients map their current and future states and information flows.)

Figures 5.29–5.38 show portions of a large diagram that represents product and lifecycle activities and information created in each activity. Information flows left-to-right in the diagrams. Figures 5.29–5.34 are sequential. Information and the process continue from the right side of a figure to the left side of the following figure (e.g. from the right side of Figure 5.29 to the left side of Figure 5.30).

Figures 5.30–5.32 are similar, and represent conceptual, developmental, and production-ready design phases, common in large, structured design and development projects. Note that many design projects do not include all three phases, and organizations may use different terms to describe these phases.

Figures 5.35 and 5.36 present the process from a different point of view. The process is shown from a production point of view, with Figure 5.35 representing a prototyping phase and Figure 5.36 representing a production phase. The activities in these figures are not a continuation of the sequence shown in the previous figures but represent a different flow with focus on production. The top row in the figures represents a component-level view, the second row in the figures represents an assembly-level view, the third row in the figures represents an installation-level point of view, and the bottom row represents a system-level point of view. The figures show the same processes and activities in these different production phases, as the figures represent high-level idealized workflows.

Figures 5.37 and 5.38 show activities that recur many times throughout the product and information lifecycles. The activities in the diagrams are not represented in sequence or order.

Figures 5.29–5.38 illustrate that MBD information can be used and created in many lifecycle activities throughout the product and information lifecycles.

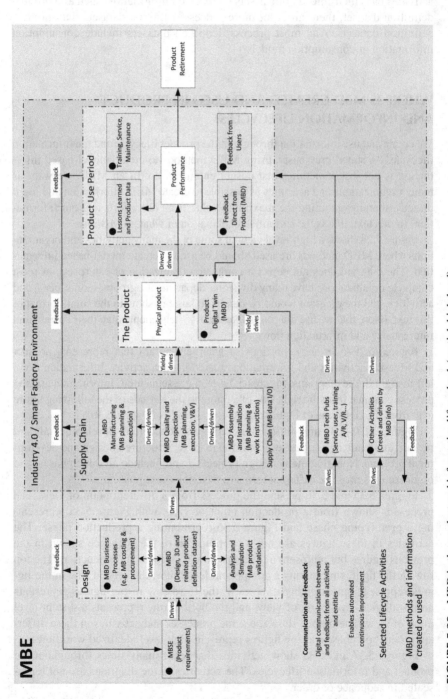

FIGURE 5.28 MBD in the product and information lifecycles 1 (partial, high-level).

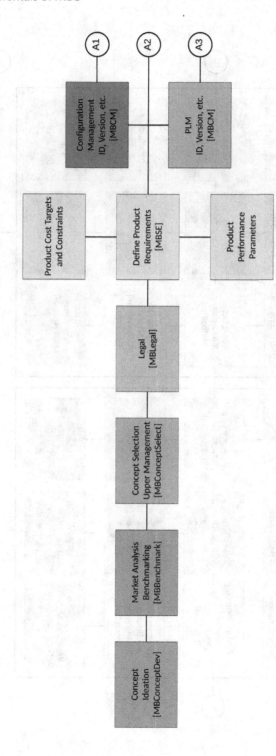

FIGURE 5.29 MBD in the product and information lifecycles 2 (pre-design).

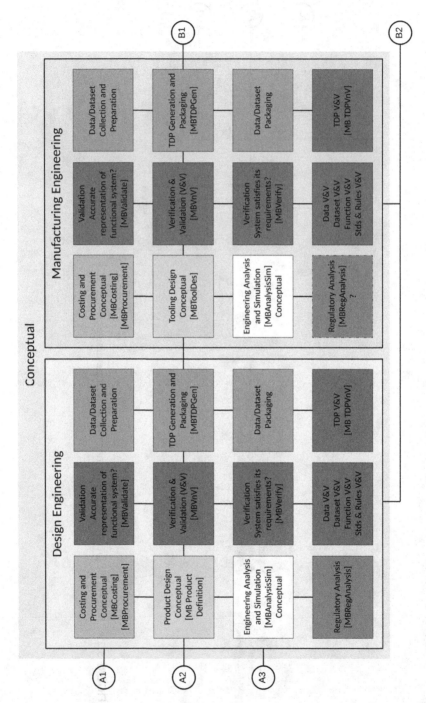

FIGURE 5.30   MBD in the product and information lifecycle 2 (conceptual design).

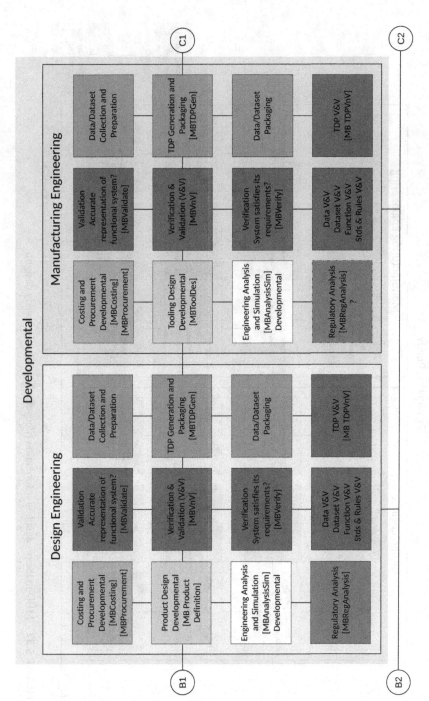

FIGURE 5.31 MBD in the product and information lifecycles 2 (developmental design).

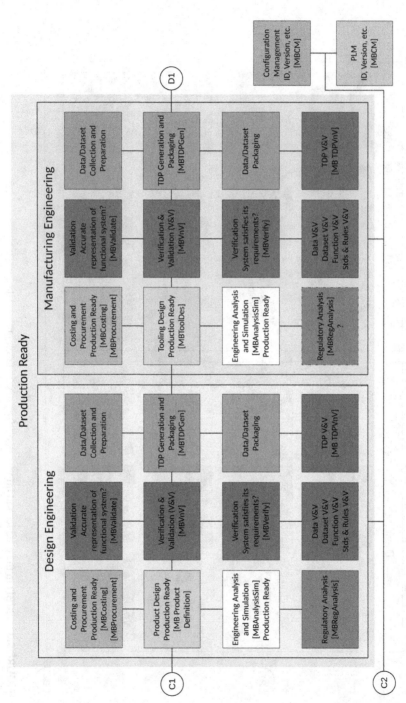

FIGURE 5.32 MBD in the product and information lifecycles 2 (production-ready design).

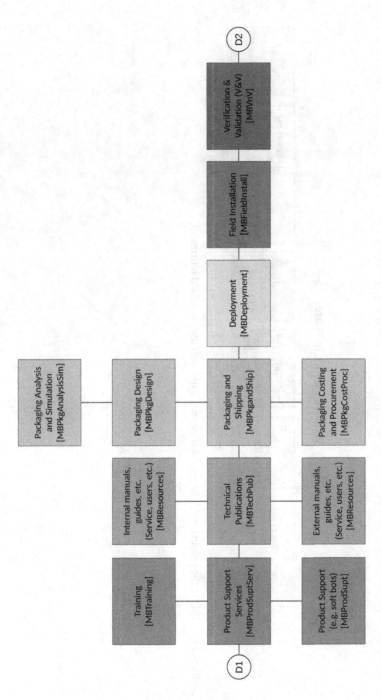

FIGURE 5.33 MBD in the product and information lifecycles 2 (support and deployment).

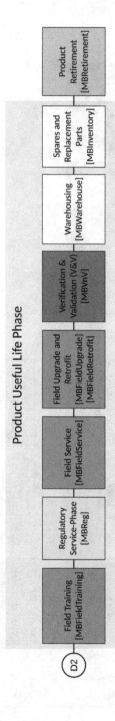

FIGURE 5.34 MBD in the product and information lifecycles 2 (product use phase and retirement).

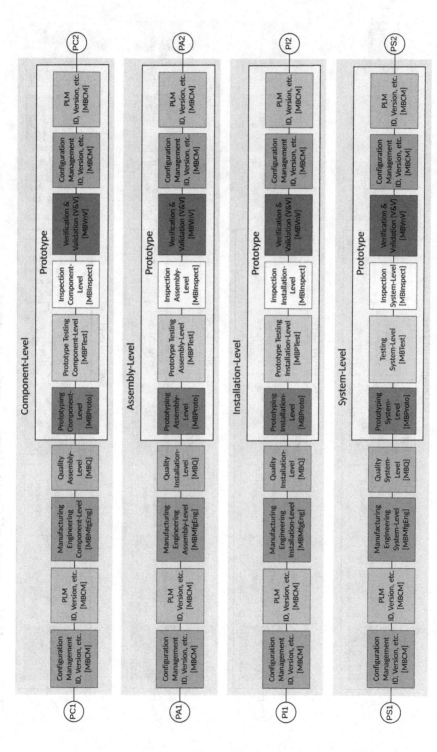

FIGURE 5.35 MBD in the product and information lifecycles 2 (production viewpoint 1).

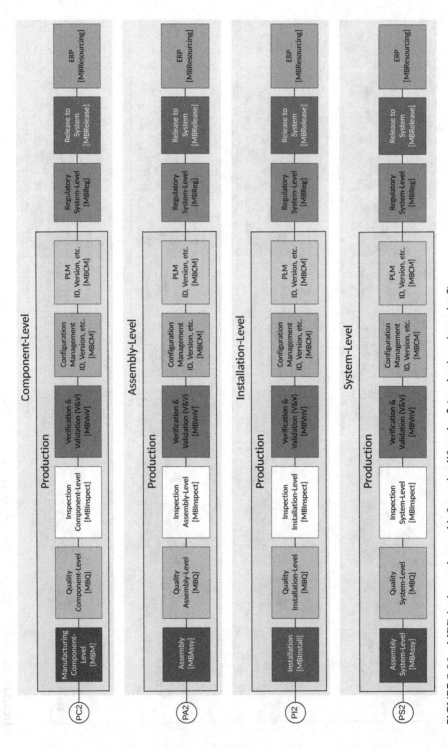

FIGURE 5.36 MBD in the product and information lifecycles 2 (production viewpoint 2).

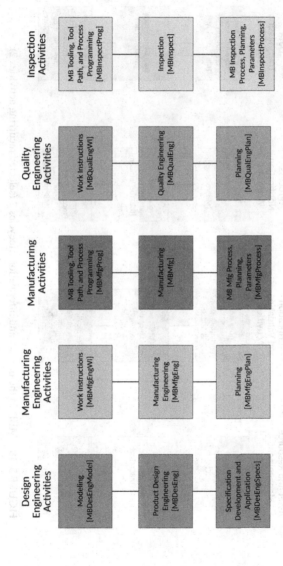

FIGURE 5.37 MBD in the product and information lifecycles 2 (selected recurring activities 1).

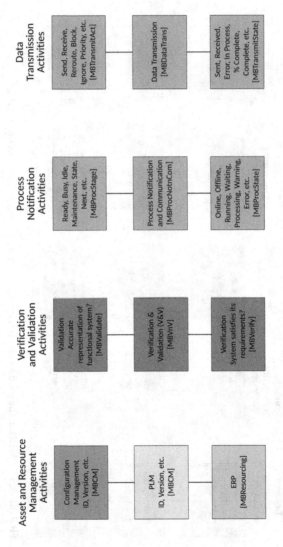

FIGURE 5.38 MBD in the product and information lifecycles 2 (selected recurring activities 2).

6 Becoming an MBE

ACRONYMS AND INITIALISMS

IT   Information Technology

The initialism "IT" is used to describe the discipline of information technology and it is used to describe the organization that is in charge of that discipline in an enterprise.

"IT is the technology backbone of any organization. It's necessary for monitoring, managing, and securing core functions such as email, finance, human resources (HR), and other applications in the data center and cloud."

Definition from Cisco.com webpage "How Do OT and IT Differ?", © 2024 Cisco (online)

https://www.cisco.com/c/en/us/solutions/internet-of-things/what-is-ot-vs-it.html#:~:text=The%20key%20difference%20between%20IT,%2Dend%20production%20(machines).

OT   Operational Technology

Hardware, software, platforms, programs, etc. used for industrial purposes. Examples are computers (workstations, servers, etc.), routers, sensors, programmable controllers, clouds, corporate, operating systems, platforms, software applications, etc.

"OT is for connecting, monitoring, managing, and securing an organization's industrial operations. Businesses engaged in activities such as manufacturing, mining, oil and gas, utilities, and transportation, among many others, rely heavily on OT. Robots, industrial control systems (ICS), supervisory control and data acquisition (SCADA) systems, programmable logic controllers (PLCs), and computer numerical control (CNC) are examples of OT."

Definition from Cisco.com webpage "How Do OT and IT Differ?" © 2024 Cisco (online)

https://www.cisco.com/c/en/us/solutions/internet-of-things/what-is-ot-vs-it.html#:~:text=The%20key%20difference%20between%20IT,%2Dend%20production%20(machines).

TERMS AND DEFINITIONS

Copycat MBE

MBE ideas, methods, goals, etc. based on what another company has done, how another company has approached MBE and implemented it. Usually, the MBE implementations being copied

DOI: 10.1201/9781003203797-6

are inadequate and not based on up-to-date ideas and methods, and thus do not provide adequate value, either to the organization being copied or the copier.

I do not recommend copying other organizations' MBE approaches. It may be okay to emulate some of the technical aspects of their approach, but the cultural aspects and overall goals will be insufficient. From what I have seen, often the organization being copied shares sanitized versions of what they did, how wonderful the results are, what they had to do to get there, the costs, pain, etc.

Opportunity Cost

The cost of not doing something, the potential loss from or value not achieved by doing something, used in the context of selecting one alternative over another.

"Opportunity cost represents the potential benefits that a business, an investor, or an individual consumer misses out on when choosing one alternative over another."

Investopedia (online) ("Opportunity Cost: Definition, Formula, and Examples") https://www.investopedia.com/terms/o/opportunitycost.asp

"The value of the next-best alternative when a decision is made; it's what is given up."

Federal Reserve Bank of St. Louis, Money and Missed Opportunities, Andrea J. Caceres-Santamaria (online) https://research.stlouisfed.org/publications/page1-econ/2019/10/01/money-and-missed-opportunities

WHAT NOT TO DO

In Chapter 5, I showed that model-based information is created and used in many places in an organization and throughout the product and information lifecycles. Most implementations I have seen do not come anywhere near achieving this ubiquity. Only a few groups create model-based information and use model-based information, confined to only a few siloed activities. Often only some of the people in these groups are using model-based methods, with others using legacy methods. The implementation is limited and inconsistent, which causes problems in the rest of the organization. Given these considerations, most implementations do not and cannot achieve much value, and the value is low compared to the costs (money, time, labor, CAPEX, OPEX, cultural, …). If we start our transformation by trying to shoehorn archaic work methods into new paradigms, we fail before we even start.

Most MBE implementations are too limited in scope. Why? The main reasons are:

- MBE development and implementation projects tend to be championed by people too low in the organization, people without adequate power, influence, responsibility, or authority
  - No true committed C-level support, ownership, and responsibility

- The people running the project tend to have technical roles, responsibilities, and authority to make technical decisions, but lack managerial and corporate-level authority to drive how the organization does business. This is a mistake and leads to conflict, chaos, and underperforming implementations.
- A dedicated team is not assigned exclusively to the project
  - People are distracted by, overly influenced by, and care more about their day job than the development work
  - The very people who should be working as champions trying to promote and advance the new methods also act as barriers right from the start.
- The work is approached in a siloed manner
  - Cross-organization and supply chain buy-in and cooperation are not achieved because the project is owned by a siloed group, thereby losing the ability to adequately influence other groups
  - Siloed approach leads to silo-centric goals (push rather than pull)
  - Engenders turf battles between siloed groups.
- The people leading MBx projects tend to focus on technical issues and technical goals rather than business issues and business goals – this also applies to people working on MBx projects, but the project leaders exert greater influence on the implementation and outcomes
  - Because MBE is usually considered as creating and using 3D CAD models instead of 2D engineering drawings, people treat MBE as a technical topic with technical goals and technical benefits
  - Technical benefits and value do not always yield benefits and value for the business – in fact, technical benefits are not equivalent to business benefits
  - Literature, engineering standards, blogs, vlogs, podcasts, and *"experts"* offering guidance offer the wrong guidance
  - People are often more focused on trying to climb the corporate ladder than doing something meaningful; thus, they aim low so they can take credit for something even if it doesn't add value.
- Many people think that MBD and MBE are IT or OT topics and driven by software; thus, they approach the project in too limited a manner.
  - If MBD and MBE are treated as an IT or OT topic, MBx development and implementation projects will be treated as IT/OT projects
  - If MBD and MBE work is an IT/OT project, it is likely that the work will be resisted by other groups, such as design, manufacturing, and inspection, as these other groups are the ones that create and use the MBD information (siloed, push rather than pull, not invented here, ...)
  - If MBD and MBE work is an IT/OT project, the scope, problem statement, solutions, and success criteria will be in terms of IT/OT (if the problem is an IT nail, the solution is an IT hammer)

- Some people think the key to migrating to MBD and becoming an MBE is software, that MBD and MBE are software disciplines and software problems, which are solved by software
  - People overemphasize the role of software in MBD and MBE (software salespeople help encourage this thinking)
  - People overemphasize the potential for software to solve their problems (software salespeople help encourage this thinking)
  - Software, IT, OT, etc. play a large role in MBD and MBE development and deployment, but they are tools in the toolbox and the environment in which MBE is implemented
  - Focusing too much on software or allowing software vendors to influence project goals and technical solutions is risky
  - If the problem is perceived in terms of software, then the solution is also perceived in terms of software. The desire to plug-and-play is compelling and valuable, but our focus must be to maximize the business value we obtain from our information.
- Pursuing MBE without a clearly understood goal of achieving maximum business value from the transformation (thinking baby steps are big steps – migration versus transformation).
- People feel they have to copy what another company has done (they don't actually want to try something new)
  - This is a problem in that no companies have really done a good job yet
    - They have not based their approaches on value propositions and methods like the ones I emphasize in this text, particularly:
      - Clearly and relentlessly focusing on replacing manual-use methods with direct-use methods
      - Understanding the many costs of information
      - Understanding that information has a lifecycle
      - Understanding that information incurs cost and drives value for an organization throughout the information lifecycle – its mere existence has a cost
      - Recognizing the lifecycle cost for ever-more-scarce information-related skills
      - Recognizing how to decrease the need for information-related skills
      - Maximizing value of information across the information lifecycle.
- Early adopters of model-based methods made many missteps, and their implementations are inefficient and hamstrung by inertia from:
  - Sunk capital
  - Getting there first
    - Cost of developing new capability in software and systems
  - People protecting their turf (carved from/built on now outdated ideas, approaches, and technology)

- Methods created years ago around cobbled-together new-at-the-time technology
- Outdated ideas about MBD and MBE baked into their business models
- Mature operating procedures for now outdated and inefficient MBx methods
- They see their implementations as gold standards, but in reality have become their inertia-laden status quo.
- Avoid Copycat MBE.
- Big consulting firms claim to know this space and this discipline, but they don't. Over many years of working with others to develop this discipline, no one from the big consulting firms has participated
  - When I meet with large clients, they often tell us about how large consulting firms are helping them in MBE
  - I am more than skeptical that they are actually "helping", as I don't think they even understand the topic.
- In summary, most MBE projects are driven by the wrong group in the organization for the wrong reasons to achieve the wrong goals with too little support.

Of the reasons listed above, it is a problem that development and implementation are driven from too low within an organization, often led by design engineers, their managers, or middle management. It is also a big problem that the problem and solution are approached from a technical or technological perspective. Related to these issues is development and implementation are constrained by siloed thinking. Note that these issues are interrelated. For example, because the work is not supported at the highest level in the organization, the project is led by technical professionals in a siloed group, the technical professionals see MBx as a technical problem with technical solutions, the siloed group has limited power and insufficient influence over other siloed groups, there are often rivalries and contradictory goals between siloed groups, upper-level management doesn't fully understand the work, so they don't fully support it, a formal dedicated team is not created to do the work, the people in the team are assigned to work on the project while still responsible for maintaining their production activities, they are more worried about their day job and see the MBx project as a side activity, thus they have a greater stake and place a higher priority on maintaining how they work today than how they could work tomorrow – people see the work primarily as a technical, CAD-related, CAD-driven activity, which is part of the reason that upper-level management thinks it is trivial and not an enterprise-wide improvement project. Whew!

Remember, MBD and MBE are not software problems. To quote the first part of the text, MBD and MBE are not about CAD, and MBD and MBE are not about software. Software plays a role in MBx, but software is the easy part. In fact, the technological aspects of MBD and MBE are the easy part. Understanding and accepting what MBD and MBE actually are, driving cultural change for the better, and overcoming organizational inertia are the hard parts.

There are many misconceptions about MBD and MBE. And many of the misconceptions have led to failed, stalled, or underperforming implementations.

ARE WE THERE YET?

What does it take to become a successful MBE? How do we use MBD and MBE to achieve value from transformation, preferably the maximum value possible? The answers to these questions are layered, and we need to bring many interrelated ideas together. The first questions we need to ask are:

Question 6.1:   What do we think an MBE is?

Question 6.2:   Why do we want to become an MBE?

I've spent a lot of time in the preceding chapters trying to address and answer Question 6.1. While many people are enthusiastic about MBE, I don't think that many have an adequate understanding of MBE or an adequate understanding of its value proposition. I think most people are excited about not having to create drawings anymore, for all the reasons I listed in previous chapters, the most pressing reasons being not having the skills needed to create adequate quality drawings, not understanding what drawing quality entails or requires, not caring much about drawing quality, not being empowered to obtain the skills or to use those skills in the design process, and not being given time to be careful and do a good job, and for people that receive, must understand, and use drawings, similar skills- and time-related problems and priorities exist.

Drawing-related skills generally don't exist in new hires, as they are under-emphasized in education and apprenticeships, and thus have become underdeveloped, undervalued, and atrophied in the workplace. There is inadequate staff and inadequate time to use the skills if they are present at all. This trend has been growing since 1990, as solid modeling CAD software has made it easy to create a drawing. Drafters and designers are extinct in many organizations, and the skills they had and emphasis they placed on drawings and drawing quality went extinct along with them. Refer to Figures 2.4 and 2.5 for more information about how the design engineering workforce, skillsets, and priorities have changed.

Fast forward to today and current ideas about MBD and MBE, and it seems that the main reason people want to pursue MBD and MBE is because they don't want to make drawings anymore. And that desire is driven by the problems described in the previous paragraph. Many people end up creating annotated models for manual use, and thus migrate rather than transform. I think of process migration as a baby step, a horizontal shift in how we work. I think of process transformation as a significant order-of-magnitude change in how we work that is intended to yield significant benefits. See Figure 6.1 for a graphical description of transformation versus migration.

In Figure 6.1 I include the idea that transformation offers significant rewards, such as 10X improvements in one or more business characteristics. "10X" is a buzzword and part of a long history of claims about programming, business

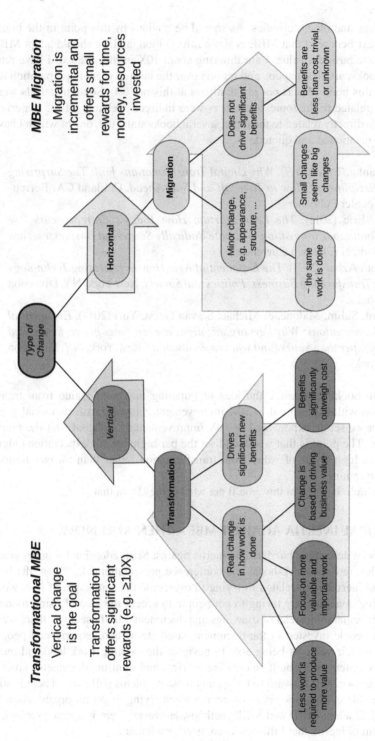

FIGURE 6.1 MBE transformation versus migration.

operations, and other activities. As should be obvious by this point in the book, in this text I contend that MBE is about information, and we should adopt MBE to increase business value. I am thinking about 10X in this context. I have read many books, articles, papers, and reports over the last ten years in preparation for writing this book (this is not my first pass at this project). Some materials were directly related to the topic, and others were indirectly related. Of the materials that were directly related to the topic, several books stand out, one of which I have already mentioned and quoted.

Saldanha, Tony (2019). *Why Digital Transformations Fail: The Surprising Disciplines of How to Take Off and Stay Ahead*. Oakland CA: Berrett-Koehler Publishers.

Ries, Eric (2011). *The Lean Startup: How Today's Entrepreneurs Use Continuous Innovation to Create Radically Successful Businesses*. New York, NY. Crown Currency

Azeem Azhar (2021). *The Exponential Age: How Accelerating Technology is Transforming Business, Politics and Society*. New York, NY. Diversion Books.

Ismail, Salim, Malone, & Michael S., van Geest, Yuri (2014). *Exponential Organizations: Why new organizations are ten times better, faster, and cheaper than yours (and what to do about it)*. New York, NY. Diversion Books.

These books emphasize the idea of pursuing maximum value from transformation with exponential or 10X improvement being a transformational goal. Whether we set an exponential or 10X improvement as our goal isn't the point, however. The point is that we should set the bar high, set our expectations high, and drive high levels of value from transformation rather than meager returns from migration.

Aim high. If you aim low, you'll get what comes from that.

CULTURAL INERTIA AGAINST MBE – THEN AND NOW

I have been describing barriers and inertia against MBE adoption for many years in articles, reports, workshops, and conference presentations. Early on, the barriers and inertia were related to trying to overcome the status quo of how work was being done, namely trying to get people to accept the idea of using model-based information instead of drawings and documents. In those days, there were a lot of people invested in the document-based status quo, experienced people who saw their value as being able to navigate the existing workflow, and most business systems were built on creating, storing, and shuttling documents around the workplace. Fast forward to today, and these problems still exist. There is still considerable cultural inertia to overcome when trying to get an organization to accept and adopt MBD and MBE methods. However, there is a new problem, a new form of inertia that I think is even more problematic.

As the concepts of MBD and MBE are no longer nascent, some companies have migrated to model-based methods, and there are now a few engineering standards that describe model-based methods, there is a new cultural inertia to overcome. This is the inertia built on the old idea that MBD and MBE are CAD topics and that the goal of MBD and MBE is to use 3D information manually. People want to help you adopt these outdated methods, software companies want you to buy their software that supports these outdated methods, and people like to present their accomplishments in these outdated methods at conferences and as influencers, and so on. There is a lot of hype in this space.

Today, we not only have to get over the old hump of "we work with drawings, not models", we also have to get over a new hump, the inertia from sunk capital and reputations built on faulty, misguided, limited, and outdated ideas of MBD and MBE. Now we also have to overcome the inertia of existing MBE implementations, the software businesses profiting from them, and people who want to keep their software businesses, reputations, and consulting businesses afloat. We must generate enough momentum with the ideas presented in this text to overcome the inertia of current MBE thinking, which I sum this up in Figure 5.2, the figure with the horses pulling the car. I find it difficult to help companies advance beyond inefficient approaches to MBD and MBE and Copycat MBE.

There is a better way.

Okay. Enough of that. Let's focus on what we should do.

BECOMING AN MBE – DIGITAL TRANSFORMATION AND MBE

We need to rethink how we work and how we create and use information, to untangle the current state of how we work and how we think we should work from what is possible. Companies struggle with these activities, as people are hard-wired into how they currently work, current priorities, what they are rewarded for, and what constitutes doing a "good job". As stated elsewhere in the text, many companies operate chaotically, and people are expected to respond to and excel in managing chaos, working as firefighters, expediting orders, and responding to unnecessary-but-common crises. In such an environment, processes are out of control, and people are expected to adapt to and manage the chaos. I hope it is obvious that this is not the best way for an organization to operate. I first learned about out-of-control processes in 1989 or 1990. I've seen a lot of chaos ever since. It is not always easy for people to accept that the way they are working is inadequate. This is true even in cases where they are very aware of their environmental shortcomings and complain about it with their coworkers. As a consultant, I often embed with clients and work as an advisor to their staff. It is different to commiserate with your peers than to have someone from outside tell you that you're operating inefficiently.

While it is not absolutely necessary, I strongly recommend that companies clean up their operations before transforming how they work by adopting MBE methods. Most companies have formal processes in place. Maybe the processes are not 100% complete, but there are formal processes, often in the form of SOPs.

Some companies take things further and map their processes and procedures using Unified Markup Language (UML), Business Process Modeling Notation (BPMN), or similar methods. Environments like these are easier to transform, as they already have a culture (or a semblance of a culture) of working formally and a model that represents how they work (or are supposed to work). It is best to transform formal processes. One of the main differences between manual processes and automated processes is that manual processes are usually very flexible, so flexible that the path followed from start to finish can vary greatly. This variation is due to human intervention and intermediation. People assess inputs and outputs during process steps and make adjustments accordingly.

I'm reminded of a longtime friend, a machine designer, explaining how he had worked for a small company that didn't always make formal engineering drawings to document their products. My friend said he'd make an informal sketch of equipment on green engineering paper and then spend the next week or two in the shop completing and adjusting the design as part of the production team during production. While such an approach may work for a small company putting out custom products, the process is informal, it is not scalable, it is not repeatable, it is not reliable, it is inadequately recorded (or documented), and by no means should it be replicated as a model-based process. One of the problems that results from informal processes is that we don't really understand what we end up with, what the end product is, or why it is the way it is (e.g. we don't understand the product, why it works or doesn't work, what the problem is, what causes the problem, what should be done to correct the problem, what process was actually used, etc.). Informal processes do not yield trustworthy results, as uncertainty and variability are too great. Informal processes are by no means optimal.

Informal processes and the problems they cause are not minor issues. Informal processes may work where the status quo provides adequate support, in terms of time, expertise, experience, expectations, resources, and skills. Informal processes rely on experienced staff, often very experienced staff, who can make judgments on the fly and adapt to a chaotic workplace and chaotic workflows. Usually, such an environment is not subject to regulatory oversight or much oversight of any kind. People may be thrust into environments that do not provide the right tools to do the job as it is formally described – the worker has to make do and *wing it*, and do the best they can, given the constraints. During a recent interview process for an MBE implementation, I asked a skilled inspector to show me how he evaluated the position of holes on parts. A stack of parts that needed to be inspected sat on a bench in the inspection lab. The work wasn't being done in the company's main inspection lab, which had an assortment of inspection tools and equipment. Very few tools were available to this inspector, so he had to inspect the position of the holes visually using an engineering scale/ruler. The scale and the process were not precise enough to adequately evaluate the holes, and evaluating the relationship between holes and a datum reference frame wasn't possible with the tools available…. The manager expected the inspector to inspect the parts with the tools available, without complaints or statements that the tools were inadequate. People make do. We are adaptable. We're good for that. However, we should not try to automate such a poorly executed process. The reason the process was executed poorly was that the worker was inadequately supported and was expected to make do and use the wrong tools in too short of a time to adequately evaluate the parts. A checkmark on a document was more important than actually ensuring the parts conformed to their specifications. The moral of that story is that the parts may "pass" inspection, but they might not fit or align properly during the assembly process. This is another side effect of informal processes, that information needed by later steps doesn't exist or is inadequate, e.g. just in time manufacturing requires knowing that part geometry is within specification before we try to assemble the parts. So the assemblers ream out the holes to provide more clearance, the fasteners fit through the holes, but now the parts have inadequate bearing surface for the fasteners, the fit is too loose, the quality of the assembly is diminished, and the customer buying the product suffers. This gets back to Dr. Taguchi's quality loss function. Such a scenario is a loss for society. Everyone loses. But we met the inspection schedule …

We can separate projects into two broad types: *greenfield* projects and *brownfield* projects. I'm using *greenfield* and *brownfield* to describe organizations or companies. Simply, *greenfield* is a construction term meaning that a site is undeveloped. From the point of view of MBE transformation, *greenfield* describes a new organization, one that isn't burdened by preconceived notions of how they work, so there is less organizational inertia that must be overcome during transformation. *Brownfield* in this context is an organization that exists, has history, has a status quo, has a work culture, and has strong ideas of "this is how we work, why we think we have to work this way, why we can't change, …", and thus is burdened by significant organizational inertia. From this point of view, transformation for a greenfield organization should be much easier than for a brownfield organization. However, given that even the newest company has some inertia, as mission statements have been written, C-suite executives have made promises to stakeholders, and as many new hires come with experience working in other environments, the overall organization may not be overly burdened with baggage, but there will be pockets of resistance. Resistance. There is always resistance, even when the value of doing something is blatantly obvious. It may be as simple as grumbling about "who messed with my turf?" to worrying that "my turf is no longer relevant". The latter is part of the impending changes we will see from broader AI adoption.

To undergo a transformation:

- We need to understand an organization's formal procedures or SOPs, if they exist (e.g. are documented or modeled)
- We need to map how the organization actually works, as the formal procedures might not be closely followed in practice
- We need to understand:
  - What information is created/authored
  - Why the information is created
  - Who creates the information (e.g. which department, activity, group)
  - What the information is based on (what is the basis of the information)
  - The intended purpose of the information
  - How does the authoring activity think the information is used
  - How the authors think the information will be used
  - The content of the information
  - If the adequacy and quality of the information is evaluated
  - If the information is evaluated, how is it done, how is it recorded, who does it, etc.
  - What form and format the information is released
  - How the information is released
  - How the information is transmitted (which system(s), formally, informally, …)
  - How the information is revised
  - Who revises the information
  - If the information released is official, binding, unofficial, …
  - Who uses the information (e.g. which department, activity, group)
  - If the information changes state in the use case (e.g. converted to other units, translated, changed from one thing to another such as hole dimensions and tolerances converted into a drilling process and tool selection, to a different language)
  - How the information is actually used (by which department, activity, group, for what tasks, for which use cases …)
  - What the users are trying to accomplish in the activity
  - Which tasks use the information in the use case
  - How the information is actually used correlates to its expected use (actual use versus intended use)
  - If the information and the activities it supports are necessary
  - How, what, why, and when information flows through the organization, supply chain, clients, regulators, etc.
- We may find that information is created but not required by anyone, perhaps because processes have changed, but the authoring activity hasn't been notified or freed from the task of creating the information, or stubbornly clings to the status quo and keeps creating the information (*I'm talking to you, explicit dimensions...*)

- We need a clear understanding of the organization's objectives, what the organization is trying to accomplish, stripped of distraction and outdated burden
  - For example, if we are an OEM designing and making a product, our goal is to produce products that meet our clients' needs for a competitive price, with adequate quality, and to maximize profitability or stakeholder value (e.g. minimize scrap, rework, time to market, cycle time, material usage, CAPEX and OPEX burden, etc., maximize productivity, quality, CAPEX and OPEX utilization, throughput, efficiency, perceived value to customers, perceived brand value, etc.)
  - Okay, that isn't stripped down. How about this instead?
  - Our goal is to produce products in a manner that generates the greatest value for the organization, our stakeholders, and our customers over the life of the organization (likely in that order)
    - E.g. We want to generate maximum value and return for the organization while staying in business over the intended lifespan of the organization
- We need to understand what information is actually required to achieve the objective
- We need to understand the cost required to create the information (all types of costs, e.g. money, time, labor, skill requirements, etc.)
- We need to understand the cost of using the information (all types of cost, e.g. money, time, labor, skill requirements, etc.)
- We need to understand how to optimize the workflow so that the organization requires the least amount of information to obtain the maximum benefit while meeting their objectives
- We need to achieve our goals in an ethical manner that serves our stakeholders and society.

Aside from the last bullet, the items above may seem mercenary. From an economic perspective, and paraphrasing quality master Dr. Genichi Taguchi and his quality loss function, inefficiency can be thought of as an organizational quality problem, and reduced quality means a loss in value for society. Inefficiency at work equates to a cost for society, a tax if you will. Inefficient creation, use, and management of information impart a cost on an organization, its stakeholders, and its customers. Those costs affect society. I am playing fast and loose with the ideas of the quality loss function here. But it provides a basis for the ideas in this text. There is a high cost to using documents and document-based workflows. Manual-use information and manual workflows have a high cost. There is a high cost to using archaic inefficient methods. These costs not only affect the organization but also society as a whole.

Thus, mercilessly stripping down an organization's information creation, use, and management to bare essentials and rebuilding processes to optimize the value obtained from information is ethical. As stated earlier, MBE is efficient, it is lean, and it is green.

Starting the Model-Based Journey

To start a model-based transformation process, we must ask a lot of questions, so we understand the current state, where we want to get to, and how we want to get there (Footnote 6.1). First are questions 6.1 and 6.2, "What does the organization think an MBE is?" and "Why does the organization want to become an MBE?". We also need to ask enough questions to gain the understanding described in the bulleted list in the previous section.

In the context of MBE transformation rather than migration, we also need to understand the prevalence and extent of document-based processes that use information manually. We need to understand the systems, methods, justifications, behavior, influences, culture, etc. built to support those processes. We need to understand:

- How much we use documents and visual information
- Why we use documents and rely on visual information
- Why we think we need documents and visual information
- How we use documents and visual information
- The cost of our manual activities that use documents and visual information
- Associated metrics and milestones.

From this information, we can understand the current state of information. We can establish a baseline and from there determine transformation goals and chart the future state. Once the current state is understood, we can begin to distill and condense our workflows to extract the true goals and core information needed to achieve those goals. We can condense 50-step processes into 5-step processes.

FOOTNOTE

6.1 *Understanding and mapping the current state should be a separate activity from defining and mapping the future state. These activities should not be combined. It is likely that different people will be involved in each activity, and at the early stage of determining the current state, project team members will not yet have an adequate understanding of MBE to define the future state.*

As we progress through understanding and mapping how we work today and deciding how we should work tomorrow, we have many options and decisions to make. We need courage, vision, ownership, management support, and the right environment to transform optimally. To succeed, we need a strong understanding of our core mission, and we need to determine the minimum information needed to achieve it. We must find and replace activities that require low-value information, and we must find and replace activities that require high-skill levels to use the low-value information (low-value high-skill activities). We need to stop wasting our time and resources.

ESTABLISHING AN MBE TRANSFORMATION PROJECT

The first activity in an MBE transformation project is to hold one or more MBE workshops. The workshops serve a dual purpose:

1. Train: Teach staff about MBD, MBE, the MBx ecosystem, and MBx value propositions
2. Establish a baseline: Ensure staff has a common up-to-date understanding of MBD, MBE, and MBx value propositions.

People need training because few people have actually been trained in MBD, MBE, and especially the information-management-related aspects of the topics provided in this book. Some people, many people in fact, think MBE is an IT or OT discipline, that IT and OT systems are the answer. Many others think MBE is a software issue, that software is the answer. While IT, OT, and software play a role in MBE, they are not the main challenges, not by a long shot.

People need to be elevated to a common baseline because MBD and MBE have been around for a few decades, there are still a lot of outdated narrow ideas being promoted, people have mixed ideas of what MBD and MBE are and why they should adopt them, and how to go about it. What, why, and how. Big questions. If we don't establish adequate understanding and get everyone aligned to the same baseline, we are in for a chaotic transformation, or more likely, a migration that causes a lot of pain, costs a lot, but yields little real value for the organization. Most people today who talk about MBD and MBE have outdated ideas, describing MBD and MBE as CAD topics, as replacing drawings with models. If needed, refer back to the beginning of Chapter 1 that explains what MBD and MBE are and what they are not. MBD and MBE are not merely about replacing drawings with models. Another recent trend is that MBE is about inspection. This is misguided and based on examples of Copycat ideas about MBD and MBE.

Following the initial workshops, the next step is to set the goals, scope, parameters, budget, and operating constraints for the work. How much does the company want to do? Do they want to do enough? The answer to the latter question is often driven by how much they understand and their answers to questions 6.1 and 6.2. The purpose of the workshop is so the organization can provide the right answers to questions 6.1 and 6.2. I don't want to waste my time, and I don't want my clients to waste their time, money, and resources pursuing the wrong goals for the wrong reasons.

After establishing the scope of the transformation project, one of the big questions is how do we get there, how do we get to the end state? Do we do it in one step, two steps, or more steps? We will discuss this in a few sections after this one. Suffice to say that there is not a one-size-fits-all answer. Every organization is different. We may all end up with the same goals, but how we decide to get there depends on a lot of factors. Eventually, we'll all get there because there will no longer be a reason or value to work the way we currently work. As I said, someday people will look back to today and chuckle at what seems to be primitive and inefficient work methods.

PUSH AND PULL

The concepts of *push* and *pull* were introduced in Chapter 2. Our goal should be to ensure we are developing our new workflow with greater emphasis on *pull* than *push*. Remember, *push* and *pull* refer to how ideas and information spread through an organization, which matters for transformation activities. In a push scenario, an author/creator, generally in an upstream activity, is driving, as they push something to users, regardless of whether the recipients want what's being pushed on them or not. This scenario is a failure waiting to happen, ready to foment a revolt. Often the response to *push* is *pushback*. Not an effective business strategy. In a *pull* scenario, the users/recipients are driving. The users request something from the authors/creators, upstream activities. I am describing supply-and-demand. If we have to choose *push* or *pull*, *pull* is stickier than *push*, it is more likely to persist if recipients want something than if they don't.

In this scenario, it's best if we have mutual interest and willingness to do something, that what the authors/creators in upstream activities want to create corresponds to what the users/recipients want to use. From an MBE perspective, this equates to the supply side aligning with the demand side, authors/creators in upstream activities wanting to create and provide model-based direct-use information and users/recipients in downstream activities wanting to use the model-based information directly. This is classic win-win. Another purpose of baselining staff in the initial MBE workshops is to establish enough understanding, enthusiasm, and momentum so that we have both *push* and *pull* for creating and using MBD and becoming an MBE.

BUILDING A PROJECT TEAM

It's important that we establish a team that is dedicated to the project, whose main priority, preferably their sole activity, is the MBE transformation project. The project team should consist of a broad cross-section of the organization. Some organizations are too small and staffed too lean to dedicate enough staff to the project, so their implementation teams will be distracted. In other cases, due to factors listed at the beginning of this chapter, the team leading the work might not have adequate management support to pull people from other departments or activities. Regardless, we want to include people representing every affected activity or department in which relevant information is created and used. Authors/ creators, users, customers, and stakeholders.

In addition to ensuring that project participants prioritize the transformation work, they should be shielded from undue pressure from production or from people trying to block transformation and retain the status quo. I cannot emphasize enough that this is a very real problem that can derail transformation without adequate intervention. The values, key performance indicators, metrics, rewards, and priorities of manufacturing organizations are to get work done, which often includes whatever it takes to get it done. As discussed earlier, schedule is often the main priority regardless of what it says on our motivational posters.

We are trying to achieve several goals with our project team participants. We want:

1. To ensure success by selecting key staff members for the team
2. Project team participants to work on the phases of the transformation project
3. People who know how we are supposed to work versus how we actually work, who have technical competence, who create and use information, and who represent as many relevant activities as possible
4. A balanced team of staff members, supported by advisors and vendors
   a. Staff including developers, authors/creators/upstream activity representatives, users/recipients/downstream activity representatives, stakeholders, internal and external organizations for all affected activities across the organization and the product and information lifecycles
5. People who are willing to try new ideas and accept process improvement ideas over maintaining the status quo
6. People who are respected and trusted within the extended organization
7. To protect the transformation project team from hostile or obstructive intervention from people
   a. Who are opposed to transformation
   b. Who are overly invested in the status quo
   c. That see transformation as a threat to production.
8. To convert key members of our staff into MBE champions, people who will act as mentors, guides, and leaders who motivate others to accept, adopt, implement, and optimize the work.

UNDERSTANDING AND MAPPING THE CURRENT STATE

After scoping, baselining, selecting the transformation team, outlining the project, securing funding and management support and sponsorship, the next phase of MBE transformation is understanding and mapping an organization's current state. This activity should include peripheral organizations, such as customers, regulators, partners, contractors, and suppliers. This phase requires planning and a lot of information gathering. Some information is obtained from company documents, some information is obtained from interviews and elicitation, and some information is obtained through observation. Don't underestimate the importance of observation.

The first steps of information gathering should be discussions or meetings, preferably with key staff members.

Many companies operate formally and have published Standard Operating Procedures (SOPs). This is a great place to continue the elicitation process.

Remember, MBE is about information and the information lifecycle, and the information lifecycle may be different than a corresponding product lifecycle. We should focus on information and the information lifecycle aspects of MBE and digital transformation. Thus, understanding and mapping the current state is an exercise in understanding and mapping information and information lifecycles, how and where information is authored, how it flows through the extended organization, how and where it is used, for what reasons, in what forms, and activities related to products and processes used to design, develop, manufacture, evaluate, and deliver products. Our maps and flowcharts representing the current state are focused on information, information state, information flow, how information is used, the information lifecycle, and processes and activities in which this occurs.

Note that this section merely provides a brief overview of understanding and mapping the current state. The full process is long and iterative, as it takes a lot of effort to unravel how companies work and how they create information, how they store it, what they think is necessary and important, who owns the information, how the information is revised and why, how the information is managed, how the information is used, and what users do with the information. If we don't understand clearly how a company is operating and creating, using, and maintaining information today, our chances of a successful and valuable transformation are diminished.

To close this section, I assume that every OEM and similar organization that should be interested in MBD and MBE are operating today as document-based organizations that primarily use information manually – I assume this is true even for organizations that think they've already transformed to become an MBE. Organizations that use CAD, CAE, CAI, CAM, CMMs, etc. tend to be primarily document-based and focused on using information manually. We have a lot of computing power, systems, platforms, hardware, and software, but we like to use them to do old-fashioned stuff digitally. It makes sense, as we've been working that way for hundreds of years.

Figure 6.2 shows a current state workflow with 54 steps. Note the repeating occurrence of the RIDICT sequence throughout the workflow, which is represented in the boxes simply as *Parse and Convert Input*.

While we might not recognize it, because the RIDICT sequence is part of how we work today (the status quo), using RIDICT requires a high level of skill, and each time we use the RIDICT sequence, we incur significant expense, burden, risk, and inefficiency. The process steps with *italicized text* and marked with "M" in the figure represent manual activities where information is created for manual use or used manually. Note that there are 41 manual-use-related steps in this simplified workflow. There is a lot of room for improvement.

Understanding an organization's culture is critical. Cultural factors are much more important than technical factors. Many people think MBD and MBE are technical topics. They see technical problems that require technical solutions. Unfortunately, this thinking is widespread, and it leads to solving the wrong problems. The most important aspects of an organization that need to be understood are its culture, its priorities, and its approach to information. We can develop and

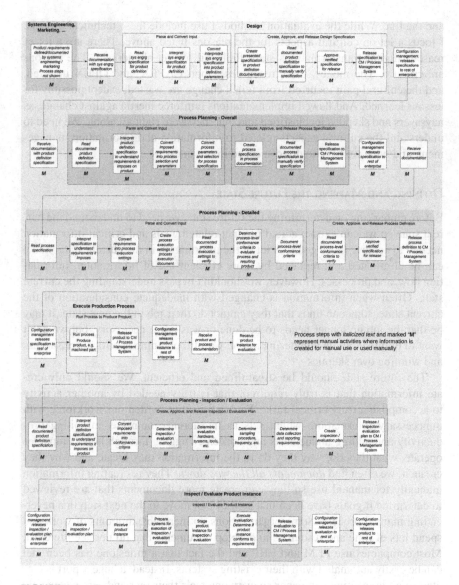

FIGURE 6.2   Current-state manual-use workflow example.

implement whatever technical solution we want, but if it is implemented in an environment that is focused on the wrong things or operates too informally, the solution won't solve the real problem.

Early stages of MBE transformation include careful analysis of an organization from technical and cultural perspectives. Several MBE capability and maturity matrices and assessment methods have been developed. They focus on technical aspects, such as CAD, drawings, models, BOMs, PDM, PLM, infrastructure, and the like. While the evaluation methods I use include these technical aspects, I focus a lot on cultural aspects for the reasons listed above. There are a lot of reasons digital transformations stall, flail, fail, or get abandoned. The reasons are cultural. The money, time, resources, labor, CAPEX, and OPEX invested, and the cultural pain incurred from failed change are wasted, except possibly for lessons learned. And flawed implementations reinforce inertia against change, as naysayers and skeptics feel justified and that they were right all along. We need to understand the cultural aspects related to information and information management, and what we have to do to succeed in our quest.

Defining and Mapping the Future State

Before embarking on determining the future state, we need to understand the current state so we can extract actual product-related goals and critical information needed to achieve those goals. A lot of the information in the current state model can be eliminated and replaced by more efficient use of information, information flow, and processes. However, we shouldn't bypass understanding the current state. Often when information is changed with inadequate consideration of the current state, someone finds that they cannot do their job in the new state. It may be that the job no longer needs to be done or was never necessary, but we don't want to find that out by blundering into it. Careful planning is critical for success and winning over skeptics.

The future state should be streamlined and efficient. We should only create information that is truly required, that is truly needed by another activity to accomplish a truly critical task. We don't want to burden our workforce with unnecessary information, which acts as interference or noise, inhibiting productivity and adding undue costs and burdens. Digging through how companies operate, what they do, why they do it, the information they create and why they create it, we often find *documents all the way down*. They're creating information manually for manual use so they can create more documents that are reviewed to determine if goals have been met so they can move to the next step in a never-ending manual process. Some organizations do incredible amounts of work and spend lots of time and resources to develop documents that do not need to exist. Most companies use PLM and ERP and the fact that a milestone has been met can be easily automated with their existing systems. Instead, a huge report is created so a big meeting can occur so everyone feels their rear ends are adequately covered and they can move to the next manual step.

If we scrape our processes hard enough, if we ask *why* enough times and answer honestly with a clear understanding of what we are truly trying to accomplish, we will end up with very streamlined processes. Mind you, we still need complete and accurate product definition information, complete and accurate process definition information, and to validate and verify all information and processes. Those requirements do not change when we transform our processes. We change our processes to eliminate wasted steps, unnecessary information, information that is difficult to create and requires specialized skills to create, information that

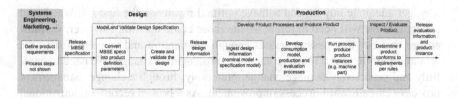

FIGURE 6.3 Transformed workflow: future-state direct-use workflow example.

is difficult to use and requires specialized skills to use, and information that we just don't need. All information has a cost, a cost that is incurred by the organization and whoever else uses or owns the information. Our future state should be optimized to maximize the value we obtain from our information, from our IP. The formulas in Chapter 5 illustrate how we can understand and quantify the cost and value of our information, and hopefully help us realize the potential of MBE.

As with the previous section on the current state, this section will not delve too deeply into mapping the future state. My goal is for industry to envision, accept, adopt, and achieve a future state in which information is value-optimized, in which direct-use information is created as much as possible and used directly as much as possible.

If we can envision a future where we've been operating for a long time as an efficient streamlined MBE, and someone suggests that we revert back to how we work today, it will be a very funny joke, like suggesting we go back to the days before computers.

Figure 6.3 shows a transformed future state workflow of the workflow shown in Figure 6.2. Whereas Figure 6.2 had 54 steps, the workflow in Figure 6.3 has 7 steps. All instances of the RIDICT sequence have been eliminated, no manual-use information is created, and there are no manual activities that require manual-use information.

As stated in the previous section, cultural aspects must be addressed before transformation. We must understand our culture, and we must define our success strategy before we roll out our transformed processes (kudos to Tony Saldanha for this idea). We need to understand what the cultural barriers are and how to overcome them *before* we encounter them. Our approach must include recognizing barriers to success and how to overcome them, and this must occur early in the transformation process. Success must be baked into the transformation process.

How to Get There – The Transformation Process

Digital transformation from a current state into an MBE is a long journey. We like to think of discrete events or projects in which we transform something, but there are many somethings to transform, and there are lessons to be learned, and obstacles and barriers to overcome, and we can and should be ready to continuously improve the results. We should aim high right from the start, but we need to be ready for the long haul. It is rare that a company would do everything the right way, get adequate buy-in, and achieve optimal results on the first try. We should expect that we will need to refine whatever we come up with. However, we must ensure that our first pass generates enough momentum and value to justify the cost and pain incurred to overcome the inertia against change. We need to carefully and proactively manage change, preferably through a change management process geared toward process improvement. As a broken record, organizational culture presents the greatest challenges to successful transformation.

I see the transformation process occurring in four phases or stages. Each phase consists of several activities with specific goals and milestones. I envision a largely serial or waterfall process, but many activities can and should be run in parallel. Some things have to happen before others can begin in a meaningful way.

Earlier, I recommended several books about Lean and Lean Startup methods. There are many great ideas in these texts. However, the ideas of minimum viable product (MVP) and pivoting may be detrimental in MBE implementation. We may inadvertently *prove* the wrong thing for the wrong reasons. For example, if we build an MVP for MBD information assets and MBE use cases, and then evaluate these MVP assets and use cases in an inadequate manner, with inadequate understanding, or in an inadequate environment, we may *prove* that our goals are not realistic or attainable. That is, if we are not careful, we may portray partial success as a failure. This will occur if we do not run our MBE development project in a sandboxed environment, if the work is not separated from production systems, or if it is executed by people who are unfamiliar with the relative immaturity of the work or who are hostile to the transformation objectives. An MVP approach is great if we are careful to understand what we are trying to accomplish, why we are doing it, what we are trying to prove and evaluate, and what the success criteria are. One of my main concerns about the MVP approach is the pivot, the idea that we learn from the MVP and alter course based on the results. I've seen this play out wrong many times in industry with prototyping. A few prototypes are produced to early-stage conceptual or developmental specifications, or maybe just to a model, the prototypes are tested, and some things are *proven* by the results. And by *proven*, I mean cast in stone. I've seen many cases where, because prototypes performed adequately, design engineers decided they didn't need to specify GD&T on production drawings or perform tolerance analysis on the design. Often, the only thing a prototype *proves* is that the prototype met some objectives. In cases like this, the prototype may do more harm than good.

A model-based example of an MVP failure is where a product definition dataset that does not include explicit dimensions (DCC 5 + DUC 5.2) is sent to a machine shop, but the shop is unaware of the project goals and the full purpose of the MVP, and the shop is inadequately prepared to run the pilot.

TABLE 6.1
The Ten Steps of Transformation

Steps		Technical	Cultural
1.	Understanding	x	x
2.	Accepting		x
3.	Adopting		x
4.	Evaluating	x	x
5.	Planning	x	x
6.	Developing	x	x
7.	Testing	x	x
8.	Optimizing	x	x
9.	Implementing	x	x
10.	Continuously improving	x	x

MBE TRANSFORMATION PROJECT

I break the transformation process into four phases in the following sections. The first two phases occur pre-launch, the third phase is launch, and the fourth phase occurs post-launch. A project can be structured differently if desired; however, this approach minimizes risk and helps to expedite the work. Some steps can be run in parallel, and others must be completed enough before other steps can make meaningful progress.

I repeat Table 2.1 here as Table 6.1, which shows the ten steps of transformation. I will repeat portions of the table in the following transformation project phases. Note that some steps are grayed-out in Tables 6.1a–d. The grayed-out steps are not major parts of that phase. However, some of the grayed-out steps play a role in all activities.

PHASE 1 – ANALYSIS [PRE-LAUNCH]

Engage, Understand, Envision, Roadmap

1. Immersion

   It is important to have a deep understanding of and familiarity with a company's environment and culture. We can review flow charts, corporate standards, SOPs, drawings, documents, product definition datasets, etc., but that doesn't tell us how they actually work, what motivates them, what they are rewarded and penalized for, and what happens between start and end of shift. Immersion is mainly important for consultants and other participants who are not part of the organization being transformed.

2. Training and Baselining

Everyone involved in MBE transformation must have an adequate and consistent understanding of MBD and MBE and their value proposition, from C-level senior management to the workers participating in the transformation development project. If training and baselining don't occur, the project will probably fail.

Note: Training and baselining should happen *before* setting transformation scope and goals. If training and baselining occur after setting the scope, the team will be stuck trying to achieve ill-advised goals or will have to start over and rescope the work.

3. Evaluation and Planning

An organization's and its peripheral organizations' MBE maturity and MBE capability must be evaluated (assessed) to determine their current state and what it will take to get to their desired future state. Evaluation is a detailed iterative process involving a lot of interviews, reviewing procedures, documents, flow charts, process maps, how information is created, how information is used, how work is performed, who owns what, CAD practices, CAx practices, PDM/PLM/ERP/CM ... infrastructure, culture, how culture affects information, how work is performed, etc. Once the current state is adequately understood, and stakeholders and project participants are baselined, we can plan for a future state (or future states (plural) if a stepwise approach is desired).

Once the plan is mature enough, it is important to get buy-in from senior management and a champion is selected from the senior management staff to own the project. The transformation is unlikely to succeed without adequate high-level support.

Estimates for short-term and lifecycle savings in terms of cycle time, development time, time to market, schedule, labor, CAPEX, OPEX, skill

TABLE 6.1.a

Ten Steps of Transformation – Phase 1

	Steps	Technical	Cultural
1.	Understanding	x	x
2.	Accepting		x
3.	Adopting		x
4.	Evaluating	x	x
5.	Planning	x	x
6.	Developing	x	x
7.	Testing	x	x
8.	Optimizing	x	x
9.	Implementing	x	x
10.	Continuously improving	x	x

load, training requirements, ROI, and other metrics are developed as part of project justification and funding. In many cases, ROI estimates and similar justifications are very much estimates. Usually, actual ROI is lower than estimates because of extraneous factors, such as inadequate management support, undue influence and barriers erected by unaffiliated groups, people with a stake in the status quo, skeptics, etc. This is a cultural problem, ultimately because a senior staff member inadequately supported the transformation. Or to say it differently, underperforming transformation is often due to inadequate support for transformation, often with short-term goals overriding longer-term opportunities. We must recognize this potential and guard against it in our early-stage work.

PHASE 2 – DEVELOPMENT AND PILOTS [PRE-LAUNCH]

Develop, Validate, Optimize

4. Pilot MBE Development

MBE pilots are developed to achieve, prove, and optimize specific things. Goals should be discrete and lead to the next stage of development. For example, converting from a drawing-based DCC 2 + DUC 2.1 product definition dataset to a model-based DCC 5 + DUC 5.2 product definition dataset to evaluate if model geometry can be used directly in pilot production processes. Iteration in processes and procedures will likely be required.

Pilots may use informal or less formal information, methods, processes, systems, etc. to obtain needed information without undue burden or expense, akin to minimum viable products (MVPs). As stated in the Sidebar/Info box above, we must be careful how we use MVPs and rigorously adhere to defining pilot goals and success criteria. There is a very real danger of pivoting too soon or for the wrong reason in this phase. MVPs must be seen as a means to refine pilots, not as a means to test overall project goals (such as if becoming an MBE is a good idea).

5. Pilot MBE Execution

Pilots should be executed in stages, and it is a good idea to run distinct product types and processes in separate pilots (e.g. pilots for discrete parts, separable assemblies, inseparable assemblies, custom COTS parts, machined parts, sheet metal parts, additive manufactured parts, molded parts, etc.).

Separate pilots should be run to evaluate internal processes, such as parts manufactured in-house, and processes that include external activities, such as parts manufactured by external suppliers. The goal is to build methods to achieve optimal results using pilot methods for all business cases that are in the project scope. The difference between Phase 2 and Phase 3 is that in Phase 2, the methods, processes, and information used are not required to be production ready. Mock-up simulations

TABLE 6.1.b
Ten Steps of Transformation – Phase 2

	Steps	Technical	Cultural
1.	Understanding	x	x
2.	Accepting		x
3.	Adopting		x
4.	Evaluating	x	x
5.	Planning	x	x
6.	Developing	x	x
7.	Testing	x	x
8.	Optimizing	x	x
9.	Implementing	x	x
10.	Continuously improving	x	x

may be used instead of full production software, such as running activities from spreadsheets rather than from full production software (Note: doing this is only okay if the purpose of the pilot is not to evaluate the software represented by the spreadsheet).

It is important that pilots are extensively tested and validated. In the pilot phases, users evaluating the pilots can be from the pilot project or a select team of power users or super users.

Different people will likely be brought into this phase. It is likely that this phase will include people who transition from earlier phases and people new to the project. Additional training and baselining is required. The training will be similar to the training presented to the initial team. Rebaselining is a good idea at this point.

PHASE 3 – ROLLOUT [LAUNCH]

Release, Transform, Realize

6. Production MBE Development

Based on success and lessons learned in phase 2 pilots, the project moves to the production phase. Pilot information assets, methods, processes, tools, activities, systems, etc. are transformed to or replaced by production-ready information assets, methods, processes, tools, activities, systems, etc.

Similar scopes, testing, and objectives to Phase 2 occur here, except we are using formal official-equivalent information assets, methods, processes, tools, activities, systems, etc. We are developing, evaluating,

TABLE 6.1.c

Ten Steps of Transformation – Phase 3

	Steps	Technical	Cultural
1.	Understanding	x	x
2.	Accepting		x
3.	Adopting		x
4.	Evaluating	x	x
5.	Planning	x	x
6.	Developing	x	x
7.	Testing	x	x
8.	Optimizing	x	x
9.	Implementing	x	x
10.	Continuously improving	x	x

and optimizing production-ready information assets, methods, processes, and systems.

It is important that the production-ready information assets, methods, processes, tools, activities, systems, etc. are extensively tested and validated. Initially, testers will be from the project team or a select team of power users or super users. Once their feedback has been addressed, additional testing by representative users is required.

7. Production MBE Deployment

Once production-ready information assets, methods, processes, tools, activities, systems, etc. are ready for deployment, a larger base of users outside of the project team should be made aware of the project. Selected representatives from outside the project team in all affected activities/groups/departments should be trained and baselined in MBD and MBE. At this time, the training should be customized for the MBE implementation, and people should be guided from how-we-do-it-today to how-we-will-do-it-using-these-new-methods-tomorrow. It is important that we assuage their concerns about the new methods and that everything for success has been included, and what they think is important or necessary may be rendered moot by the new methods and processes. Champions play an important role at this stage.

Different people will likely be brought into this phase. It is likely that this phase will include people who transition from earlier phases and people new to the project. Additional training and baselining is required. The training will be similar to the training presented to the initial team but customized for the actual methods implemented. Rebaselining is a good idea at this point.

TABLE 6.1.d
Ten Steps of Transformation – Phase 4

	Steps	Technical	Cultural
1.	Understanding	x	x
2.	Accepting		x
3.	Adopting		x
4.	Evaluating	x	x
5.	Planning	x	x
6.	Developing	x	x
7.	Testing	x	x
8.	Optimizing	x	x
9.	Implementing	x	x
10.	Continuously improving	x	x

PHASE 4 – AUDITING [POST LAUNCH]

Manage, Continuously Improve

8. Auditing

The auditing phase is different than the previous phases and probably will be considered as a separate project. Auditing is used to evaluate conformance to the transformed work methods and to prescribe corrective action. On-going training may be included in the auditing phase as well. I modeled this idea on ISO 9000 and similar auditing that is used to ensure that work methods continue to adhere to a stated norm. Auditing may be done for workflows for internal activities/groups/departments and for workflows that include external activities/groups/departments.

HOW MANY STEPS?

Several critical decisions are needed when considering transformation:

1. Determining the desired or expected end state
2. Determining how many steps or states are necessary to get to the end state.

Let's assume that an organization buys into the idea of a full digital transformation and decides to become an optimized MBE as defined in this text. Should they transform to the end state in one step, take one big leap? Should they transform in a few large steps? Or should they transform in many small steps? These are business decisions with business consequences. Some consequences are predictable, and others may be surprising.

Many times companies want to take small steps. They want to take small steps, affected by concerns about risk and change aversion, and thinking that small steps will eventually get them to the desired end state. There is indeed greater risk with greater change, if the reasons for the change and the value the change brings are poorly planned, poorly understood, inadequately supported by management, or inadequately staffed, developed, or executed. There are many ways for transformations to fail. Notice that I am not limiting my comments here to MBE. MBE transformation is no different from other types of corporate transformation in this regard. There is not a one-size-fits-all answer to these questions. Business decisions must be made in the context of an organization's environment and constraints. However, I recommend transforming in the fewest steps possible. If an organization is serious about obtaining value from MBE and transforming successfully and optimally, its chance of success is greatest if it transforms in the fewest steps. The main reasons for my sentiment are what I have seen in industry, that cultural inertia is usually inadequately addressed by partial steps because small steps do not adequately provide the value needed to overcome the inertia of the status quo.

I recommend transforming in the fewest steps possible. If an organization is serious about obtaining value from MBE and transforming successfully and optimally, its chance of success is greatest if it transforms in the fewest steps.

Some companies try implementing MBE methods in a single team, division, product line, silo, etc., hoping that the limited implementation will justify greater adoption by proving the value to the rest of the organization. The desired result may occur, but the more direct result from one part of an organization operating one way and other parts of the organization operating a different way is that staff working in these areas require different skills and must be familiar with different workflows, rules, standards, SOPs, etc. Thus, from an overall organizational perspective, cases where significantly different work paradigms are implemented in different parts of an organization mean that the skills required and what I call the "cognitive load", increase, and the overall organizational costs increase and overall efficiency and productivity decrease. Some of this is because we lose interchangeability of staff, as they become adept and specialized in the different workflows, and averse to working in the other paradigm. This aversion to the other paradigm, and particularly toward the new work paradigm (e.g. MBE), adds to the cultural inertia against change.

Success with MBD and MBE comes down to having a clear understanding of MBE and its true value proposition, having the support and initiative to transform to become an optimized MBE, and to make sure that the results are so good and beneficial that cultural inertia will be overcome.

Figure 6.4 shows four optional transformation paths to become an MBE. As I stated, I recommend transforming in fewer steps, as the organization's chances of success and not ending up with a stalled or failed transformation are greater. Figure 6.5 shows a variation of Figure 6.4, which is focused more on transformation based on which product definition dataset classes are used. This is how I used to think of transformation, which is more from an MBD point of view and more from an ASME

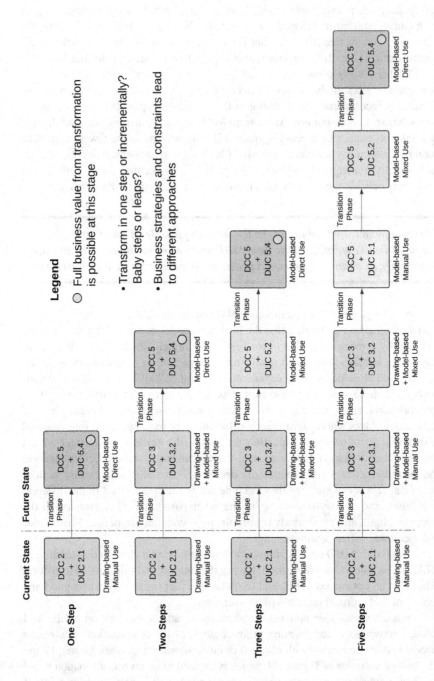

FIGURE 6.4 Steps to becoming an MBE – DCC + DUC Focus.

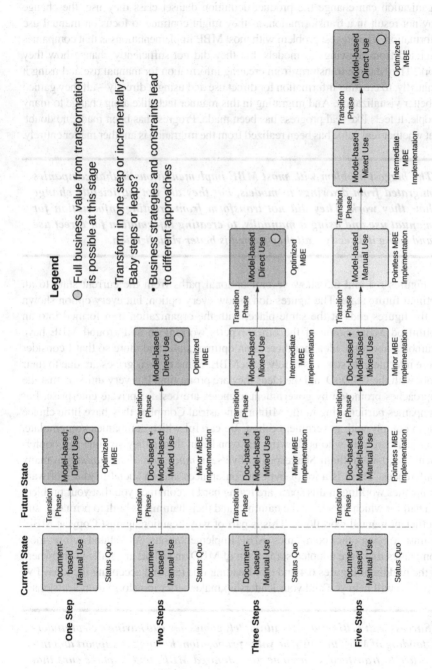

FIGURE 6.5 Steps to becoming an MBE – information focus.

Y14.41/ISO 16792 point of view. The problem with this approach is that while an organization can change the product definition dataset class they use, the change may not result in a transformation, as they might continue to focus on manual use information. The biggest problem with most MBE implementations is that companies migrated from drawings to models, but they did not sufficiently change how they work. They did not transform from creating information for manual use and using it manually, to creating information for direct use and using it directly – all they gained is better visualization. And migrating in this manner feels like a big change to many people. It feels like real progress has been made. Progress has been made, no doubt. But whether real value has been realized from the migration is another matter entirely.

The biggest problem with most MBE implementations is that companies migrated from drawings to models, but they did not sufficiently change how they work. They did not transform from creating information for manual use and using it manually, to creating information for direct use and using it directly – all they gained is better visualization.

Figures 6.4 and 6.5 show several optional paths from the current state to an optimal future state. The figures don't show every option, but every option shown in the figures ends at the same place, with the organization transformed into an optimized MBE. Some of the other experts working in and around MBE have grumbled about the idea that there is an optimal state, and more so that I consider these intermediate states as barely even MBE. Some of their gripes are due to their work with the US DoD and the ideas they are promoting. It is very unlikely that the approaches promoted by government agencies are best for private enterprise. For companies participating in the Military Industrial Complex, they have little choice but to comply with government diktat. Be careful which companies you emulate. Consider carefully who is showing the wonderful things they've done at a conference. To quote William Shakespeare, I've seen *much ado about nothing* at many conferences. There is a lot of hype at conferences and on social media. Software companies working in this space are trying hard to convince you that your problem is a nail for which they have the hammer, and their hammer is built to primarily suit inefficient manual workflows. This is part of why I caution against Copycat MBE, against copying other companies' MBE implementations. As I stated before, most companies that have adopted some form of MBD and MBE are stalled somewhere in the middle of Figures 6.4 and 6.5. Reading this text and accepting the ideas I've provided will help you and your leadership make better transformation decisions.

Success with MBD and MBE ultimately comes down to having a clear understanding of MBE and its true value proposition, having the support and initiative to transform to become an optimized MBE, and to make sure that the results are so good and beneficial that cultural inertia will be overcome.

STANDARDS AND MBE

I included some commentary on standards in Chapter 4. Engineering and industrial standards have been around for a long time. Here I will discuss the effect of standards on other standards, on MBD and MBE, and on MBE transformation.

Figures 6.6 and 6.7 show a hierarchical relationship between various standards and their effect on how we work today and how we do business today. The standards in the figures relate to engineering drawing practices, product definition standards, modeling standards, data format standards, and MBE standards. Note that many of the standards in Figures 6.6 and 6.7 are US national standards, many from ASME, which I helped develop. I have alternate versions of these figures focused on ISO engineering standards. Both sets of figures tell the same story, so I am only showing the US versions here, although the information modeling and PDQ standards are ISO standards.

Starting at the top left corner of the figure, I show what I call 2D and PMI-Defining Standards. These are legacy standards, standards that are or have evolved from industrial drafting standards. The first US mechanical engineering-related drawing standard I have is from 1935. The overall purpose and approaches used in these standards haven't changed much since then. Standards are primarily created to support how we work today, which is based on how we worked in the past. In some ways, engineering product definition standards provide a self-reinforcing loop in which we codify and cement how we work today. These legacy drafting standards form the top tier of the hierarchy of standards shown in Figures 6.6 and 6.7. In the context of MBE, these standards act primarily as impediments to MBE. Nearly everything in drafting standards is defined in the context of manual use – drafting standards are designed to support visualization and to facilitate using information visually – manual use. The legacy standards flow down to, affect, and constrain the next tier of standards.

I call the next tier of standards Model Construction and Usage Standards. ASME Y14.41 and ASME Y14.47 are shown. There are a few other standards not listed, such as ASME Y14.46, which is relatively new and focuses on additive manufacturing. ASME Y14.41, as discussed earlier in the text, was created to codify rules, methods, and terminology for 3D model-based product definition practices. As I've stated, I think ASME Y14.41 supports the wrong workflows and tells the wrong story. After reading that standard, many people have told me they thought it is about applying GD&T to models, that it is a GD&T standard (it is not). Many ideas in the standard are in terms of manual use, such as query and highlight being used almost synonymously with associativity, which both require a logical link between digital elements in a product definition dataset. The current state of ASME Y14.41 and ASME Y14.47 provides a framework and guidance for defining manual use MBD information assets and for using that information manually. There are a few tidbits in those standards to support direct use, but they are outliers and not

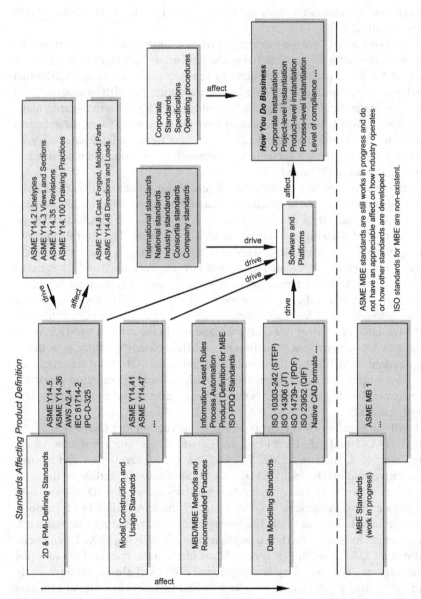

FIGURE 6.6 Tiers of standards: hierarchical relationship between standards and how we work 1

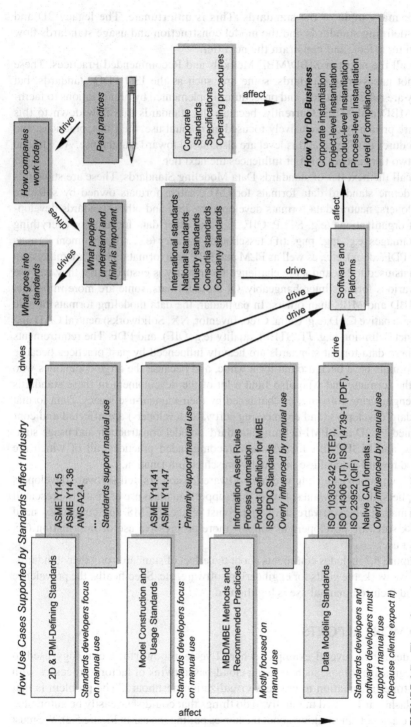

FIGURE 6.7 Tiers of standards: hierarchical relationship between standards and how we work 2

the primary topic of the standards. This is unfortunate. The legacy 2D and PMI-defining standards and the model construction and usage standards flow down to, affect, and constrain the next tier.

I call the next tier MBD/MBE Methods and Recommended Practices. These are not necessarily standards, some are, such as the ISO PDQ standards, but many are rules, practices, and procedures implemented by organizations to facilitate MBD and MBE. Currently, because the standards that flow down to this tier are primarily or exclusively focused on manual use, the rules, practices, and procedures developed at this level are also biased toward manual use. As with the first two tiers above, this tier influences the next tier.

I call the next tier of standards Data Modeling Standards. These are standards that define standard data formats for CAD, native formats owned by software developers, neutral data formats developed by ISO and other standards development organizations (e.g. STEP, QIF, JT), and other data formats for everything from images (e.g. .jpg, .png, .tif), tessellated geometry (e.g. .stl), document formats (e.g. .PDF, .docx), etc., as well as PLM and ERP data formats. There are many data formats used today and a big challenge for industry is ensuring that they can use the various formats interchangeably. Of these formats, some are more important in MBD and MBE than others. In particular, the data modeling formats of note here are native CAD (e.g. Catia, Creo, Inventor, NX, Solidworks) neutral CAD and product definition (e.g. JT, STEP), quality (e.g. QIF), and PDF. The requirements for these data format standards are heavily influenced by past practices because the formats have been around for a while, and because the large companies who use the formats, and who also fund a lot of the development of these standards and engineering software, are burdened by their manual-use legacy. Data format standard developers (and engineering software developers) are affected and constrained by 2D and PMI-defining standards, model construction and usage standards, and MBD/MBE methods and recommended practices, all of which are biased toward manual use of visually-presented information.

If companies using engineering software (customers) tell software developers (suppliers) that they want software that supports manual use of visually-presented information, the software developers must oblige. OEMS and companies need to ask software developers to support more progressive use of information (i.e. direct use).

Figure 6.7 includes comments about the effect of standards on other standards, how we work, the goals for engineering software, etc. Specifically, the prevalence of and focus on manual use is highlighted.

NO MORE REPORTS

I've described several examples where I've seen powerful technology used to do regressive things, such as using cloud-based twins of factory processes with real-time information to create a visualization dashboard. The problem is that the dashboard is used manually to do things that could very easily be automated (Once again, I am thinking about the horses pulling the car in Figure 5.2). We must seek out unnecessary manual activities that require manual-use information and replace them. Another example was the tremendous effort of many people to create a huge report so the team could decide if they had passed a milestone and could move to the next step. Determining if a milestone has been met could easily have been automated in the company's PLM software. They didn't need to spend all the time and effort on the report. They didn't even need the report. Or at least they shouldn't need the report. If the environment was operating out of control, if there was inadequate trust between the teams, groups, and individuals in the organization, then such a report and all the manual scrutiny is a way to make sure that no one is pulling a fast one, that a milestone has really been met. A much better alternative is to get the organization's act together, to stop relying on a very laborious and inefficient method, to automate milestone compliance/achievement, and get on with it.

No more reports.

The report is not the point, however. The purpose of this report was to provide the information needed to determine if a milestone had been met. In the end, this is equivalent to checking many boxes, and those boxes could be built into the activity management system.

In my GD&T work, I help companies with inspection, understanding engineering specifications, the requirements they impose upon the finished product, what the conformance criteria are for the requirements, and how to evaluate conformance (e.g. inspect). Most companies create some kind of inspection report, which includes the results of an inspection. This may be a formal activity, such as using SAE form AS-9102 for first article inspection. Using a form is old-fashioned (by now I assume you expected me to say that), as forms and reports are intended for manual use. For MBE, we should be using model-based (MBD) methods to represent inspection data. Rather than an inspection report, we should create an inspection record. An inspection report is designed to be read and acted upon, which is a classic RIDICT activity. It does not need to be done today. We should be able to easily automate a process in which the inspection criteria and boundary conditions are represented in an inspection plan, the inspection is executed, the record populated with the results, and the system should be able to determine if the item conforms to its requirements or not.

No more reports.

We don't need to read them. We don't need to write them. Our organizations don't need to own them. We don't need to be burdened with learning all the specialized skills needed to read and write them. A few people do. Not everyone in the workflow. We certainly don't need to have reports and RIDICT use cases to determine if it's okay to move to the next step in a process. That information is already in modern machinery and facilitated by standards such as MTConnect (ANSI/MTC1.4-2018). Break the loop. Free your and your employees' time to focus on productive activities.

Recognize, replace, and eliminate low-value and low-value high-skill activities. Eliminate low-value legacy manual activities, including the information needed to drive them and the information they create, which in turn drives other inefficient manual activities.

No more reports.

MBE VALUE PROPOSITION REVISITED

Figure 6.8 shows a histogram with comparisons of relative cost, value, and other criteria related to pre-MBE work methods, Copycat MBE implementations, and optimized MBE implementations. Note that in some areas on the graph higher values are better for the organization and in other areas lower values are better for the organization. The moral of the story depicted in the histogram is that regardless of whether a company decides to implement MBE optimally or if they

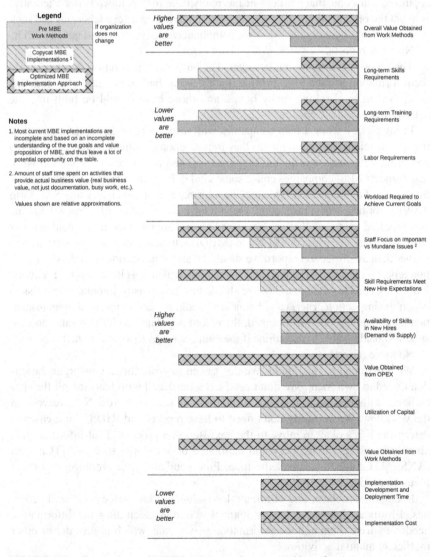

FIGURE 6.8   Comparison of MBE implementation cost and benefits.

decide to take small iterative steps, such as copying other inefficient or ineffective implementations, the implementation costs are about the same, but there are big differences in the value the organization obtains from transformation, as well as the problems of increased cultural inertia that comes from small steps that are painful for the organization but yield little real value. As stated in the figure, the values shown are approximations and not on a particular scale – the values represent relative costs and values for the criteria shown.

Figure 6.9 shows the relative opportunity cost or loss from choosing to do nothing (choosing not to invest in MBE transformation), implementing MBE

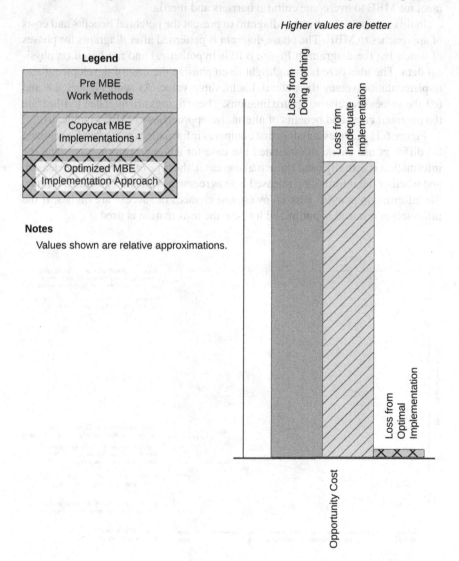

FIGURE 6.9 Comparison of MBE implementation opportunity cost.

inadequately, or implementing MBE optimally. I use the concept of *opportunity cost* to explain why it may be detrimental to take too small of a step toward becoming an MBE, to partially implement MBE methods, or to copy another organization that did so. Not understanding the full value of MBE and its true purpose will lead to inadequate MBE implementation. While inadequate MBE implementations and optimal MBE implementations have similar development and implementation costs, inadequate MBE implementations have a much greater opportunity cost (a much greater potential loss), as inadequate MBE implementations do not yield adequate value, and thus will not make a strong enough argument for MBE to overcome cultural barriers and inertia.

In this section, I use a phase diagram to present the potential benefits and costs of approaches to MBE. The phase diagram is patterned after diagrams for phases of water, but the diagram in Figure 6.10 is hypothetical and not based on physical data. The idea here is to highlight or emphasize the cost of development and implementation versus the potential achievable value. As with Figures 6.8 and 6.9, the values are relative approximations. These figures are intended to illustrate the potential costs and benefits of alternative approaches to MBE graphically.

Figure 6.11 includes a matrix that compares information created and optimized for different use cases, the intended use case for authored information, how the information is actually used (its actual use case), the type of information released, and whether the information released is in agreement with and optimized for how the information is used. Risk is lowest and chances of success are highest if the information released is optimized for how the information is used.

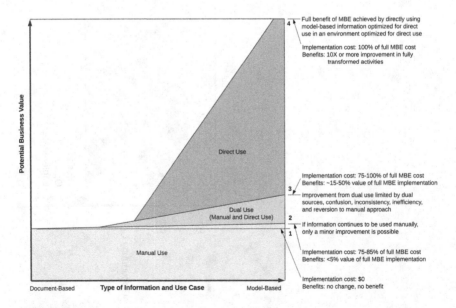

FIGURE 6.10   MBE phase diagram.

	Information optimized for	Actual Use: Used manually	Actual Use: Used directly	Type of Information Released: Presented information	Type of Information Released: Direct-use information	Info/Use/Type in Agreement?	
1	Manual use	X		X		Yes	✓
2	Manual use	X			X	No	
3	Manual use	X		X	X	No	
4	Manual use		X	X		No	
5	Manual use		X		X	No	
6	Manual use		X	X	X	No	
7	Direct use	X		X		No	
8	Direct use	X			X	No	
9	Direct use	X		X	X	No	
10	Direct use		X	X		No	
11	Direct use		X		X	Yes	✓
12	Direct use		X	X	X	No	
13	Dual use [1]	X		X		No	
14	Dual use [1]	X			X	No	
15	Dual use [1]	X		X	X	No	
16	Dual use [1]		X	X		No	
17	Dual use [1]		X		X	No	
18	Dual use [1]		X	X	X	No	
19	Dual use [1]	X	X	X		No	
20	Dual use [1]	X	X		X	No	
21	Dual use [1]	X	X	X	X	Yes	✓

FIGURE 6.11 Correspondence between information created and how information is used.

BEST CASE AND RECOMMENDED PRACTICE:

Official information should be optimized for its intended use, and it should be used as intended. These are critical criteria for obtaining value from MBD and MBE. As shown in the matrix in Figure 6.11, only three rows conform to the best case (rows 1, 11, and 21).

Official information should be optimized for its intended use, and it should be used as intended.

Notes

1. Best practice:
   Create information optimized for one use, manual use or direct use.
2. Dual use should only be used if one information type is official information, and the other information type is for reference only.

Goals

- Designate direct use as primary use case.
- Use direct-use information as official information.

RECOMMENDED PRACTICE

- Designate and use direct use information as official information as much as possible.
- Presented information may be made available on demand in cases where business constraints require information to be used manually.
- If direct-use information is designated as official information and presented information is available, people will circumvent the process and use manual methods regardless of which information is designated as official.

MBE QUIT POINTS

Over the years, I've had many discussions with people about MBD and MBE. Many discussions were with other people working in this space, and others were with people unfamiliar with design, manufacturing, inspection, MBE, etc. A few years ago as I was explaining the value proposition and failure modes of MBE to a friend from a different field. I was explaining how companies get stuck, give up, and settle for incomplete or inadequate MBE implementations. He said I was talking about "quit points". *Quit point* is a common concept in education, smoking cessation, weight loss, process improvement, ... for any activity that is difficult. I hadn't heard the term before, but I got it. My friend used the term in a gaming context, where a *quit point* was a point in a game where people gave up or stopped playing. I've played a lot of computer games and many times I've hit the point where it was difficult, seemed hopeless, or even pointless. Maybe it was time to give up, cave in, call it good, and move on and do something else. Usually, gamers persevere, as do students, athletes, etc.

I have seen many companies hit quit points in their MBE implementation. There are many places to stumble that occur throughout the process, from initial concept to deployment, through all 10 steps in Table 6.1. However, the most common quit point is near the beginning of the process. Early in the process, someone presents the idea of becoming an MBE to other parties in the organization and supply chain to get feedback and to set the scope. The conversations often occur in the context of limited information, with one or both people having a limited understanding of the value proposition, and with the recipient bolstered by considerable cultural inertia. Often the recipient is completely focused on production and maintaining the status quo, as workers are hard-wired to protect the status quo of meeting production goals from any possible threat, and they view MBE as a threat, as *crazy talk*. (In some cases, both sides of the conversation may think it is a bad idea.) Often, the proponent is too low in the organization's hierarchy to influence the other party. Regardless of the reason, in these early decision-making phases, people often decide to pursue a watered-down version of MBE. This is unfortunate.

Thus, many companies undermine their MBE implementation at the very beginning by aiming so low that they won't generate enough value to overcome cultural inertia and change how they work. This is why I have companies start their transformation with MBE workshops. They need to understand what MBD and MBE are, what their true value propositions are, and the consequences of not doing enough to make MBD and MBE stick. Note that the idea of stickiness or making something stick is an idea I got from a book, "Made to Stick: Why Some Ideas Survive and Others Die (Footnote 6.2)." Not sticking means not enduring or being remembered, thus not being sticky enough to endure or be remembered. A slowly dying process improvement initiative obviously isn't sticky enough.

OVERCOME BY EVENTS

It is a shame that most MBE implementations are undermined so early in their development, often at the earliest scoping stages. The goals are set too low for the reasons listed above, and other reasons as well. Sometimes people running the project or higher up want to hedge their bets and make sure they don't try anything too audacious. They want to make sure they look good. Climbing the corporate ladder or trying to stay on top of it affect what organizations do and their culture. I think this is a common reason Copycat MBE is popular. I've gotten the feeling that some clients were more concerned about the ladder than actually generating value for the organization. The quit point here is that the champion decides to quit focusing on building organizational value and focus on their image instead, and it occurs before they even start the project.

FOOTNOTE

6.2 *Heath, Chip, Heath, Dan (2007). Made to Stick: Why Some Ideas Survive and Others Die. New York, NY: Random House.*

CONCLUSION

While I didn't cover everything about MBD and MBE in this text, I addressed many topics that I think are critical to understanding MBD and MBE, their value, their costs, the opportunities they present, and what we can do to realize those opportunities. I repeated some things many times. While I think it is a good idea to read the text from front-to-back, I assume many people won't read it that way. I assume they'll search for a specific topic and read that portion (or get frustrated when they reach a quit point!). Thus, I tried to make some sections a bit more self-contained. Many topics are repeated for emphasis. I've found some of these topics hard sells – I've found it difficult to get people with outdated or muddy ideas about MBD and MBE to accept the different spin I put on MBD, MBE, what they are, why they matter, what their value is, what the end state is, and how to get there. Some folks don't seem to want to think about this from an economic or business-value perspective. Note that I am absolutely convinced that the reason most MBE implementations flail, stall, or fail is because they are not considered from economic or business points of view. For process improvement initiatives like MBE to succeed and provide maximum value, or even provide meaningful value, upper C-level management must understand the topic and what we are trying to achieve. For some reason, executives in manufacturing have embraced Lean, Six Sigma, ISO 9000, and related process improvement initiatives and quality management systems, taking the time to learn about them and to prioritize them in their operations. MBE needs similar attention, priority, and widespread adoption. It's time to show the love!

A lot of the MBD and MBE material on the Internet, presented on social media, at conferences, in blogs and vlogs is focused on the wrong things. The focus is on technology and technical solutions, on software, menus, data formats, CAD, BOMs, PDM, PLM, ERP, etc. People want an easy button, a technical solution they can plug-and-play to be an MBE (see Figure 6.12). Of course, technical aspects are important, as they enable or impede success, but we shouldn't start our MBE journey with a technical discussion. We should start with a discussion about what MBE really is, what it really brings to the table, and why it is so important. As I've stated, MBE is lean. MBE is about information, information management, optimizing information, how information is used, and recognizing and focusing on the information lifecycle. It is about stripping away the noise, burden, and soul-crushing tedium inherited from legacy document-based workflows

FIGURE 6.12 The MBE easy button.

and using information manually. MBE is about aligning and optimizing how we work, the information we create, and how we use information with our current technological state. MBE is about increasing value for the organization and for society. MBE is about making what we do at work more rewarding by increasing the value of what we do and the value of time we spend doing it. MBE is about automation, automating tasks that no longer need to be done manually, and eliminating manual-use information needed to support those manual activities.

Doing something that provides real value to the organization is more important than doing what's easy.

Remember, success with MBE primarily depends on organizational culture and how well change is managed. How well MBE is understood and accepted, how well it is supported by management, and how it is implemented matter, but MBE implementations fail, flail, or stall for cultural reasons. There is not an easy button for changing corporate culture. And trying to copy what another organization says they did will not yield the results you expect, for every organization has a different culture, and it is likely that the organization being copied didn't get the results they expected either. Think it through. Make sure you have high-level buy-in and support. Aim high.

MBE is about information, information management, optimizing information, how information is used, and recognizing and focusing on the information lifecycle. It is about stripping away the noise, burden, and soul-crushing tedium inherited from legacy document-based workflows and using information manually.

Over the years, as I realized the end goal of MBE is automation, I discussed this idea with other experts, and some are uncomfortable with automation. By automation, I am not describing a bunch of robots bouncing around in lights-out factories (although industry will do whatever it does whether we like it or not).

The US manufacturing industry has been pursuing automation for decades, and even with our reduced manufacturing workforce we maintain very high productivity. As I have clarified, many of our processes are already automated, and like all good automation, we don't even notice it anymore.

To get started on our journey, we need to break activities down and ask enough questions. Once we ask *why* enough times, the answers will guide us to higher productivity. Some of the questions are:

- Why are we using documents?
- Do we need the documents?
- What are the documents used for? (I know, they're used to make more documents...)
- Why are we doing this manually?
- Why are we doing this at all?
- Why are we creating this information?
- Do we really need to do this manually?
- Do we really need this information?
- Do we really need this to be visual information?
- Can we use direct-use information instead?
- What is the lifecycle cost of creating information?
- What is the lifecycle cost of using information?
- What is the lifecycle cost of maintaining information?
- What is the lifecycle value obtained from information?
- Why do we work this way?
- How well do the skills we require align with the skills of people entering the workforce?
- Are we forcing staff to focus on low-value tasks and activities?
- Are we forcing staff to focus on low-value high-skill tasks and activities?
- Are we increasing churn and turnover because our staff must develop skills and do tasks they think are low value, archaic, unnecessary, or overly complex (e.g. read engineering drawings)?
- What are the cascade of benefits driven by converting a task from manual use to direct use?

I hope this text has sparked new ways to think about how we work, the information we create, how we use information, and the cost of and value we obtain from information over its lifecycle.

My primary goals in this text have been:

- To inform and elevate readers' understanding of MBD and MBE and their value proposition
- To describe the disciplines of MBD and MBE with terms that clarify and lay bare their purpose and value propositions
- To clarify how much we focus on creating manual-use information and using information manually

- To clarify that MBE is about information
- To clarify that information has a lifecycle
- To clarify that we create and use information assets, they are not merely artifacts or datasets, and we must think of them in terms of their cost and their value
- To clarify that to obtain meaningful value and ROI from MBE, MBE must be considered from economics and business points of view
- To clarify that MBE is an important productivity improvement methodology
- To clarify that MBE is worthy of upper-management commitment and ownership, which is essential for successful MBE implementation
- To clarify that MBE increases organizational efficiency, productivity, quality, and competitiveness
- To energize readers and to instill enthusiasm about MBE
- To establish a new elevated baseline for all involved
- To help people achieve the potential of MBE.

To maximize the value we obtain from MBD and MBE, we need a common understanding. We must focus on increasing the value of our information, our information assets, our IP, CAPEX, OPEX, and our time. We need to think about information differently. We need to understand that every information element and information asset has lifecycle value and lifecycle cost, and our goal is to increase the value/cost ratio. We need to break free of the inefficient burden we inherited from our document-based past and take advantage of the incredible tools and systems available today. We need to change how we work in manufacturing to match the capabilities of the fourth industrial revolution in which we operate.

Digital transformation is not just a buzzword. Becoming an MBE requires that we transform how we work, the information we create, and how we use that information. MBE requires digital transformation and information transformation. And, most importantly, it requires that we transform our ideas about how we work and how and why we create and use information. There's a better way. At our fingertips.

Bibliography

1. Clarke, Arthur C. (1997). *3001: The Final Odyssey*. NYC, NY: Random House.
2. Echeberria, Ana Landeta (2020). *A Digital Framework for Industry 4.0: Managing Strategy*. 1st ed. Cham: Palgrave Macmillan.
3. Appian (2022). *A Productivity eBook from Appian: 6 Key Advantages of Eliminating Manual Document Processing*. McLean, VA: Appian.
4. Freeman, Joshua Benjamin (2018). *Behemoth: A History of the Factory and the Making of the Modern World*. NYC, NY: W. W. Norton & Company.
5. ARK Invest Management LLC (2023). Big Ideas 2023: Annual Research Report. St Petersburg, FL: ARK Invest Management LLC.
6. ARK Invest Management LLC (2024). Big Ideas 2024: Annual Research Report. St Petersburg, FL: ARK Invest Management LLC.
7. Krum, Randy (2013). *Cool Infographics: Effective Communication with Data Visualization and Design*. 1st ed. Hoboken, NJ: John Wiley & Sons, Inc.
8. Moore, Geoffrey A. (2014). *Crossing the Chasm: Marketing and Selling Disruptive Products to Mainstream Customers*. 3rd ed. NYC, NY: Harper Business, Collins Business Essentials.
9. Saini, Kavita, et al. (2021). Deloitte Insights. Deloitte Development.
10. Ismail, Salim; Diamandis, Peter H.; Malone, Michael S. (2023). *Exponential Organizations 2.0: The New Playbook for 10x Growth and Impact*. 1st ed. Powell, OH: Ethos Collective.
11. Roser, Christoph (2016). *Faster, Better, Cheaper in the History of Manufacturing: From the Stone Age to Lean Manufacturing and Beyond*. 1st ed. Boca Raton, FL: CRC Press; Productivity Press; Taylor & Francis Group.
12. Littman, Michael L., et al. (2021). *Gathering Strength, Gathering Storms: The One Hundred Year Study on Artificial Intelligence (AI100) 2021 Study Panel Report*. Stanford, CA: Stanford University.
13. Ury, William (1993). *Getting Past No: Negotiating in Difficult Situations*. NYC, NY: Bantam Dell.
14. Huff, Darrell (1954). *How to Lie with Statistics*. NYC, NY: W. W. Norton & Company.
15. Daugherty, Paul R.; Wilson, H. James (2018). *Human + Machine, Updated and Expanded: Reimagining Work in the Age of AI*. 1st ed. Boston, MA: Harvard Business Review Press.
16. Ward-Dutton, Neil (2020). *IDC Technology Spotlight: The Future of Work in the Age of Cloud and Low-Code Automation (Report)*. Framingham, MA: IDC.
17. Halpern, Marc (2021). Innovation Insight for Model-Based System Engineering (Report). Stamford, CT: Gartner.
18. Carreira, Bill (2004). *Lean Manufacturing That Works: Powerful Tools for Dramatically Reducing Waste and Maximizing Profits*. 1st ed. NYC, NY: Amacom Books.
19. Heath, Chip; Heath, Dan (2007). *Made to Stick: Why Some Ideas Survive and Others Die*. NYC, NY: Random House.
20. MIT Technology Review Insights (2024). *MIT Technology Review: Taking AI to the Next Level in Manufacturing (Report)*. Cambridge, MA: MIT Technology Review Insights.

21. Hughes, Andrew; Murray, Diane (2018). MOM and PLM in the IIoT Age: A Cross-Discipline Approach to Digital Transformation (Report). Cambridge, MA: LNS Research.
22. Siemens (2022). PCB Sourcing and Quoting: Overcoming Complex Challenges with Intelligent Software Solutions (Report). Plano, TX: Siemens Digital Industries Software.
23. Winton, Brett (2024). Platforms Of Innovation: How Converging Technologies Should Propel A Step Change In Economic Growth (Report). St Petersburg, FL: ARK Invest Management LLC.
24. Stuart, David (2016). *Practical Ontologies for Information Professionals.* 1st ed. Chicago, IL: ALA Neal-Schuman.
25. Madison, Dan (2005). *Process Mapping, Process Improvement and Process Management: A Practical Guide to Enhancing Work Flow and Information Flow.* Chico, CA: Paton Professional.
26. Stark, John (2015). *Product Lifecycle Management (Volume 1): 21st Century Paradigm for Product Realisation (Decision Engineering).* 3rd ed. Geneva: Springer.
27. Schwab, Klaus (2018). *Shaping the Future of the Fourth Industrial Revolution.* Geneva: World Economic Forum.
28. Schwab, Klaus (2021). *Stakeholder Capitalism: A Global Economy that Works for Progress, People and Planet.* 1st ed. Hoboken, NJ: John Wiley & Sons, Inc.
29. Delligatti, Lenny (2013). *SysML Distilled: A Brief Guide to the Systems Modeling Language.* 1st ed. Boston, MA: Addison-Wesley Professional.
30. Ross, Phillip J. (1995). *Taguchi Techniques for Quality Engineering.* 2nd ed. NYC, NY: McGraw-Hill Professional.
31. Schedlbauer, Martin (2010). *The Art of Business Process Modeling: The Business Analyst's Guide to Process Modeling with UML & BPMN.* 1st ed. Sudbury, MA: The Catharsis Group.
32. Davidow, William H.; Malone, Michael S. (2020). *The Autonomous Revolution: Reclaiming the Future We've Sold to Machines.* Oakland, CA: Berrett-Koehler Publishers.
33. Suleyman, Mustafa; Bhaskar, Michael (2023). *The Coming Wave: Technology, Power, and the Twenty-first Century's Greatest Dilemma.* NYC, NY: Crown.
34. Walton, Mary M. (1986). *The Deming Management Method.* NYC, NY: Perigee Books.
35. Gazzaley, Adam; Rosen, Larry D. (2016). *The Distracted Mind: Ancient Brains in a High-Tech World.* Cambridge, MA: The MIT Press.
36. Azhar, Azeem (2021). *The Exponential Age.* 1st ed. NYC, NY: Diversion Books.
37. Reis, Eric (2011). *The Lean Startup: How Today's Entrepreneurs Use Continuous Innovation to Create Radically Successful Businesses.* 1st ed. NYC, NY: Crown Currency.
38. Ullman, David G. (2010). *The Mechanical Design Process.* 4th ed. NYC, NY: McGraw-Hill.
39. Tufte, Edward R. (2001). *The Visual Display of Quantitative Information.* 2nd ed. Cheshire, CT: Graphics Press.
40. Blum, Andrew (2019). *The Weather Machine: A Journey Inside the Forecast.* NYC, NY: Ecco.
41. Stotz, Andrew (2015). *Transform Your Business with Dr. Deming's 14 Points.* N Charleston, SC: CreateSpace Independent Publishing Platform.

42. Fisher, Adam (2018). *Valley of Genius: The Uncensored History of Silicon Valley (As Told by the Hackers, Founders, and Freaks Who Made It Boom)*. NYC, NJ: Twelve, Hachette.

43. Martin, Karen; Osterling, Mike (2013). *Value Stream Mapping: How to Visualize Work and Align Leadership for Organizational Transformation*. 1st ed. NYC, NY: McGraw Hill.

44. Wolfram, Stephen (2023). *What Is ChatGPT Doing and Why Does It Work?* Champaign, IL: Wolfram Media Inc.

45. Saldanha, Tony (2019). *Why Digital Transformations Fail: The Surprising Disciplines of How to Take Off and Stay Ahead*. 1st ed. Oakland, CA: Berrett-Koehler Publishers.

46. Schwab, Klaus (2016). *World Economic Forum*. 1st ed. Geneva: World Economic Forum.

Index

Printed in the United States
by Baker & Taylor Publisher Services

Printed in the United States
by Baker & Taylor Publisher Services